T0342193

HIGH-DENSITY AND DE-DENSIFIED SMART CAMPUS COMMUNICATIONS

HIGH-DENSITY AND DE-DENSIFIED SMART CAMPUS COMMUNICATIONS

Technologies, Integration, Implementation, and Applications

Daniel Minoli

DVI Communications
New York, NY, USA
Red Bank, NJ, USA

Jo-Anne Dressendofer

Slice Wireless Solutions
New York, NY, USA

Registered Office
John Wiley & Sons, Inc., 111 River Street, Hoboken, NJ 07030, USA

Editorial Office
111 River Street, Hoboken, NJ 07030, USA

For details of our global editorial offices, customer services, and more information about Wiley products, visit us at www.wiley.com.

Wiley also publishes its books in a variety of electronic formats and by print-on-demand. Some content that appears in standard print versions of this book may not be available in other formats.

Library of Congress Cataloging-in-Publication Data:

Names: Minoli, Daniel, 1952– author. | Dressendofer, Jo-Anne, author.
Title: High-density and de-densified smart campus communications : technologies, integration, implementation and applications / Daniel Minoli, Jo-Anne Dressendofer.
Description: Hoboken, NJ : Wiley, 2022. | Includes bibliographical references and index.
Identifiers: LCCN 2021050372 (print) | LCCN 2021050373 (ebook) | ISBN 9781119716051 (hardback) | ISBN 9781119716068 (adobe pdf) | ISBN 9781119716082 (epub)
Subjects: LCSH: Wireless communication systems. | Smart materials.
Classification: LCC TK5103.2 .M5665 2021 (print) | LCC TK5103.2 (ebook) | DDC 621.384–dc23/eng/20211110
LC record available at https://lccn.loc.gov/2021050372
LC ebook record available at https://lccn.loc.gov/2021050373

Cover design by Wiley
Cover image: © enjoynz/Getty Images

Set in 10/12pt TimesTenLTStd by Straive, Pondicherry, India
10 9 8 7 6 5 4 3 2 1

In loving memory of my wife Anna (Dan)

Era una santa e completò la sua missione con passione, pur giovane.

"E se dal caro oggetto, Lungi convien che sia, convien che sia, Sospirerò penando, Ogni momento" (from a stanza in Vivaldi's "Vedrò con mio diletto")

In loving memory of my mother Helene (Jo-Anne)

Who was there for every tear along my not-so-easy career and pushed me to dream even bigger

CONTENTS

PREFACE

High-density campus communications have traditionally been important in many environments, including airports, stadiums, convention centers, shopping malls, classrooms, hospitals, cruise ships, train and subway stations, evangelical megachurches, large multiple dwelling units, boardwalks, (special events in) parks, dense smart cities, and other venues. These communications span several domains: people-to-people, people-to-websites, people-to-applications, sensors-to-cloud analytics, and machines-to-machines/device-to-device. While the later Internet of Things (IoT) applications are generally (but not always) low speed, the former applications are typically high speed. In many settings, people access videos (*a la* Over The Top [OTT] mode) or websites and applications that often include short videos or other high data-rate content. Deploying optimally performing high-density campus communication systems is desired and required in many cases, but it can, at the same time, be a complex task to undertake successfully.

High-density campus communications play a role in the evolution of Smart Campuses but also drive the Smart City and Smart Building use cases. Connectivity is now considered a fourth utility (in addition to gas, water, and electricity). In fact, massive-type communication is a recognized requirement of 5G, even if just in the machine-type communication environment. In the campus applications just cited, people-to-people, people-to-websites, and people-to-applications connectivity is increasingly important, given that nearly everyone now carries a smartphone and many apps entail high-throughput transmissions.

There are unique requirements and unique designs required for high-density communications, particularly because of the relative scarcity of available spectrum. In addition, there has been and continues to be a set of transitions, even transformations, of the underlying technologies. The world has moved to IP for all data, voice, and video communications. Additionally, there is a trend toward the use of Wi-Fi-based hotspot communication in all practical situations, due to near ubiquity of service, lower end-user costs, higher bandwidth, technical simplicity, lower infrastructure costs, decentralized administration, regulation relief, and non-bureaucratic delivery of service (without the reliance of large institutional providers). While 5G promises to deliver a set of new capabilities, neither 3G nor 4G displaced Wi-Fi as a common access technology in the office, in the campus, on the street, and in travel. The technologies per se used for high-density communications are not new (perhaps with the exception of 5G), but the requirements, as well as the design and system synthesis, are relatively unique.

As the second decade of the twenty-first century rolled along, however, a new requirement presented itself due to the worldwide pandemic: physical/desk distancing in support of Office Social Distancing (OSD) and Office Dynamic Cluster Monitoring and Analysis (ODCMA). Wireless technologies have been harvested to address and manage these pressing issues. Real-Time Locating Systems (RTLS) have been employed for a number of years to automatically identify and then track the location of objects or people in real-time, within a building, or in other constrained locations are seeing renewed interest and applications. Even if effective vaccines are found and distributed globally, the common opinion is that many (but not all) societal and workplace changes driven by the pandemic may become permanent.

This book assesses the requirements, technologies, designs, solutions, and trends associated with High-Density Communications (HDC). We believe this to be the first book that specifically synthesizes the topic of applied high-density communications. Chapter 1 looks at the functional requirements for high-density communications. Chapter 2 discusses the traditional data/Wi-Fi Internet access, including OTT video. Chapter 3 addresses the traditional voice/cellular design for campus applications, especially the Distributed Antenna System (DAS). Chapter 4 peruses the traditional sensor networks/IoT services approaches. Chapter 5 is the core of this text and examines evolved Wi-Fi hotspot connectivity and related technologies (Wi-Fi 5, Wi-Fi 6, spectrum, IoT, VoWiFi, DASs, microcells issues, 5G versus Wi-Fi issues), as well as intelligent integration of the discrete set of campus/venue networks into a cohesive platform usable in airports, stadiums, convention centers, classrooms, hospitals, and the like.

Chapter 6 starts the discussion on de-densification, using the same kind of technologies discussed in part one of the book; it considers the topic of office social distancing and discusses one of the available technologies. Chapter 7 covers the use of Ultra-Wideband (UWB) technologies. Chapter 8 addresses the office social distancing challenge using Wi-Fi, Bluetooth, and cellular/smartphone methodologies. Chapter 9 provides a use case for HDC systems, and Chapter 10 offers a pragmatic view for some of the economics of broad deployment of HDC.

The book is targeted to networking professionals, technology planners, campus administrators, service providers, equipment vendors, and educators. It is not a research monograph, but rather it aims at integrating the real-world deployment of technologies, strategies, and implementation issues related to delivering an actual working HDC environment in any of the key venues listed above. It is important to note that the composition of this book started in February 2020. While social distancing in the office and public venues was a crucial short-term goal at press time, the business- and public-venue density requirements will likely resurge over time, likely with some yet to be foreseen modifications.

Many books delve extensively on general technologies of all types; however, they fall short in terms of the economics of such technologies, deployment challenges, associated security issues, and most lack tangible case studies. This book addresses these key aspects, based on actual deployment by the team associated with this writing, at a top US airport.

Some portions of this text make use of patent material filed with the United States Patent Office. All inventors cited are implicitly acknowledged for their contribution to this synthesis.

DANIEL MINOLI
DVI Communications

JO-ANNE DRESSENDOFER
Slice Wireless Solutions
30 December 2020

ABOUT THE AUTHORS

DANIEL MINOLI

Mr. Minoli is the principal consultant at DVI Communications. He has published 60 technical telecom and IT books, many are the first in their field (e.g., the first-ever book on VoIP, the first-ever on outsourcing of telecom services, the first-ever book on metro Ethernet, the first-ever book on green networks, the first-ever book on IPv6 security, the first book on public hotspots, and the first book on IPv6 support of IoT, among others); he has also published 340 other papers (the majority of which are peer-reviewed). Many books focus on raw technologies and fail to address Return on Investment (ROI), deployment, security considerations, and to provide case studies; Mr. Minoli's books aim to address these key issues when documenting the applicability of the underlying technologies.

Mr. Minoli started to work on wireless LANs in the late 1970s as part of ARPANet-sponsored R&D and continued wireless work in the form of Geo/Meo satellite transmission, microwave, free space optics, mmWaves/"wireless fiber," cellular, Wi-Fi WLANs, sensor networks, wireless IoT, crowdsensing, 900 MHz SCADA, BMSs, UltraWideband, and 5G. He has written two books on LANs and several long book chapters on WLANs in other books; and, as noted, he has written a book on public hotspots and a book on metroEthernet/VPLS. At press time, over 225 published US patents, as well as 38 US patent applications, cite his work. Additionally, 5917 academic researchers cite his work in their own publications, according to Google Scholar, including 1887 citations of his books on Wireless Sensor Networks, 569 of his books/papers on IoT, 344 of his books on enterprise architectures, 262 of his books on video, and 259 of his books on VoIP. Mr. Minoli is a reviewer for several publishers, including Elsevier, Springer, IEEE, and Wiley. He has taught (adjunct) over 75 college graduate/undergraduate courses at New York University, Stevens Institute of Technology, and Rutgers University. He has been affiliated with Nokia, Ericsson, AT&T, SES, Prudential Securities, Capital One Financial, and AIG, and has been an expert witness/testifying expert in about 20 patent lawsuits. He has undertaken Intellectual Property (IP) work related to patent invalidity, infringement/non-infringement analysis, breach-of-contract, dispute of equipment functionality, and IP portfolio valuation in the area of packet video/IPTV, packet voice/VoIP, networking, imaging (scanned checks), IoT, and wireless. He has provided Court testimony, sustained numerous depositions, and produced numerous Expert Reports, Rebuttal Reports, and Post Grant Review Declarations.

JO-ANNE DRESSENDOFER

Jo-Anne (Josie) Dressendofer is the founder of SliceWiFi. The firm was launched in 2016 to address the rapidly expanding need for fast, reliable Wi-Fi service in permanent and temporary locations. What started as a goal to become the first "Managed Wi-Fi Brand" ended up becoming the first company to compete with the goliath cellular companies, with Wi-Fi and an all-inclusive technology, turning SliceWiFi into a telecommunications company overnight. SliceWiFi initially achieved market recognition in New York City, as one of the leading Wi-Fi providers in the NY metro area, after successfully supporting difficult, densely populated networking environments such as the Javits Center and downtown Brooklyn rebuilding after Hurricane Sandy; NY Fashion Week's many simultaneous event locations; many hackathons with over a thousand users; the Staten Island Ferry during peak travel over the Hudson River; and the parks at Hudson Yards where no fiber was to be had. In 2017, SliceWiFi won *CIO* magazine's category award for "Top Wireless Solution Providers."

Ms. Dressendofer has led a 25-year career in the tech industry, competing aggressively and winning repeatedly against larger, better-financed multi-billion-dollar competitors. Her firms have a record of being more creative with leading-edge technology deployment and networking engineering than all the legacy providers in play. The recent win at BWI Thurgood Marshall Airport (BWI) against major players in the telecommunications industry was transcendent and proof that the SuperNetwork concept (Chapters 9 and 10) is not only a trendsetter but a victory for all women in technology.

ACKNOWLEDGMENTS

In addition to the inventors cited in this work, Mr. Minoli wishes to warmly thank Mr. Benedict Occhiogrosso, President, DVI Communications, for the continued support and input in all the bleeding-edge technologies discussed in this text. DVI Communications, Inc. is a leading and highly respected Information Technology, ICT consultancy, and systems engineering firm with core competencies in IT, ICT, IoT, M2M, wireless, telecom, security, and audiovisual systems. Throughout its 40+ year history, the firm has supported many organizations deploying traditional and emerging technologies, serving both large enterprises and smaller organizations in numerous vertical markets with complex, state-of-the-art systems often working alongside legacy systems, supporting several generations of technology simultaneously.

Ms. Dressendofer wishes to credit and thank the staff of Slice Wireless Solutions, Inc. (SliceWiFi) for the support of this initiative, as described in Chapter 9 and further synthesized in Chapter 10, in the context of designing and deploying a reimagined Thurgood Marshall Airport (BWI) SuperNetwork and the development of WiSNET. The complete redesign and the initial redeployment of the entire BWI Airport terminal-side and some portions of the operations wireless communication infrastructure, amid the COVID-19 pandemic and the span of 12 months, all while maintaining reliable, uninterrupted airport service, was an enormously complex task. Much has been learned at the practical level and is documented in the last two chapters of this book. John Hutzler, COO, and Ed Wright, CTO, have been instrumental in the successful design and completion of this SuperNetwork redeployment mission, even more so as evinced by the relatively small size and the recent debut of SliceWiFi, and this win against the competition backed by billions faced during the RFP process. Without their labor, there would be no SuperNetwork and no chapters to document herewith. Thanks to Cheryl Beck, CMO and Jeffrey Forester, our legal council.

Lastly, to those who were there before SliceWiFi and who without their contribution would never had led down the path of this incredible development. I especially owe that to Morris Williams, Jiamini Erskine, and Ricky Smith of BWI for having the courage to choose a better way not the old way and stay by our side during the tough times, our Nashville investors and investment team, Eddy Wong, my former partner and mentor, Irwin Cohen whose inspiration and endless contacts led me to the incredible support of Jason Zuckerbrod and Jody Westby, and my six nieces who inspired me every day to do more to open doors and make the world a better place for them. Thank you will never be enough for your help in creating a dream this big, against such odds and see it actualized. Dan Minoli you stand alone in genius and my admiration.

1 Background and Functional Requirements for High-Density Communications

This introductory chapter covers two topics: (i) a basic introduction to the underlying technologies and principles that apply to High-Density Communications (HDC), but not high-density specifics, which are covered in the chapters that follow, and (ii) a discussion of the main requirements for HDC in the context of key use cases. Use cases include airports, stadiums, convention centers, classrooms, amusement parks, train and subway stations, large multiple dwelling units, open air special events, and other venues.[1]

As the second decade of the twenty-first century rolled along, however, a new requirement presented itself due to the worldwide pandemic: physical/desk distancing in support of Office Social[2] Distancing (OSD) and Office Dynamic Cluster Monitoring and Analysis (ODCMA). A "de-densification" effort was established at the time. The de-densification effort in the workplace impacts a large number of factors, including network connectivity services and architectures. Propitiously, wireless technologies have been harvested to address and manage these pressing distancing issues. Even if effective vaccines are found and distributed globally, many agree that some of the societal and workplace changes driven by the pandemic may become permanent. One change likely to remain is the increased reliance on Work From Home (WFH) and along with it, are the implications of greater utilization of a global workforce in what might be called Outsourcing 2.0 (with the 1.0 version having taken place in the 1990s and 2000s). However, "the sun will rise again," and in a few years, people-based HDC may yet again become the norm; in the meantime, a large population of Internet of Things (IoT) devices may indeed require HDC support, and during the pandemic, the e-commerce warehouse use case continues to need HDC support. Thus, while "social distancing" was a short-term goal at press time, the business- and public-venue high-density requirements are expected to resurge and/or continue over time. Further discussion of these issues is provided in the latter part of the chapter.

1.1 BACKGROUND

The principal ways people currently communicate (especially when away from home) are via 4G/Long-Term Evolution (LTE) cellular access, for both voice and data, and/or via a public, institutional, or corporate Wi-Fi™ hotspot. In less populated areas and while in motion, cellular access is typically the norm, rather than Wi-Fi access. In large business and commercial

---

[1] The composition of this book started in February 2020. While "social distancing" was a short-term goal at that juncture, the business and public venue high-density requirements will resurge and/or continue over time.
[2] Some (more properly) use or prefer the term "spatial distancing."

High-Density and De-Densified Smart Campus Communications: Technologies, Integration, Implementation, and Applications, First Edition. Daniel Minoli and Jo-Anne Dressendofer.
© 2022 John Wiley & Sons, Inc. Published 2022 by John Wiley & Sons, Inc.

buildings (e.g. skyscrapers, hospitals, hotels), internal systems known as Distributed Antenna Systems (DASs) may be used to provide better signal quality to cellular users; these systems interoperate with the public cellular network in a number of ways. When stationary, both choices may be available.

Cellular services are offered by carriers using specific carrier-allocated Radio Frequency (RF) spectrum. Relatively high monthly fees are incurred; additionally, there may be both physical and administrative limits to the amount of bandwidth and interval-accumulated throughput. Wi-Fi makes use of bands that are freely allocated; services could be free or could be nearly free based on some account subscription arrangement.

There are plusses and minuses with both technologies: a signal associated with a cellular service such as 4G/LTE reaches longer distances and is often the best choice in sparsely populated areas (assuming the service is available); high-speed mobility is supported and roaming between towers (cellular access points) is seamless; the service is typically provided by well-established carriers that have experience with availability and Quality of Service (QoS) metrics; large portions of the United States are covered, and; the session bandwidth is often guaranteed for the session's duration once the session is established. Conversely, the service costs for 4G/LTE are relatively high and there are limits to the user throughput; there is relatively limited practical competition among carriers; large base-station antennas are needed to cover large geographic areas; the technology is complex; indoor reception of voice and data can be problematic, creating the need for more indoor antennas; and 5G will require smaller (therefore, a larger number of) cells. Wi-Fi is often perceived to be free; the technology is simpler; the hardware and infrastructure are cheaper; it is a consistent technology between the office and the home; there is more competition in the sense that various establishments (e.g. stores, coffee shops, malls, libraries, institutions) make Wi-Fi service available. However, the technology is subject to interference; the distance is limited; roaming does not work across different providers and may not even work for a given provider, even within limited geography; congestion can occur, and; QoS is not guaranteed. Nonetheless, both technologies fill a role, and both technologies are clearly needed.

There are several Wireless Local Area (WLAN) standards that have evolved over time, including Institute of Electrical and Electronics Engineers (IEEE) standards 802.11a, 802.11b, 802.11g, 802.11n, 802.11ac, 802.11ax. The new standards have been developed to accommodate the evolving requirements for higher speeds. Some protocols and wireless routers provide backward compatibility with older Wi-Fi systems. The Wi-Fi Alliance (an industry group) has announced a banding "generation" designation, as follows:

- Wi-Fi 4 is 802.11n, released in 2009
- Wi-Fi 5 is 802.11ac, released in 2014
- Wi-Fi 6 is the new version, also known as 802.11ax (scheduled for release in 2019)

Earlier versions of Wi-Fi have not been officially branded, but one could label the previous generations as follows:

- Wi-Fi 1: 802.11b, released in 1999
- Wi-Fi 2: 802.11a, released in 1999
- Wi-Fi 3: 802.11g, released in 2003

Radio technologies in cellular communications have grown rapidly. They have evolved since the launch of analog cellular systems in the 1980s, starting from the First Generation (1G) in the 1980s, Second Generation (2G) in the 1990s, Third Generation (3G) in the 2000s, and Fourth Generation (4G) in the 2010s (including LTE and variants of LTE). Fifth Generation (5G) access networks, which can also be referred to as New Radio (NR) access networks, are currently being deployed and are expected to address the demand for exponentially increasing data traffic and are expected to handle an extensive range of use cases and requirements. Basic use cases include, among others, Mobile Broadband (MBB) and Machine-Type Communications (MTC), for example, involving IoT devices – Machine-to-Machine (M2M) communication is a specific IoT niche. The IoT refers to the network of physical objects with Internet connectivity (connected devices) and the communication between them; these connected devices and systems collect and exchange data. The IoT has been defined as "the infrastructure of the information society"; it extends Internet connectivity beyond traditional devices such as desktop and laptop computers and smartphones to a range of devices and everyday entities that use embedded technology to communicate and interact with the external environment [1]. Massive Multiple Inputs and Multiple Outputs (MIMO) designs, new multiple access methods, and novel channel coding approaches are being assessed for use in 5G and HDC environments [2–7].

The upcoming 5G access networks may utilize higher frequencies (i.e. > 6 GHz) to support increasing capacity by allocating larger operating channels and bands, although some lower frequencies can also be used. Millimeter wave (mmWave), the band of spectrum between 30 and 300 GHz, have shorter wavelengths that range from 10 to 1 mm. Currently, much of the mmWave spectrum is underutilized; thus, it can be used to facilitate the deployment of new high-speed services. While it is known that mmWave signals experience severe path loss, penetration loss, and fading, the shorter wavelength at mmWave frequencies also allows more antennas to be packed in the same physical dimension, which allows for large-scale spatial multiplexing and highly directional beamforming [8].

Some observers have predicted the "death of Wi-Fi" at various points in the recent past. To quote Mark Twain (as told by his biographer Albert Bigelow Paine), "the report of my death has been grossly exaggerated." Ignoring the ALOHAnet of the late 1960s/early 1970s, wireless LANs started to appear in the late 1980s/early 1990s (e.g. with the WaveLAN system originally designed by NCR Systems Engineering/Wireless Communication and Networking Division, available commercially in 1990 and for several years, some concepts eventually making their way into the 1997 IEEE 802.11 standard[3]). The generic technology has thus been around for 30 years. When (some form of) 3G/4G/LTE was starting to be deployed, some predicted that it would be the death knell of (public hotspot) Wi-Fi, but it did not happen. In fact, many devices developed the capability of transferring connectivity and roaming seamlessly between the local Wi-Fi (corporate, public, residential) and cellular service – some users even use their cellular-based smartphone to create a small local hotspot to support traditional Wi-Fi elements in their environment. Now with 5G on the horizon, some are offering the same (questionable) prediction about the future of Wi-Fi [9]. As is the case with many pairs of technologies, one technology moves ahead, the other lagging; then at some point, the second technology makes a quantum leap forward, and the original one lags; then again, the original technology makes a new advancement and leapfrogs the other technology, and so on. One can apply this idea to cellular and Wi-Fi in terms of speed/throughput as well as cost and end-device capabilities. In broad terms, Wi-Fi generally offers higher data rates and service can be cheaper; however, large-geography coverage and large-geography roaming are more "natural" in the cellular context. Another observation is that 5G will often require small cells, implying both a similarity with a Wi-Fi

---

[3]Classic WaveLAN (a pre-802.11 protocol) operated in the 900 MHz or 2.4 GHz ISM bands – pursuant to the publication of the IEEE 802.11 standard in 1997 WaveLAN IEEE, supporting the standard was introduced to the market. In WaveLAN, the radio modem section was hidden from the OS, making the WaveLAN card appear to be a typical Ethernet NIC. WaveLAN laid important foundation for the formation of IEEE 802.11 working group and the resultant creation of Wi-Fi. *Wikipedia, WaveLAN, retrieved 27 January 27 2020.*

hotspot and increased infrastructure and deployment cost. 5G is advocated from the perch of higher speeds, higher density, and reliable connectivity; however, it remains to be seen if these features can be achieved on a large scale (i.e. over a large geographic, national, or international geography) and in a cost-effective manner. The global standard could in theory benefit dispersed IoT sensor support, in a smart city setting, for example, but until recently, the cost of the cellular interface for the sensor tended to be fairly expensive (e.g. in the $20–40 range); thus, the use of other Low Power Wide Area Network (LPWAN) technologies such as LoRa or Sigfox have taken hold. This interface cost must decrease substantially if the use of 5G cellular in IoT applications is to become ubiquitous.

1.2   REQUIREMENTS FOR HIGH-DENSITY COMMUNICATIONS

HDC can be characterized by several (requirement) metrics. Basic metrics include, but are not limited to, user connection density, traffic volume density, experienced data rate, and peak data rate. Many venues require ultra-high connection density and ultra-high traffic volume density; applications that entail M2M and may typically (but not always) require very low end-to-end latency. For example, 5G systems aim at the following key performance indicators: (i) connection density: one million connections per square kilometer; (ii) traffic volume density: tens of Gbps per square kilometer; (iii) user experienced data rate: 0.1–1 Gbps; (iv) peak data rate: tens of Gbps, and; (v) end-to-end latency: 1–10 ms. See Figure 1.1. In addition, there is a need for scalability: it is one thing to have high density in a small area (say, a classroom), and it is another matter to be able to sustain that over a large venue (for example, a stadium or airport). For this discussion, it is assumed that the mobility speed is not a factor: pedestrian rates (≤10 km/h) are assumed.

One million connections per square kilometer (also definable as 1 connection per m^2) equates to one connection every 10 ft^2 (1 km^2 = 10 763 910 ft^2); this is considerably higher than the connectivity goals in an office environment, where typically one has an allocated space of 130–150 ft^2 per worker, with one or two connections per worker; this is also higher than the connectivity in a classroom (say a 40×40 ft locale and 32 students, or one connection every 50 ft^2). Another example could be train cars with 200 users (perhaps not all simultaneously active) in 1000 ft^2, or one connection every 10 ft^2 if only 50% of the passengers are active at any one point in time.

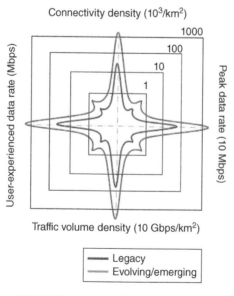

FIGURE 1.1   Requirements bouquet.

TABLE 1.1   Key Performance Indicators HDC Key Performance Indicators (KPIs)

Key Performance Indicators	Description
Connection density	Total number of connected devices per unit area (n/km^2)
User experienced data rate	Minimum data rate for a user in the actual network environment (bps)
Peak data rate	Maximum achievable data rate per user (bps)
Traffic volume density	Total data rate of all users per unit area (bps/km^2)
End-to-end latency	Time lag between the transmission of a data packet from the source and the successful reception at the destination (ms)
Scalability	The ability to retain the above-defined KPIs over large venues and/or geographic areas

In addition to traditional communications, evolving requirements for high-density environments include wearables (for example, in augmented reality applications), M2M, and vehicular traffic in Intelligent Transportation Systems (ITSs) environments. For example, densities of 1 node per m^2 have been identified for augmented reality applications, as with Personal Area Network (PAN) mechanisms [10]. For ITSs, vehicle density has been one of the main metrics used for assessing road traffic conditions: a high vehicle density usually indicates that the road or street is congested [11]; the communication traffic is comprised of beacon signals and user-generated signals. A congested road with stopped vehicular traffic might have, say, 12 cars in an area of 2500 ft^2, or a density of 1 car in about 200 ft^2 – each car could have multiple user sessions. Beyond user counts, the requirements span data rates, as highlighted in Table 1.1; some M2M and process control applications have stringent reliability and latency requirements. Applications such as Ultra HD video Streaming Over The Top (OTT), augmented reality, and online gaming impose challenging requirements on bandwidth and latency; however, these applications are not expected, in the short term at least, to have major deployment in mobile environments, but more so in stationary domiciled environments.

Additional key factors to take into consideration when deploying a state-of-the-art HDC system include spectrum utilization, energy consumption, and infrastructure and endpoint system cost [2]. Spectrum efficiency is measured as the data throughput per unit of spectrum resource per cell or per unit area (bps/Hz/cell or bps/Hz/km^2); energy efficiency is quantified in terms of the number of bits that can be transmitted per unit of energy (bits/J); infrastructure cost efficiency can be defined by the number of bits that can be transmitted per unit cost as computed from network infrastructure amortization/allocation (bits/$); endpoint system costs are clearly the endsystem costs, especially for the air interface and the protocol stack resources, to support a given maximum throughput; applicable to human devices (e.g. smartphones) and M2M systems. Improvements in these metrics of one-to-two orders of magnitude are being sought compared with legacy environments.

A number of use cases follow.

1.2.1   Pre-pandemic/Long-term Requirements for Airports

Table 1.2 identifies some target design parameters for airport applications, including voice, video, data, IoT, IoT-based security (video surveillance), IoT-based automation, and wayfinding. Two characteristics of airports are as follow: (i) people at the airport are in a "slave" situation typically with nothing to do but to use their electronic devices – this is unlike a stadium or a school where other events and occurrences take up some of the person's time, thus likely diminishing the connection time of the individuals; (ii) multiple automation M2M-like tasks may be at play in the airport including baggage handling, wayfinding/mobility/movement, and security. HDC requirements continue to be active, even, or especially, in emergency cases

TABLE 1.2 HDC KPIs for Airports

Key Performance Indicators	Key Performance Indicators	Pre-pandemic Requirements
Data/VoIP connection density, for people on smartphones, laptops, tablets	Data/VoIP connection density, for people on smartphones, laptops, tablets	1 per 20 ft^2 in terminals
	User experienced data rate	10–50 Mbps
	Peak data rate	100 Mbps
	Traffic volume density	5 Gbps per gate area (200 people per gate)
	End-to-end latency	100 ms
	Wayfinding	Throughout airport and in adjacent spaces, garages, car rental locations
	Area of coverage	Entire airport and in adjacent spaces, garages, car rental locations
Traditional telephony on DAS systems	Dialtone	50 Erlangs per gate area (200 people per gate)
	Call length	10 minutes per call
Connection density, IoT devices	Connection density, IoT devices	1 per 10 ft^2 throughout airport
	User experienced data rate	0.384 Mbps
	Peak data rate	0.768 Mbps
	Traffic volume density	100 Mbps per 1000 ft^2 throughout airport and in adjacent spaces, garages, car rental locations
	End-to-end latency	1–10 ms
	Area of coverage	Entire airport and in adjacent spaces, garages, car rental locations

(these requirements were instituted in early 2020 and continued to be active as of press time [12]) – one example of a challenging airport environment even as the pandemic was already raging, is illustrated in Figure 1.2. Typically, the visitor's public airport communication support is completely separate and walled-off from the high-security airport operations networks – the discussion and network design considered in this book focus on the former and not the latter, although similar technologies may be at play. Another characteristic is that, unlike stadiums, there is a nearly continuous requirement for connectivity, especially in large hub airports; stadiums are only used for relatively short periods a few times a week (once, less than once, or a few times a week). In addition to visitors, there are stationary concession businesses in the airport that would often make use of the same network infrastructure as the public network, although some administratively secure slice (for example, separate Virtual LANs [VLANs] would be used).

According to the National Plan of Integrated Airport Systems (NPIAS), there are approximately 19 700 airports in the United States. 5170 of these airports are open to the general public and 503 of them serve commercial flights. A typical gate area is 30 000 ft^2 (which would equate to an area of 40 × 75 ft); however, not all of that space is usable for sojourn (implying that some areas within the 30 000 ft^2 area may have a higher concentration of semi-stationary users). If the busy hour concentration of people is 150 people, then there will be 1 person per 200 ft^2 (a 10 × 20 feet area); however, there may be overcrowding situations where the concentration is comparable to the design goals depicted in Table 1.2. See Table 1.3 for the top 30 airports in the United States. Internationally, the Beijing Capital International Airport (Chaoyang-Shunyi, Beijing,

FIGURE 1.2 A gate area at Fort Lauderdale-Hollywood International Airport is crowded with travelers awaiting Delta flight 1420 to Atlanta Saturday, 14 March 2020. (Courtesy: John Scalzi, Photographer).

China) is the second largest in the world, following the Hartsfield–Jackson Atlanta International Airport, with about 50 million passengers per year as of 2018; Tokyo Haneda Airport (Ōta, Tokyo, Japan) had 41 million passengers; Dubai International Airport (Garhoud, Dubai, United Arab Emirates) had 42 million passengers; and London Heathrow Airport (Hillingdon, London, United Kingdom) had 39 million passengers.

1.2.2 Pre-pandemic/Long-term Requirements for Stadiums

For stadiums, a target of one million connections per square kilometer (also definable as 1 connection per m^2 or one connection every $10\,ft^2$) has been suggested by some researchers [2]. In the bleachers, the density could be high, even multiple individuals (say 2–3) every $10\,ft^2$. Requirements include high-capacity data and video access, IoT automation support, which also includes surveillance. The requirements are generally consistent with Table 1.2, with the coverage extending to parking lots. The services span more tightly defined time intervals (as contrasted to airports), possibly giving rise to a challenge in achieving certain goals for the Return on Investment on the infrastructure and the core-network connectivity. The communication session may span the entire sporting event and a specified interval before and after the event.

A football field encompasses $57\,600\,ft^2$ (1.32 acres) but the bleachers may extend the area of coverage to two acres; the parking lots can cover several acres, but the traffic is sparser. Indoor sporting arenas could be smaller. The largest US stadium is the Michigan Stadium in Ann Arbor, Michigan, that seats about 115 000 spectators – about 10 stadiums in the United States can seat over 100 000 people. There are about 90 football stadiums that seat between 50 000 and 99 999 people, and there are about 50 stadiums that seat between 28 500 and 49 999 people. See Table 1.4. There are many other types of sporting venues (e.g. basketball courts, baseball fields, hockey arenas, soccer fields). Soccer field dimensions are somewhat wider than the regulation American football field, being 100–110 m long and 64–73 m wide.

1.2.3 Pre-pandemic/Long-term Requirements for Convention Centers

A target of one million connections per square kilometer (also definable as 1 connection per m^2 or 1 connection every $10\,ft^2$) appears appropriate. The KPI are comparable to those of Table 1.2 for both people and M2M/IoT functionality. Connectivity is to be supported for both the booth exhibitors (which sometimes can be rather complex) as well as the visiting public. Often there

TABLE 1.3 Top US Airports – Actual and Heuristic Data Shown

Rank (2018)	Airports (Large Hubs)	Major City Served, State	2018 Passengers (in M) (Approx.)	Ave Daily (365 days)	Busy Hour (0.05,0.1,0.1,0.2,0.1,0.2,0.1,0.2,0.05)	Gates	Ave People per Gate at BH
1	Hartsfield–Jackson Atlanta International Airport	Atlanta, GA	52	142100	28420	192	148
2	Los Angeles International Airport	Los Angeles, CA	43	116786	23357	128	182
3	O'Hare International Airport	Chicago, IL	40	109246	21849	191	114
4	Dallas/Fort Worth International Airport	Dallas, TX	33	89865	17973	182	99
5	Denver International Airport	Denver, CO	31	85928	17186	111	155
6	John F. Kennedy International Airport	New York, NY	31	83675	16735	128	131
7	San Francisco International Airport	San Francisco, CA	28	76148	15230	115	132
8	Seattle–Tacoma International Airport	Seattle, WA	25	68204	13641		
9	McCarran International Airport	Las Vegas, NV	24	64809	12962		
10	Orlando International Airport	Orlando, FL	23	63520	12704		
11	Newark Liberty International Airport	Newark/New York, NJ	23	62461	12492		
12	Charlotte Douglas International Airport	Charlotte, NC	22	61051	12210		
13	Phoenix Sky Harbor International Airport	Phoenix, AZ	22	59243	11849		
14	George Bush Intercontinental Airport	Houston, TX	21	57967	11593		
15	Miami International Airport	Miami, FL	21	57603	11521		
16	Logan International Airport	Boston, MA	20	54823	10965		
17	Minneapolis–Saint Paul International Airport	Minneapolis/St. Paul, MN	18	50311	10062		
18	Fort Lauderdale–Hollywood International Airport	Fort Lauderdale, FL	17	48257	9651		
19	Detroit Metropolitan Airport	Detroit, MI	17	47775	9555		
20	Philadelphia International Airport	Philadelphia, PA	15	41879	8376		

21	LaGuardia Airport	New York, NY	15	41259	8252		
22	Baltimore–Washington International Airport[a]	Baltimore/ Washington, MD	13.373	36640	7328	75	98
23	Salt Lake City International Airport	Salt Lake City, UT	12	33503	6701		
24	San Diego International Airport	San Diego, CA	12	33360	6672		
25	Dulles International Airport	Washington, DC, VA	12	31858	6372		
26	Reagan National Airport	Washington, DC, VA	11	31143	6229		
27	Midway International Airport	Chicago, IL	11	29276	5855		
28	Tampa International Airport	Tampa, FL	10	28410	5682		
29	Portland International Airport	Portland, OR	10	26864	5373		
30	Daniel K. Inouye International Airport	Honolulu, HI	9	26242	5248		

Note: during 2020, most airports in the United States experienced a 60% drop in passengers. Travel was expected to improve during the second half of 2021 and beyond.
[a] *Size*: 3596.3 acres. Passenger Terminal: 2.423 million ft^2; 5 concourses (4 domestic, 1 international/swing); 73 jet gates, 2 gates dedicated to commuter aircraft; square footage per gate: 32 306 ft^2.

TABLE 1.4   Largest US Football Stadiums

Rank	Stadium	Seating Capacity	Location
1	Michigan Stadium	115 000	Ann Arbor, Michigan
2	Beaver Stadium	111 000	University Park, Pennsylvania
3	Kyle Field	111 000	College Station, Texas
4	Ohio Stadium	110 000	Columbus, Ohio
5	Neyland Stadium	109 000	Knoxville, Tennessee
6	Rose Bowl	107 000	Pasadena, California
7	AT&T Stadium	105 000	Arlington, Texas
8	Darrell K Royal–Texas Memorial Stadium	104 000	Austin, Texas
9	Tiger Stadium	102 000	Baton Rouge, Louisiana
10	Bryant–Denny Stadium	102 000	Tuscaloosa, Alabama

TABLE 1.5   Top Convention Centers in the United States

Center	Location	Exhibition Space, Approx. (ft^2)	Total Space, Approx. (ft^2)
McCormick Place	Chicago, Illinois	2 700 000	9 000 000
Orange County Convention Center	Orlando, Florida	2,100,000	7 000 000
Georgia World Congress Center (GWCC)	Atlanta, Georgia	1 500 000	4 000 000
Las Vegas Convention Center	Las Vegas, Nevada	2,200,000	3,200,000
New Orleans Morial Convention Center	New Orleans, Louisiana	1,100,000	3,100,000
America's Center	St. Louis, Missouri	500 000	2 700 000
San Diego Convention Center	San Diego, California	600 000	2,600,000
TCF/Cobo Center	Detroit, Michigan	720 000	2 400 000
Walter E. Washington Convention Center	Washington, DC	700 000	2 300 000
Sands Expo and Convention Center	Las Vegas, Nevada	940 000	2 300 000

is also a video broadcasting function among specialized media outlets that may need to be supported. Since visitors are engaged with the goings-on in the exhibit, the connectivity requirements may be somewhat diffused during those time slots. Connectivity may coincide with extended business hours.

Some events comprise both a set of lecture sessions and exhibit sessions. When lecture sessions are underway, the connectivity requirements (specifically, the traffic volume density) may be low or lower; however, when the sessions wrap up, there may be a pulse-shaped traffic requirement where a large number of participants all want to make phone calls or access the Internet.

There are about 310 convention centers in the United States of various sizes, 50 of which have more than 200 000 ft^2 of total space. See Table 1.5 for the top 10 convention centers in the United States. For example, the largest US convention center is the McCormick Place in Chicago, Illinois, with 9 million ft^2 of space and 2.7 million ft^2 of exhibition space. The exhibit space generally tends to be one-half to one-third of the total space.

1.2.4   Pre-pandemic/Long-term Requirements for Open Air Gatherings and Amusement Parks

Networks for public parks are typically designed around public safety and the availability of cellular service; first responder access is important (e.g. in the context of E911). For data and multimedia services, users will typically utilize their smartphones and 4G/LTE cellular

TABLE 1.6   Top Amusement Parks in the United States

Site	2017 Visitors
1. Magic Kingdom, Lake Buena Vista, Florida	20 450 000
2. Disneyland, California	18 300 000
3. Disney's Animal Kingdom, Florida	12,500 000
4. Epcot, Florida	12,200 000
5. Disney's Hollywood Studios, Florida	10 722 000
6. Universal Studios, Florida	10 198 000
7. Disney California Adventure	9 574 000
8. Universal's Islands of Adventure, Florida	9 549 000
9. Universal Studios, Hollywood	9 056 000
10. Knott's Berry Farm, California	4 034 000

connections; however, in some instances, Wi-Fi is available, as in the latter case, and is employed to move users toward food and merchandize concessions, or for geo-fencing applications. A target of one million connections per square kilometer (also definable as 1 connection per m^2 or 1 connection every $10 ft^2$) has been suggested by some researchers [2]. Open air gathering tends to be more "pop up" operations with short-lived operational timeframes; however, the density could be high, even multiple individuals (say 2–3) every $10 ft^2$. Requirements include high-capacity data and video access, and perhaps video surveillance.

A lower target seems appropriate for amusement parks, given that people go to these parks (usually with high entrance fees) for entertainment and less for spending time on personal communication devices. There are about 430 parks and amusement parks in the United States; Table 1.6 identifies the 10 top parks.

1.2.5   Pre-pandemic/Long-term Requirements for Classrooms

Classrooms are in session only for certain hours of the day, of the week, of the seasons. Students may toggle between being online and listening to the teachers. In broad terms, a classroom (say of 40×40 ft and 32 students) would require one connection every $50 ft^2$.

There were 132 853 K-12 schools in the United States in 2015, according to data from the National Center for Education Statistics (NCES). The average public school size is as follows: city: 591 students; suburban: 656 students; and rural: 358 students. Table 1.7 depicts the enrolment in the top 10 districts in the United States.

TABLE 1.7   Enrolments at Largest US Districts

Rank	District Name	State	Enrollment (K)
1	New York City	NY	1100
2	Los Angeles Unified	CA	634
3	Chicago	IL	378
4	Miami-Dade County	FL	357
5	Clark County	NV	327
6	Broward County	FL	272
7	Houston	TX	216
8	Hillsborough County	FL	214
9	Orange County	FL	200
10	Palm Beach County	FL	193

TABLE 1.8 Example of School Demographics (NYC)

Size Category	Number of Classrooms	Number of Offices	Total Building Area (ft^2)	Approximate Number of Sites
Small	50	10	100000	250
Medium	100	15	175000	650
Large	140	25	300000	275
Campus	200	40	450000	100
				1275

A school may have a large number of classrooms, in addition to administrative offices. For example, New York City's Department of Education (DOE) is the largest school system in the United States, serving over 1.1 million children across 1800 schools with 140000+ employees at 1300+ school buildings and 29 administrative sites across New York City. Many sites have multiple schools or administrative offices per building. While individual schools vary greatly in size, a standard set of LAN/WAN equipment, including switches, routers, servers, firewalls, and access points is deployed throughout individual school organizations and shared spaces. These networks provide e-mail, administrative and instructional applications for both wired and wireless devices. Additionally, administrative networks are typically wired and are kept in separate VLANs from instructional networks. Table 1.8 illustrates the approximate size and demographics for New York City DOE School buildings.

In addition to content traffic, there is an increasing need to provide IoT-based functionality such as bathroom sensors for smoking or vaping of substances, Heating, Ventilation, and Air Conditioning (HVAC) operations, and video surveillance.

1.2.6 Pre-pandemic/Long-term Requirements for Train and Subway Stations

While some quote a figure of 6 persons per km^2 in subway stations [2], it is our pragmatic observation that the densities at rush hour are more in line with the parameters of Table 1.2, with concentration of 1 per 10 ft^2 or 1 per 20 ft^2. Table 1.9 provides some information on the subway and rapid transit systems in the United States (about 15 systems in total).

1.2.7 Pre-pandemic/Long-term Requirements for Dense Office Environments

Office space represents a major environment where work is accomplished in the United States and around the world. Data from the Commercial Buildings Energy Consumption Survey indicates that there were 5.6 million commercial buildings in the United States in 2012 (the most

TABLE 1.9 Top Subway and Rapid Transit Systems in the United States

System	Annual Ridership (2018) (M)	Avg. Weekday Ridership (K)	Stations (Approx.)
1. New York City Subway	2629	8765	470
2. Washington Metro	226	764	90
3. Chicago "L"	226	720	145
4. MBTA, Boston	156	510	50
5. BART, Bay Area Rapid Transit	126	417	46
6. SEPTA Philadelphia	94	328	75
7. PATH NJ/NY	92	310	13
8. MARTA, Atlanta	65	206	38

Number of buildings in U.S. (000) by square footage

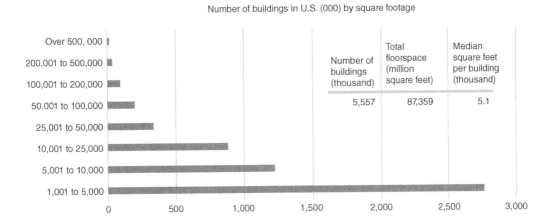

Principal building activity (000)

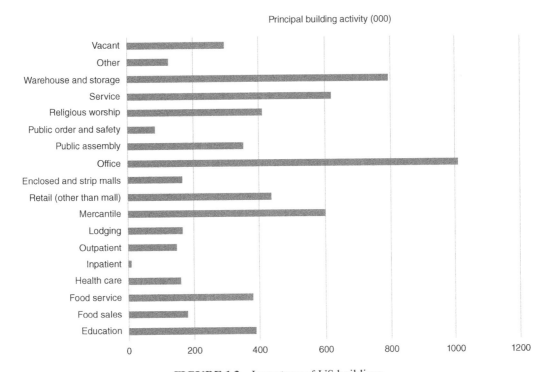

FIGURE 1.3 Inventory of US buildings.

recent year for which data are available), spanning 87.4 billion ft^2 of floorspace (see Figure 1.3) [13, 14]. The typical space allocation per employee is 130–150 ft^2, although some lower-end industries (e.g. retail) allocate less and some higher-end industries (e.g. law firms) allocate more.

Observers call out the need for tens of Tbps per square kilometer for traffic volume density. Using the data just cited, the requirement would be 3 connections per 100 ft^2 (a VoIP connection, a LAN data connection, and a wireless connection) – this equates to 1 per 33 ft^2. Conference rooms typically have higher concentration, say 20 connections in a 400 ft^2 conference room, or 1 per 20 ft^2; this motivates the need for space optimization and tools to address occupancy (for example, using occupancy sensors under desks, utilization sensors in meeting rooms, and comfort sensors in rooms and desk areas). Figure 1.4 depicts a heuristic model for office and conference room space allocations and costs. Requirements span voice, video, data, and IoT applications.

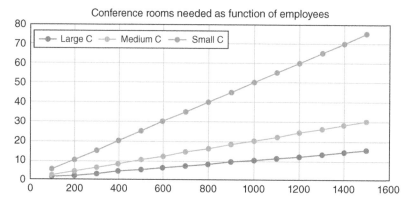

FIGURE 1.4 Heuristic model for office space allocation.

It is expected that the number of IoT sensors will increase with time for smart lighting, smart building, and other automation functions. Thus, it would not be a stretch to reach a point in the near future, where a density of 1 connection per 10 ft² is required for office environments.

1.2.8 Ongoing Requirements for Dense Smart Warehouses and Distribution Centers

As noted in Figure 1.3, there are about 800 000 warehouses and storage buildings in the United States of which, according to data from the U.S. Bureau of Labor Statistics, 18 182 are private warehousing establishments. Warehouses are increasing in size: the average size of warehouses was 182 000 ft² as of 2017. While many of these spaces are "lights off," an increasing number of high-capacity warehouses to support e-commerce are appearing. An increasing number of warehouses and distribution centers are investing in automation and robotics. The communication needs of these venues are very high, some calling these venues "hyper-connected warehouses": wireless technology and real-time inventory tracking are being adopted by nearly all operators using Radio Frequency Identification (RFID) tags attached to each inventory item, barcoding, IoT, and Global Positioning System (GPS) to transmit real-time data to and from the warehouse floor and inventory management applications [15, 16].

1.2.9 Pre-pandemic/Long-term Requirements for Dense Smart Cities

Cities around the globe are investing in IoT-based smart city technologies to undertake data-driven management aimed at improving the administrative and operational processes while ameliorating the quality of life for the residents [17–20]. In 2008, the world's population reached a 50–50 split in the distribution of populations between urban and nonurban environments. At this juncture, one is witnessing an expansion of cities, as populations accelerate the transition from rural and suburban areas into urban areas driven by economic opportunities, demographic shifts, and generational preferences. Seventy percent of the human population is expected to live in cities by the year 2050.

The recent pandemic may impact some of the urbanization trends and dynamics, both on a short-term basis as well as on a longer-term basis. For example, on a short-term basis, ridership in New York City public transportation in March 2020 was down 95%; in August 2020, subway ridership was still down 70% and bus ridership was down 30%. No change in ridership was expected by the authorities at least until early 2021, dependent on the administration of vaccines; face mask mandates and discouraged phone use on public transportation systems. The New York City subway system was shut down 1:00 a.m. to 5:00 a.m. daily for cleaning; bridges and tunnels vehicular traffic into New York City was down 20% from a normal baseline; for NJ Transit, as of beginning of August 2020, ridership was down 65% and authorities were unable to predict when ridership will be back to even 50% [21]. On a longer-term basis, some have taken a completely pessimistic view about city living (perhaps for political reasons) [22–24], while other have taken a more balanced view, as noted in this 2020 quote [25]:

> Tales of Americans fleeing cities in droves, however, are likely overstated. According to … the online real estate service Zillow, 64% of prospective homebuyers on the site are looking at suburban areas – a figure that has barely budged from previous years – while searches for property in rural and urban areas likewise represent about the same percentage as before.

Historically, the largest growth in urban landscapes has been occurring in developing countries. There are now more than 400 cities with over 1 million inhabitants and there are 20 cities with over 10 million people. In most instances, especially in the Western World, cities have aging infrastructures, such as roads, bridges, tunnels, rail yards, and power distribution plants. Some locations have experienced tremendous real estate development in recent years, yet the roads, water mains, sewers, power grids, and sometimes even communication links have seen no, or extremely limited, upgrades. The physical infrastructure that is in place in many cities is aging, and going forward, the services provided by such infrastructure may be subject to temporary rationing as necessary, even emergency upgrades are made. Sometimes, just closing a lane for a few days creates chaotic and dangerous traffic conditions and national headlines.

New technological solutions are being developed to manage the increasingly scarce infrastructure resources, especially in view of the challenges imposed by population growth, limited financial resources, and perennial political inertia. IoT technologies and principles hold the promise of improving the resource management of many assets related to city life, including the flow of goods, the movement of private and public vehicles, and the greening of the environment. Smart Cities application areas include but are not limited to ITSs (including Smart Mobility, vehicular automation, and traffic control), smart grids, smart building, goods and products, logistics (including smart manufacturing), sensing (including crowdsensing and Smart Environments), surveillance/intelligence, and smart services. Cities have been incorporating new technologies over the years, but recently the rate of technology adoption has increased, especially for, but not limited to, surveillance, traffic control, energy efficiency, and street lighting. Many of these IoT technologies require extensive communication infrastructures supporting high density, low latency, and high reliability. For example, ITSs require end-to-end latency in the order of milliseconds as well as high density in congested locations. Dense residence neighborhoods require Gbps-level user experienced data rate [2].

1.3   PANDEMIC-DRIVEN SOCIAL DISTANCING

A novel coronavirus, Severe Acute Respiratory Syndrome Coronavirus 2 (SARS-CoV-2), was identified in December 2019 as the cause of respiratory illness and other morbidities and was designated Coronavirus Disease 2019, or Covid-19.

The Covid-19 pandemic was raging worldwide in 2020. Figure 1.5 depicts illustrative infection and casualty data from the *COVID-19 Dashboard by the Center for Systems Science and Engineering (CSSE)* at Johns Hopkins University (JHU), retrieved on 7 December 2020. Airline travel has been dramatically impacted; Figure 1.6 is a personal photo at Newark Airport, 21 October 2020, at 1:09 p.m.

1.3.1   Best Practices

While medical solutions in the form of diagnostics, prophylactics, therapeutics, treatments, and vaccines were eagerly sought worldwide, two basic mechanisms have been adopted as preventive measures: mask-wearing and social distancing. Social distancing has been applied to a large number of social settings, including work locations, schools, travel, sports events/arenas, entertainment events, church attendance, cruises, and political events, to list just a few.

Social distancing can be seen as being related to the concept of "social density," which is defined in the field of Psychology as "density that can be changed by altering the number of individuals per given unit of space. Spatial density is density that can be changed by altering the amount of space while keeping the number of individuals constant." Social density is a major determinant of crowding when individuals feel that the amount of space available to them is insufficient for their needs [26].

As regions stabilized from the COVID-19 pandemic and stay-at home and WFH restrictions were eased and/or lifted, firms sought to bring workers back into the physical workplace. Many

FIGURE 1.5   Infection and casualty data from the COVID-19 Dashboard by the Center for Systems Science and Engineering at Johns Hopkins University – As of 7 December 2020. At galleys review time (August 27, 2021) the total cases were 214,647,607 and the total deaths were 4,474,716 (Johns Hopkins COVID-19 Dashboard).

FIGURE 1.6 Pandemic impact on airline travel in the Fall of 2020.

organizations endeavored to institutionalize current best practices and protocols to facilitate an orderly return of tenants to buildings. While these measures do not necessarily guarantee the safety or protection from COVID-19, they generally complied with current government regulations and also selectively adopted Industry Best Practices, including suggestions from the World Health Organization (WHO), the Center for Disease Control (CDC), the Building Owners and Managers Association (BOMA), as well as local Department of Health (DOH) (e.g. [27–46]). Practical considerations include:

1. Preparing the site or building: cleaning plans, prereturn inspections, HVAC and Mechanical checks;
2. Control Access: protocols for safety and health checks, building reception, shipping, and receiving, elevators, visitor policies;
   - Occupancy monitoring – building, tenants, visitors;
   - Occupancy monitoring – utilization of escalators, elevators;
   - Pre-authorization coordination with tenants for tenant access (per day);
   - Thermal scanning of visitors and staff.
3. Create a Social Distancing Plan: decreasing density, schedule management, office traffic patterns;
   - Mask – Social Distancing Enforcement – tenants, visitors, maintenance staff, and servicing vendors;
   - Floor markings – 6-ft distancing;
   - (IoT-type) sensor deployment to monitor occupancy, bathroom use, air quality, wellness-deployment options.
4. Reduce touch points and increase cleaning: open doors, clean desk policy, food plan, cleaning common areas.

In the work environment, social distancing takes the form of physical/desk distancing in support of OSD and ODCMA. A 6-ft/2-m goal becomes a *defacto* standard for a large number of venues. In addition, a number of firms have installed higher cubicle partitions, told employees to wear masks when not at their desks, and set up one-way aisles in the office that force people to walk the long way around to get to the kitchen or the bathroom; other firms have instituted staggered shifts (e.g. half the staff comes in on Mondays, Wednesdays, and Fridays, the other half on Tuesdays and Thursdays), deployed spaced-apart desks in open spaces, eliminated break rooms, placed sanitizer in many prominent office locations, and have been subjecting employees to daily questions about their health [47]. The "de-densification" impacts many factors (e.g. policies, economics, profitability, occupancy, real estate utilization, sustainability). For example, as of late Summer 2020, a real estate trade group estimated the occupancy rate for many office towers in downtown Boston or New York at around 5–7%, and 10–30% in the suburbs. Of course, "de-densification" also impacts network connectivity services and architectures; however, wireless technologies, including wireless sensor networks and the IoT, have been advantageously harvested to address and manage these pressing distancing issues (even contact tracing).

Real-Time Location Systems (RTLSs)[4] are wireless systems employed to automatically identify and possibly track the location of objects or people in real time, within a building, or in other constrained or defined spaces (for example, in campuses, in wide-area/regional spaces, or worldwide). These signals are typically electromagnetic: RF systems are common, but some systems use optical/infrared mechanisms (other signals can also be used in confined spaces). Position is derived by measurements of the physical properties of the radio link.

Automatic wireless monitoring is prevalent in a variety of industries, where location and other characteristics of a product or environment must be tightly monitored for safety or regulatory reasons, such as hospitals; other environmental parameters may be significant due to the nature of the items or processes (e.g. manufacturing processes) monitored. Factors such as humidity, exposure to light, motion sensors, positional location within a predetermined area, and orientation of the product (e.g. whether or not it is upright) may be important. RTLSs have thus gained popularity and are now seen as a "mainstream product" [48]; these systems are especially popular in the healthcare industry for applications ranging from asset tracking through patient and staff tracking, environmental or patient sensing (e.g. temperature), hygiene compliance, elopement (i.e. a patient leaving a facility without authorization), and theft prevention. They also have applications in de-densification of the office for OSD/ODCMA use cases.

Conceptually, there are four classifications of RTLS [49]:

- Locating an asset in a controlled area, e.g. warehouse, campus, airport – area of interest is instrumented – accuracy to 10 ft;
- Locating an asset in a very confined area – area of interest is instrumented – accuracy to tens of centimeters;
- Locating an asset via satellite – requires Line-of-Sight (LOS) – accuracy to 30 ft;
- Locating an asset over a terrestrial area using terrestrial mounted receivers over a wide area, cell phone towers, for example – accuracy 600 ft.

RTLSs enable one to identify the location of objects in real time using (i) "tags" attached to or embedded in the objects tracked, and (ii) "readers" that receive and process the wireless

---

[4]Some define the term as "Real-time locating system" while a few others define the term as "Real-time Location Service" or as "Real-Time Location Solution." An exact Google search on the terms depicts the following "Real-time Location System": 407 000 results; "Real-time Locating System": 63 100 results; "Real-time Locating Service": 1360 results; "Real-time Locating Solution": 2260 results.

signals from these tags to determine their locations. Figures 1.7 and 1.8 provide a pictorial view of an RTLS environment [51]. RTLSs may perform passive or active (that is, automatic) collection of location information, and several wireless technologies can be used to establish the

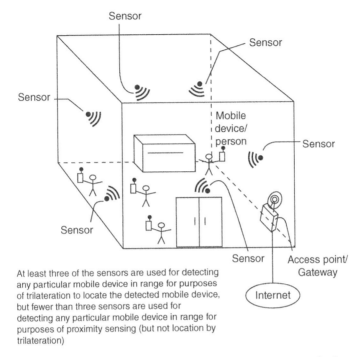

At least three of the sensors are used for detecting any particular mobile device in range for purposes of trilateration to locate the detected mobile device, but fewer than three sensors are used for detecting any particular mobile device in range for purposes of proximity sensing (but not location by trilateration)

FIGURE 1.7 A pictorial view of an RTLS environment [50].

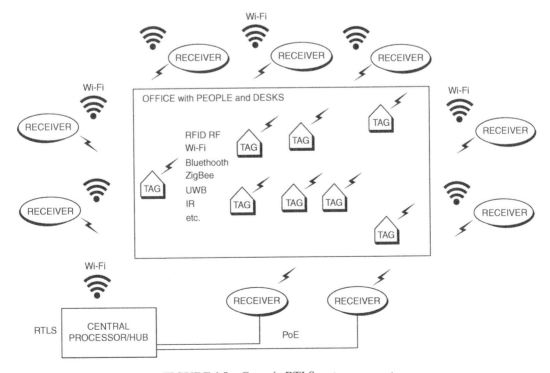

FIGURE 1.8 Generic RTLS system concept.

communication between tags and readers [50]. RTLSs can employ various types of wireless technology, including optical, RF, bar code scanning, ultrasound, global positioning (GPS), WLANs, Ultra-Wideband (UWB), Ultra High Frequency (UHF), Bluetooth™, ZigBee™, Wi-Fi, cellular-based positioning, infrared (IR), or combination thereof.

The method of location is through various types of multilateration, including:

- Angle of Arrival
- Time Difference of Arrival
- Time of Flight Ranging Systems
- Amplitude Triangulation
- Cellular Triangulation
- Satellite Multilateration (for GPS)

When RFID tags are used, two methods of locating an object or people wearing a tag include:

- Locating an asset given that the asset has passed point A at a certain time and has not passed point B;
- Locating an asset by providing a homing signal whereby a person with a handheld can find an asset.

These approaches are discussed later, starting in Chapter 6.

1.3.2   Heuristic Density for the Pandemic Era

HTC in tight spaces pre-pandemic followed an $(L/3)^2$ order of magnitude, where L is the side length of a hypothetical (square) space; post-pandemic, density follows an $(L/6)^2$ model, with a difference of the order of $(L^2/4+L)/3$. See Figure 1.9.

The relative "requirement relaxation" is $(L/3+1)^2/(L/6+1)^2$, or about one-quarter (0.25) the previous requirement (asymptotically with area size). Table 1.10 reinterprets the short-term (2020–2023) requirements for airports; the other venues, except warehouses and IoT sensors, follow the 0.25 heuristic reduction.

1.4   THE CONCEPT OF A WIRELESS SuperNetwork

Institutional (corporate/campus) networks (INETs) have evolved considerably in the past 60 years, which spans the era of data communications. Figure 1.10 depicts a view of this evolution. As the decade of the 2020s dawned, one can assert that we have reached a stage where "super-integrated," feature-rich INETs have emerged (what might be labeled as INET-v6). Such "super integration" in the corporate/campus now covers (i) voice services in the form of VoIP, VoWi-Fi, DASs, and private (virtual) cellular networks (in addition to the public cellular networks edging toward 5G); (ii) extreme reliance of Wi-Fi access in the form of Wi-Fi 6 and also new usable bands; (iii) Virtual Private Networks as the wide-area connectivity of choice; (iv) integration of Building Management Systems (BMS) to support smart building/smart campus functionality along with RTLS and more general IoT functionality; (v) cloud-based services and analytics, also based on the concepts of Software Defined Networks (SDNs) and Network Function Virtualization (NFV); (vi) massive use of videoconferencing as well as (corporate reception where/as needed) of OTT/IPTV video feeds (e.g. business TV); and the widespread use of Artificial Intelligence (AI), Machine Learning (ML), and Deep Learning (DL).

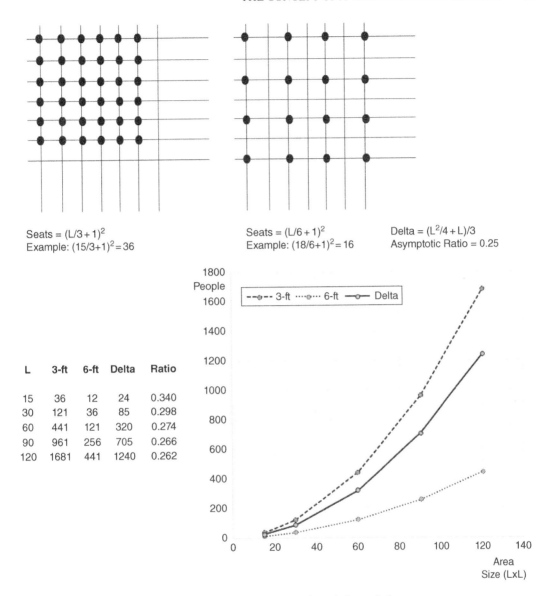

FIGURE 1.9 De-densified (heuristic model).

Herewith we introduce the concept of a Wireless SuperNetwork (WiSNET)™. A super-integrated service-rich INET network where the preponderance of the local and edge connectivity is wireless; such a WiSNET, in addition to the underlying technologies just cited, enjoys, in fact, requires a unified, highly flexible, cost-effective management and administration apparatus. The SuperNetwork embodies a unified, comprehensive, scalable architecture for high-density, high-throughput multimedia communications, supporting open, scalable, and inexpensive technology for secure, QoS-enabled, high-mobility services for a plethora of users having a variety of connectivity and access requirements that span multiple use cases.

This SuperNetwork concept is elaborated at some length in Chapter 10.

The rest of the text will discuss many (but not all) of the constituent technologies that support the realization of such WiSNETs.

TABLE 1.10   HDC KPIs for Airports

Key Performance Indicators	Key Performance Indicators	Pre-pandemic Requirements	Post-pandemic Requirements
Data/VoIP connection density, for people on smartphones, laptops, tablets	Data/VoIP connection density, for people on smartphones, laptops, tablets	1 per $20\,ft^2$ in terminals	1 per $80\,ft^2$ in terminals
	User experienced data rate	10–50 Mbps	Same
	Peak data rate	100 Mbps	Same
	Traffic volume density	5 Gbps per gate area (200 people per gate)	1.25 Gbps per gate area (50 people per gate)
	End-to-end latency	100 ms	Same
	Wayfinding	Throughout airport and in adjacent spaces	Same
	Area of coverage	Entire airport and in adjacent spaces, garages, car rental locations	Same
Traditional telephony on DAS systems	Dialtone	50 Erlangs per gate area (200 people per gate)	12 Erlangs per gate area (50 people per gate)
	Call length	10 minutes per call	Same
Connection density, IoT devices	Connection density, IoT devices	1 per $10\,ft^2$ throughout airport	Same
	User experienced data rate	0.384 Mbps	Same
	Peak data rate	0.768 Mbps	Same
	Traffic volume density	100 Mbps per $1000\,ft^2$ throughout airport and in adjacent spaces, garages, car rental locations	Same
	End-to-end latency	1–10 ms	Same
	Area of coverage	Entire airport and in adjacent spaces	Same

REFERENCES

1. Corbett, J.J., Kjendal, D.L., Woodhead, J. R. et al. (2020). System and method for low power wide area virtual network for IoT. US Patent 10,778, 752, 15 September 2020; Filed 6 May 2019.
2. Liu, G. and Jiang, D. (2016). 5G: vision and requirements for mobile communication system towards year 2020. Chinese Journal of Engineering 2016: 5974586. https://doi.org/10.1155/2016/5974586.
3. Larsson, E.G., Edfors, O., Tufvesson, F. et al. (2014). Massive MIMO for next generation wireless systems. IEEE Communications Magazine 52 (2): 186–195.
4. Nikopour, H., Yi, E., Bayesteh, A. et al. (2014). SCMA for downlink multiple access Of 5G wireless networks. *Proceedings of the IEEE Global Communications Conference (GLOBECOM'14)*, Austin, Texas (December 2014).
5. Li, B., Shen, H., Tse D. et al. (2014). Low-latency polar codes via hybrid decoding. *Proceedings of the 8th International Symposium on Turbo Codes and Iterative Information Processing (ISTC '14)*, Bremen, Germany (August 2014).
6. Zhang, C. and Parhi, K. (2013). Low-latency sequential and overlapped architectures for successive cancellation polar decoder. IEEE Transactions on Signal Processing 61 (10): 2429–2441.

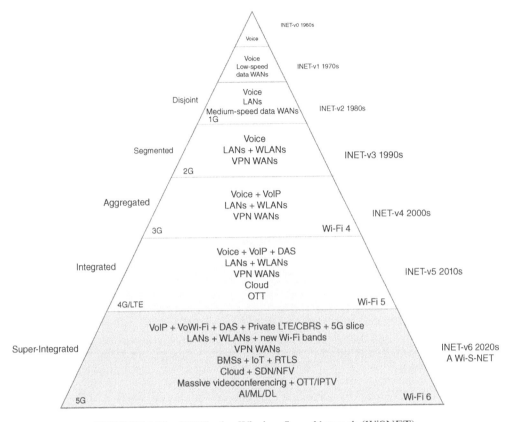

FIGURE 1.10   INET-v6: a Wireless SuperNetwork (WiSNET).

7. Yuan, B. and Parhi, K.K. (2014). Low-latency successive-cancellation polar decoder architectures using 2-bit decoding. IEEE Transactions on Circuits and Systems 61 (4).

8. Ghosh, A. and Nammi, S. (2019). Adapting demodulation reference signal configuration in networks using massive MIMO. US Patent 10,397,052, filed 8 October 2017. Uncopyrighted. https://patents.google.com/patent/US10397052B2/en?oq=10397052.

9. Higginbotham, S. (2020). The long goodbye of Wi-Fi has begun – local 5G networks could replace the familiar wireless standard. IEEE Spectrum, 22 January 2020, https://spectrum.ieee.org/telecom/wireless/the-long-goodbye-of-wifi-has-begun (accessed 5 July 2020).

10. Pyattaev, A., Johnsson, K., Andreev, S. et al. (2015). Communication challenges in high-density deployments of wearable wireless devices. IEEE Wireless Communications 22 (1): 12–18. https://doi.org/10.1109/MWC.2015.7054714.

11. Sanguesa, J.A., Fogue, M., Garrido, P. et al. (2013). An infrastructureless approach to estimate vehicular density in urban environments. Sensors (Basel) 13 (2): 2399–2418. https://doi.org/10.3390/s130202399.

12. Spencer, T. and Crawford, T. (2020). CDC recommends no gatherings of 50 or more for 8 weeks. *The Associated Press*. https://www.cp24.com/world/cdc-recommends-no-gatherings-of-50-or-more-for-8-weeks-1.4853774 (accessed 17 August 2020).

13. *Commercial Buildings Energy Consumption Survey (CBECS), Energy Usage Summary*, U.S. Energy Information Administration, Washington, DC, USA, 2012. http://www.eia.gov/consumption/commercial/reports/2012/preliminary (accessed 17 August 2020).

14. Minoli, D., Sohraby, K., and Occhiogrosso, B. (2017). IoT considerations, requirements, and architectures for smart buildings – energy optimization and next generation building management systems. IEEE Internet of Things Journal 4 (1): 269–283. https://doi.org/10.1109/JIOT.2017.2647881.

15. Rogal, B.J. (2018). Warehouses are getting bigger, and going high-tech. https://www.globest.com/2018/10/01/warehouses-are-getting-bigger-and-going-high-tech/?slreturn=20200205110541 (accessed 17 August 2020).

16. Gomez, J. (2020). 50 Warehouse automation stats you should know. https://6river.com/warehouse-automation-statistics (accessed 17 August 2020).

17. Minoli, D. and Occhiogrosso, B. (2018). Internet of things applications for smart cities, Chapter 12. In: Internet of Things A to Z: Technologies and Applications (ed. Q. Hassan). IEEE Press/Wiley ISBN-13: 978-1119456742.

18. Minoli, D. and Occhiogrosso, B. (2020). Blockchain-enabled fog and edge computing: concepts, architectures and smart city applications. In: Blockchain-Enabled Fog and Edge Computing: Concepts, Architectures and Applications (eds. M. Rehan and M. Rehmani). Boca Raton, FL: CRC Press/Taylor & Francis Group.

19. Minoli, D. and Occhiogrosso, B. (2019). Practical aspects for the integration of 5G networks and IoT applications in smart cities environments", Special Issue titled "Integration of 5G Networks and Internet of Things for Future Smart City". Wireless Communications and Mobile Computing 2019, Article ID 5710834: 30. https://doi.org/10.1155/2019/5710834.

20. Minoli, D. and Occhiogrosso, B. (2018). Ultrawideband (UWB) technology for smart cities IoT applications. *2018 IEEE International Smart Cities Conference (ISC2) – IEEE ISC2 2018- Buildings, Infrastructure, Environment Track*, Kansas City (16–19 September 2018).

21. Saikia, P. (MTA Environmental Sustainability & Compliance), Daleo, E. (NJ Transit Capital Programs) (2020). The post-pandemic commute. CorNet Webinar, 12 August 2020.

22. Altucher, J. (2020). Opinion: New York City is dead forever. *New York Post*, 17 August 2020. https://nypost.com/2020/08/17/nyc-is-dead-forever-heres-why-james-altucher (accessed 18 August 2020).

23. Klein, M. (2020). New Yorkers keep moving out of the city to suburbs, other states. *New York Post*, 11 August 2020. https://nypost.com/2020/08/11/new-yorkers-flee-nyc-in-droves (accessed 18 August 2020).

24. Post Editorial Board (2020). Opinion Editorial, A mad rush for the exits as New York City goes down the tubes. *New York Post*, 11 August 2020. https://nypost.com/2020/08/11/a-mad-rush-for-the-exits-as-new-york-city-goes-down-the-tubes (accessed 18 August 2020).

25. Roberts, J.J. (2020). Are people really fleeing cities because of COVID? Here's what the data shows. *Fortune*, 17 July 2020. https://fortune.com/2020/07/17/people-leaving-cities-coronavirus-data-population-millennials-marriage-families-housing-real-estate-suburbs (accessed 18 August 2020).

26. APA Dictionary of Psychology. https://dictionary.apa.org (accessed 17 August 2020).

27. CDC Coronavirus. https://www.cdc.gov/coronavirus/2019-nCoV (accessed 17 August 2020).

28. CDC General Recommendations on COVID-19. https://www.cdc.gov/coronavirus/2019-ncov/community/office-buildings.html (accessed 17 August 2020).

29. CDC's Guidance For Cleaning And Disinfecting. https://www.cdc.gov/coronavirus/2019-ncov/community/pdf/Reopening_America_Guidance.pdf (accessed 17 August 2020).

30. CDC's Guidance (additional). https://www.cdc.gov/coronavirus/2019-ncov/prevent-getting-sick/cleaning-disinfection.html (accessed 17 August 2020).

31. WHO Coronavirus. https://www.who.int/emergencies/diseases/novel-coronavirus-2019 (accessed 17 August 2020).

32. OSHA Guidance. https://www.osha.gov/Publications/OSHA3990.pdf (accessed 17 August 2020).

33. OSHA Guidance (additional). https://www.osha.gov/SLTC/covid-19/controlprevention.html#health (accessed 17 August 2020).

34. NY State Dept of Health Safety Plan Template for all Businesses. https://www.governor.ny.gov/sites/governor.ny.gov/files/atoms/files/NYS_BusinessReopeningSafetyPlanTemplate.pdf (accessed 17 August 2020).

35. NY State. https://www.nysda.org/page/Coronavirus2020OfficeRe-OpeningResources (accessed17 August 2020).

36. Building/Biological Surface Testing for Coronavirus. https://www.ncbi.nlm.nih.gov/pmc/articles/PMC7141890/pdf/mSystems.00245-20.pdf (accessed 17 August 2020).

37. New York State Association of Counties, County Workforce Re-Entry Guide. 2 June 2020. (accessed 17 August 2020).

38. New York Interim Guidance for Office-Based Work During the COVID-19 Public Health Emergency. https://www.governor.ny.gov/sites/governor.ny.gov/files/atoms/files/offices-interim-guidance.pdf (accessed 17 August 2020).

39. Reopening New York Commercial Building Management Guidelines for Employers and Employees. https://www.governor.ny.gov/sites/governor.ny.gov/files/atoms/files/BuildingManagementSummary Guidance.pdf (accessed 17 August 2020).

40. New York Forward. https://forward.ny.gov (accessed 17 August 2020).

41. New York State Department of Health. Business Re-opening Safety Plan Template (accessed 17 August 2020).

42. Directory of County Health Departments. https://www.nysacho.org/directory (accessed 17 August 2020).

43. NYS Empire State Development Moving New York Forward: Business Reopening. https://esd.ny.gov/nyforward (accessed 17 August 2020).

44. NYS Department of Health COVID-19 Homepage. https://coronavirus.health.ny.gov/home (accessed 17 August 2020).

45. NYS Department of Health (2020). Interim Guidance on Executive Order 202.16 Requiring Face Coverings. https://coronavirus.health.ny.gov/system/files/documents/2020/04/doh_covid19_eo20216e mployeefacecovering_041420.pdf (accessed 17 August 2020).

46. NYS Department of Labor. https://dol.ny.gov (accessed 17 August 2020).

47. Anderson, M. Taller cubicles, one-way aisles: Office workers must adjust. *AP*, 16 August 2020. https://apnews.com/0ceae050cb6b01bd71601515b0fd0d81 (accessed 17 August 2020).

48. Amir, I. (2016). System and method of enhanced RTLS for improved performance in wireless networks. US Patent 9,298,958; 29 March 2016; Filed 2 May 2013.

49. ISO/IEC 24730:2014 Information technology — real-time locating systems (RTLS), Second edition 2014-02-15, specifically ISO/IEC 24730-1:2014(E).

50. Staff, Clarinox Technologies. (November 2009). Real time location systems. https://www.clarinox.com/docs/whitepapers/RealTime_main.pdf (accessed 17 August 2020).

51. Worsfold, G.R. and Girdler, G. (2019). System and method for locating a mobile device. US Patent 10,390,326, 20 August 2019, Filed 5 November 2018.

2 Traditional WLAN Technologies

High-Density Communication (HDC) technologies include the extension of traditional technologies (e.g. Multi-User Multiple-Input[s] And Multiple-Output[s] [MU-MIMO] approaches, also sometimes known as Massive MIMO), as well as new or evolving technologies.[1] In this basic overview chapter, some of the underlying technologies and principles that are applicable to HDC in local environments are surveyed,[2] but the discussion is not focused per se on the high-density specifics, which are covered in the chapters that follow.

2.1 OVERVIEW

The Electrical and Electronics Engineers (IEEE) developed Wireless LAN (WLAN) standards starting in the late 1990s; IEEE 802.11b was ratified in July 1999. The 802.11b Wi-Fi standard had a maximum link speed of 11 Mbps; the 2003 802.11a/g revision increased the speed to 54 Mbps with the introduction of Orthogonal Frequency Division Multiplexing (OFDM) technology. The 2009 802.11n allowed a single stream link to operate up to 150 Mbps. The 2013 802.11ac revision of the standard offered the possibility of link speeds around 866 Mbps on a single spatial stream with wider channels (on 160 MHz) and higher modulation orders (256-QAM) (433 Mbps on smaller channels). While using the maximum number of spatial streams allowed in the standard, this system could support a speed of 6.97 Gbps (these maximum speeds are only achievable in the controlled Radio Frequency [RF] lab environments[3]). 802.11ax, also called High-Efficiency Wireless (HEW), aims at improving the average throughput per user fourfold in dense user environments; additionally, this standard implements multiple mechanisms to support more users with consistent and reliable data throughput in crowded spectrum environments.

---

[1] While this text focuses on the extension of traditional technologies, newer technologies may also become important in the intermediate future (these new technologies include among others: multiple access schemes aimed at increasing the spectrum efficiency, user experienced data rate, system capacity, and connection density). Examples of new technologies include Sparse Code Multiple Access, Multiuser Shared Access, Pattern Division Multiple Access, and Resource Spread Multiple Access; new signal waveforms such as filtered OFDM, window-OFDM, Universal Filtered Multicarrier; novel channel coding such as polar coding and Low-Density Parity Check coding; and software-defined air interface also in conjunction with end-to-end network slicing [1].

[2] Some portions of this chapter are based on reference [2].

[3] These figures refer to the raw PHY speed. The cited 6.97 Gbps figure would need an AP to operate on the 160 MHz channels (e.g. channel 50 or 114); other arrangements (e.g. 80 MHz channel using 8 × 8 MIMO on channels 42, 58, 106, 122, 138, 155, or 171) provide a maximum speed of 3.4 Gbps; however, a more conservative maximum speed is 1.7 Gbps on an 80 MHz channel using 4 × 4 MIMO; observers note that although one can find 4 × 4 Wi-Fi Network Interface Cards for PCs, a much more realistic maximum PHY speed for devices using batteries and not line power is 866 Mbps for an 80 MHz channel with a 2 × 2 client.

High-Density and De-Densified Smart Campus Communications: Technologies, Integration, Implementation, and Applications, First Edition. Daniel Minoli and Jo-Anne Dressendofer.
© 2022 John Wiley & Sons, Inc. Published 2022 by John Wiley & Sons, Inc.

802.11-based WLANs include an infrastructure Basic Service Set (BSS). The BSS provides the basic building-block of the environment and typically includes an Access Point (AP) and one or more associated stations (STAs) – also known as Wireless Nodes (WNs) (or also known as User Equipment [UE] in some other contexts). The AP is a station configured to control and coordinate functions of the BSS. Obviously, a physical environment can include a large number of BSSs. The stations can transmit data to the AP, and in some environments, they can also transmit information to and receive information from another station. A station such as a personal computer or cellular phone may operate as an AP, such as when a cellular phone is configured to operate as a wireless hotspot. The AP can transmit information to a single station, selected from the plurality of stations in the BSS using a single frame, or it can simultaneously transmit information to two or more stations in the BSS using either a single OFDM broadcast frame, a single OFDM MU-MIMO transmission frame, or a single Orthogonal Frequency Division Multiple Access (OFDMA) frame [2]. An AP is a particular type of station; however, for clarity, the term STA is generally used to refer to non-AP stations.

A WLAN device incorporates a Medium Access Control (MAC) and a Physical Layer (PHY) protocol stack at the lower layers and a set of Internet protocols (e.g. as promulgated by the Internet Engineering Task Force [IETF]) at the upper layers. See Figure 2.1. An AP may include

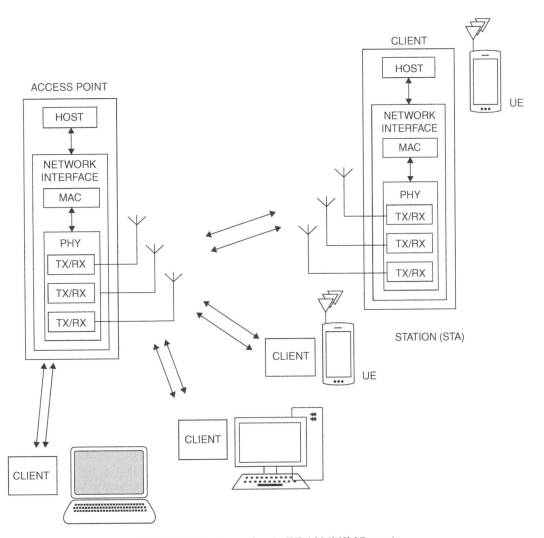

FIGURE 2.1   Example of a WLAN (MIMO case).

TABLE 2.1 Comparison of key features of WLANs/WPANs [3]

	Wi-Fi	ZigBee	Bluetooth
Type	WLAN	WPAN	WPAN
Frequency band	2.4 GHz	2.4 GHz	2.4 GHz
	5 GHz		
	Evolving 6 GHz proposals		
PHY/MAC	Direct Sequence Spread Spectrum (DSSS)	Direct Sequence Spread Spectrum (DSSS)	Frequency Hopping Spread Spectrum (FHSS)
	Multiple In Multiple Out (MIMO)		
	IEEE 802.11n	IEEE 802.15.4	
	IEEE 802.11ac		IEEE 802.15.1
	IEEE 802.11ax		• Classical Bluetooth
			• Bluetooth Low Energy (LE)
			• Bluetooth Basic
			• Rate/Enhanced Data Rate
			• (BR/EDR)
			• Bluetooth Mesh
			• Bluetooth 5.1
Upper layers	TCP/IP stack	ZigBee IP/PRO and various application profiles	TCP/IP stack and/or 6LoWPAN
	Possibly UPnP with TCP/IP, SSDP, SOAP		CoAP
Range	100 m	30 m	10 m
Data rate	High, depends on standard	250 kbps	Depends on version, 1–2 Mbps
Number of nodes in network	32/AP	High (64000)	Low (7)
Transceiver costs	Medium (but depends on Wi-Fi generation)	Very low	Low

a WLAN router function, or it can be a stand-alone AP, a WLAN bridge, or a Light-Weight Access Point managed by a WLAN controller. See Table 2.1 for some preliminary system details. In addition to the infrastructure mode, relatively new "ad hoc" modes (including a form known as Direct Wi-Fi) have emerged.

2.2 WLAN STANDARDS

Standards for WLAN technology have been developed over the years by the IEEE under the 802.11 Work Group. IEEE 802.11™ (generally "802.11") is a set of physical and MAC specifications for implementing WLAN communications. Part 11 is known as "Wireless LAN Medium Access Control (MAC) and Physical Layer (PHY) Specifications" and is maintained and enhanced by the IEEE WG802.11 – Wireless LAN Working Group. These specifications provide the basis for wireless network products using the Wi-Fi brand, managed and defined by the Wi-Fi Alliance. The specifications define the use of the 2.400–2.500 GHz as well as the 4.915–5.825 GHz unlicensed bands. Each spectrum is subdivided into channels with a center frequency and a specified bandwidth. The 2.4 GHz band is divided into 14 channels spaced 5 MHz apart, though some countries regulate the availability of these channels. The 5 GHz band is more heavily regulated than the 2.4 GHz band and the spacing of channels varies across the spectrum, with a minimum of a 5 MHz spacing dependent on the regulations of the respective country of operation [4].

The IEEE 802.11 family of standards has gone through several iterations in recent years. In the 1990s, IEEE 802.11a and 802.11b were developed. IEEE 802.11b provided a transmission rate of 11 Mbps and IEEE 802.11a provided a transmission rate of 54 Mbps. In the 2000s, the IEEE 802.11 g was developed; it provides a transmission rate of 54 Mbps by applying OFDM at 2.4 GHz. To overcome the limits of WLAN communication speed, recent WLAN standards have introduced new schemes for increasing the speed and reliability of a network and extending a management distance of a wireless network. For example, IEEE 802.11n has introduced the standard of MIMO, using multiple antennas for both transmitter and receiver to support high throughput (HT), as depicted graphically in Figure 2.1. IEEE 802.11n provides a transmission rate of 300 Mbps with four spatial streams by applying MIMO-OFDM; the standard also supports a channel bandwidth of up to 40 MHz and thus, provides a theoretical transmission rate of 600 Mbps with four spatial streams. These earlier standards have evolved into IEEE 802.11ac that can utilize a bandwidth of up to 160 MHz and supports a transmission rate of up to about 1 Gbps; it can make use of up to 8 spatial streams. The IEEE 802.11ax standard was under finalization at press time; the standard defines a high-efficiency WLAN for enhancing the system throughput in high-density environments; 802.11ax contemplates dynamically adjusting the energy level when a channel is clear, depending on whether the energy corresponds to its BSS signals or signals from another BSS [5, 6]. Such a scheme helps to promote spatial reuse between neighboring networks.

More broadly, a series of standards have been adopted as the WLAN evolved, including IEEE Std 802.11-2012 (March 2012). This standard was subsequently amended by IEEE Std 802.11ae-2012, IEEE Std 802.11aa-2012, IEEE Std 802.11ad-2012, IEEE Std 802.11ac-2013, IEEE Std 802.11af-2013, IEEE Std 802.11aj-2018, IEEE Std 802.11ak-2018, and IEEE Std 802.11aq-2018 [2, 7, 8, 9]. Table 2.2 provides a list of IEEE 802.11 active projects at press time [10].

2.3 WLAN BASIC CONCEPTS

The transmission processes operate at the PHY layer and the Data Link layer. A WLAN device typically includes a baseband processor, a RF transceiver, an antenna unit, a storage device (e.g. memory), an input interface unit, and an output interface unit. The baseband processor performs baseband signal processing and includes a MAC processor and a PHY processor.

- The MAC processor includes a MAC software processing unit and a MAC hardware processing unit. The PHY processor includes a transmitting signal processing unit and a receiving signal processing unit. The PHY processor implements a plurality of functions of the PHY layer. These functions may be performed in software, hardware, or a combination thereof according to implementation.
- The PHY processor may be configured to generate Channel State Information (CSI), according to information received from the RF transceiver. The CSI may include one or more of an RSSI; a Signal to Interference and Noise Ratio (SINR); a Modulation and Coding Scheme (MCS); and the Number of Spatial Streams (NSS). CSI may be generated for one or more frequency blocks, a sub-band within the frequency block, a subcarrier within a frequency block, a receiving antenna, a transmitting antenna, and combinations of a plurality thereof.

TABLE 2.2   IEEE 802.11 Active Projects at Press Time

802.11 Amendment	Description
P802.11ay – IEEE Draft Standard for Information Technology – Telecommunications and Information Exchange Between Systems Local and Metropolitan Area Networks – Specific Requirements Part 11: Wireless LAN Medium Access Control (MAC) and Physical Layer (PHY) Specifications – Amendment: Enhanced Throughput for Operation in License-Exempt Bands Above 45 GHz	This amendment defines standardized modifications to both the IEEE 802.11 Physical Layers (PHY) and the IEEE 802.11 Medium Access Control layer (MAC) that enables at least one mode of operation capable of supporting a maximum throughput of at least 20 gigabits per second (measured at the MAC data service access point), while maintaining or improving the power efficiency per station.
P802.11ba – IEEE Draft Standard for Information Technology – Telecommunications and Information Exchange Between Systems Local and Metropolitan Area Networks – Specific Requirements Part 11: Wireless LAN Medium Access Control (MAC) and Physical Layer (PHY) Specifications Amendment: Wake-up radio operation	This amendment defines modifications to both the IEEE 802.11 Physical Layer (PHY) and the Medium Access Control (MAC) sublayer for wake-up radio operation.
P802.11 – IEEE Draft Standard for Information Technology – Telecommunications and Information Exchange Between Systems Local and Metropolitan Area Networks – Specific Requirements – Part 11: Wireless LAN Medium Access Control (MAC) and Physical Layer (PHY) Specifications	Technical corrections and clarifications to IEEE Std 802.11 for Wireless Local Area Networks (WLANs), as well as enhancements to the existing Medium Access Control (MAC) and Physical Layer (PHY) functions, are specified in this revision.
P802.11ax – IEEE Draft Standard for Information Technology – Telecommunications and Information Exchange Between Systems Local and Metropolitan Area Networks – Specific Requirements Part 11: Wireless LAN Medium Access Control (MAC) and Physical Layer (PHY) Specifications Amendment Enhancements for High-Efficiency WLAN	This amendment defines modifications to both the IEEE 802.11 Physical Layer (PHY) and the Medium Access Control (MAC) sublayer for high-efficiency operation in frequency bands between 1 and 7.125 GHz.
P802.11az – IEEE Draft Standard for Information Technology – Telecommunications and Information Exchange Between Systems Local and Metropolitan Area Networks – Specific Requirements Part 11: Wireless LAN Medium Access Control (MAC) and Physical Layer (PHY) Specifications – Enhancements for Positioning	This amendment defines modifications to both the IEEE 802.11 Physical Layer (PHY) and Medium Access Control (MAC) sublayer that enable determination of absolute and relative position with better accuracy with respect to the Fine Timing Measurement (FTM) protocol executing on the same PHY-type, while reducing existing wireless medium use and power consumption and is scalable to dense deployments. This amendment requires backward compatibility and coexistence with legacy devices. Backward compatibility with legacy 802.11 devices implies that devices implementing this amendment shall (a) maintain data communication compatibility and (b) support the FTM protocol.

TABLE 2.2   (Continued)

802.11 Amendment	Description
P802.11bb – Standard for Information Technology – Telecommunications and Information Exchange Between Systems Local and Metropolitan Area Networks – Specific Requirements – Part 11: Wireless LAN Medium Access Control (MAC) and Physical Layer (PHY) Specifications Amendment: Light Communications	The scope of this standard is to define one Medium Access Control (MAC) and several Physical Layer (PHY) specifications for wireless connectivity for fixed, portable, and moving stations (STAs) within a local area.
P802.11bc – Standard for Information technology – Telecommunications and information exchange between systems Local and metropolitan area networks – Specific requirements – Part 11: Wireless LAN Medium Access Control (MAC) and Physical Layer (PHY) Specifications Amendment: Enhanced Broadcast Service	The scope of this standard is to define one Medium Access Control (MAC) and several Physical Layer (PHY) specifications for wireless connectivity for fixed, portable, and moving stations (STAs) within a local area.
P802.11bd – Standard for Information technology – Telecommunications and information exchange between systems Local and metropolitan area networks – Specific requirements – Part 11: Wireless LAN Medium Access Control (MAC) and Physical Layer (PHY) Specifications Amendment: Enhancements for Next-Generation V2X	The scope of this standard is to define one Medium Access Control (MAC) and several Physical Layer (PHY) specifications for wireless connectivity for fixed, portable, and moving stations (STAs) within a local area.
P802.11be – Standard for Information technology – Telecommunications and information exchange between systems Local and metropolitan area networks – Specific requirements – Part 11: Wireless LAN Medium Access Control (MAC) and Physical Layer (PHY) Specifications Amendment: Enhancements for Extremely High Throughput (EHT)	This amendment defines standardized modifications to both the IEEE Std 802.11 Physical Layers (PHY) and the Medium Access Control Layer (MAC) that enable at least one mode of operation capable of supporting a maximum throughput of at least 30 Gbps, as measured at the MAC data Service Access Point (SAP), with carrier frequency operation between 1 and 7.250 GHz while ensuring backward compatibility and coexistence with legacy IEEE Std 802.11 compliant devices operating in the 2.4, 5, and 6 GHz bands. This amendment defines at least one mode of operation capable of improved worst-case latency and jitter.

Figure 2.2 illustrates a schematic block diagram of a wireless device at some level of specificity.

Each of the stations and the AP includes a processor and a transceiver and may further include a user interface and a display device. The processor is configured to generate frames to be transmitted through the wireless network, to process frames received through the wireless network, and to execute protocols of the WLAN. The processor performs some or all its functions by executing computer programming instructions stored on a non-transitory computer-readable medium. The transceiver represents a unit that is functionally connected to the processor and designed to transmit and receive frames through the wireless network. The transceiver may be defined using a single component that performs the functions of transmitting and receiving, or two separate components, each performing one of such functions [2]. As noted earlier, a station may typically include a desktop computer, a laptop computer, a tablet PC, a

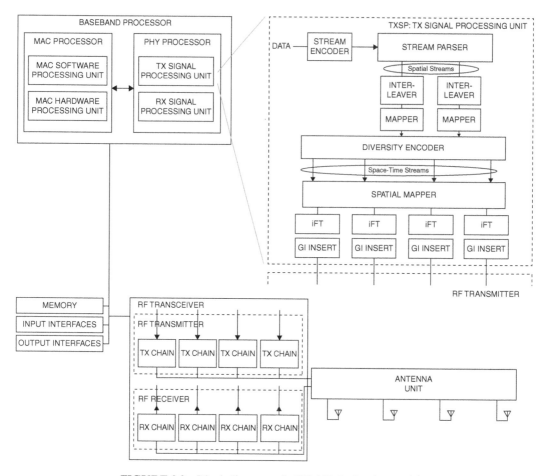

FIGURE 2.2   Block diagram of a WLAN device (example).

wireless phone, a mobile phone, a smartphone, an e-book reader, a Portable Multimedia Player, a portable game console, a navigation system, a digital camera, and so on.

WLAN devices are being deployed in diverse environments; these environments are characterized by the existence of many APs and non-AP stations in geographically limited areas; increased interference from neighboring devices gives rise to performance degradation. Furthermore, WLAN devices are increasingly required to support a variety of applications such as video, cloud access, and cellular network offloading. In particular, video traffic is expected to be a major, if not the dominant type of traffic in many high-efficiency WLAN deployments. With the real-time requirements of some of these applications, WLAN users require improved performance in delivering their applications, including improved power consumption for battery-operated devices [2].

Some of the PHY techniques are discussed first, followed by a discussion of the Data Link layer techniques.

2.3.1   PHY Layer Operation

Traditionally, at the PHY level, the 802.11 protocol uses a Carrier Sense Multiple Access (CSMA)/ Collision Avoidance (CA) channel management method: WNs first sense the channel and endeavor to avoid collisions by transmitting a packet only when they sense the channel to be idle; if the WN detects the transmission of another node, it waits for a random amount of time for that other WN to stop transmitting before sensing again to assess if the channel is free. The process is based on the AP or the WN establishing signal detection energy on a given channel; specifically, the Received Signal Strength Index (RSSI) of the received PLCP (Physical Layer Convergence Protocol) Protocol Data Unit (PPDU)[4]: if the signal detection energy is less than a Clear Channel Assessment (CCA) threshold, the AP or the WN then contends for the channel and transmits its data.

As described, a terminal in a WLAN checks whether a channel is busy or not by performing carrier/channel sensing before transmitting data. Such a process is referred to as *CCA*, and a signal level used to decide whether the corresponding signal is sensed, is referred to as a *CCA threshold*. When a radio signal is received by a terminal, it is processed to determine if it has a value exceeding the CCA threshold. When a radio signal having a predetermined or higher-strength value is sensed, it is determined that the channel under consideration is physically busy, and the terminal delays its access to that channel. When a radio signal is not sensed in the channel under consideration or a radio signal is sensed having a strength smaller than the CCA, then the terminal determines that the channel is idle.

At the PHY layer, the data frame exchanges in WLANs could be performed with a single-antenna transmission or multiple-antenna transmission (MIMO techniques). In general, a MIMO communication system employs multiple (N_T) transmit antennas and multiple (N_R) receive antennas. A MIMO channel formed by the N_T transmit and N_R receive antennas may be decomposed into N_S independent channels, with $N_S \leq \min \{N_T, N_R\}$. Each of the N_S independent channels is also referred to as a spatial subchannel or eigenmode of the MIMO channel. In some environments, MIMO exploits multipath propagation. MU-MIMO is an evolution from the single-user MIMO technology. In WLAN applications, MIMO methods allow the APs and STAs to increase the number of antennas for both transmitting and receiving, thus improving the system capacity for wireless connections. MIMO methods are utilized in IEEE 802.11n, IEEE 802.11ac, and IEEE 802.11ax; they are also used in Evolved High Speed Packet Access (HSPA+)/3G cellular, Long Term Evolution (LTE)/4G cellular, 5G cellular, and WiMAX. Older WLAN standards, such as 802.11b, 802.11g, and 802.11n, do not support MU-MIMO. Initially, only routers and APs supported the technology; now, many endpoint devices support MU-MIMO (including smartphones) support 802.11ac MU-MIMO technology. Figure 2.3 depicts a general MU-MIMO system.

In the case of a multiple-antenna, or MIMO transmission, multiple spatial streams (SS) are sent within the same frame from one station or AP, which usually is called a beamformer (BFer),

---

[4]Layer 1 has two sublayers: the Physical Layer Convergence Procedure (PLCP) and the Physical Medium Dependent (PMD) (note at the same time Layer 2 has two sublayers: the Logical Link Control [LLC] and the MAC). A description of PHY sublayers follows from [11]:

PLCP Sublayer: the MAC layer communicates with the PLCP sublayer via primitives (a set of "instructive commands" or "fundamental instructions") through a Service Access Point (SAP). When the MAC layer instructs it to do so, the PLCP prepares MAC Protocol Data Units (MPDUs) for transmission. The PLCP minimizes the dependence of the MAC layer on the PMD sublayer by mapping MPDUs into a frame format suitable for transmission by the PMD. The PLCP also delivers incoming frames from the wireless medium to the MAC layer. The PLCP appends a PHY-specific preamble and header fields to the MPDU that contain information needed by the PHY layer transmitters and receivers. The 802.11 standard refers to this composite frame (the MPDU with an additional PLCP preamble and header) as a PLCP Protocol Data Unit (PPDU). The MPDU is also called the PLCP Service Data Unit (PSDU) and is typically referred to as such when referencing PHY layer operations. The frame structure of a PPDU provides for asynchronous transfer of PSDUs between stations. As a result, the receiving station's PHY must synchronize its circuitry to each individual incoming frame.

PMD Sublayer: under the direction of the PLCP, the PMD sublayer provides transmission and reception of PHY layer data units between two stations via the wireless medium. To provide this service, the PMD interfaces directly with the wireless medium (that is, RF in the air) and provides modulation and demodulation of the frame transmissions. The PLCP and PMD sublayers communicate via primitives, through an SAP, to govern the transmission and reception functions.

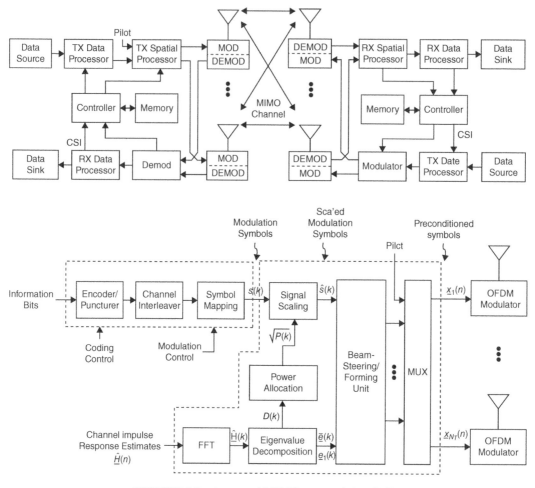

FIGURE 2.3   A general MIMO system (after [12]).

to another station or AP, which is usually called a beamformee (BFee); this type of transmission is called beamforming (BF) transmission. BF and MIMO transmissions are usually enhanced by some initial frame exchanges so that the BFer knows about the MIMO channel conditions. This initial exchange of frames before the actual data frame exchange is called a sounding procedure. The frames that might be used in a sounding procedure are (i) the HT and very high throughput (VHT) Null Data Packet (NDP) frames, (ii) the VHT MIMO Compressed Beamforming Report frame, (iii) the VHT NDP Announcement (NDPA) frame, and (iv) the VHT Beamforming Report Poll frame. Each of these frames may have various fields and subfields such as: VHT MIMO Control, VHT Compressed Beamforming Report, MU Exclusive Beamforming Report, Sounding Dialog Token, STA Info, and related fields that are utilized for exchanging information relevant to beamforming [4].

In 802.11ac standards, MU-MIMO is employed for the downlink (DL). MU-MIMO is a technology that enables a plurality of signals to be transmitted in the same time slot by space division multiplexing; with this technology, it is possible to improve utilization efficiency of the spectrum [13]. MU-MIMO is used in WLANs to support congested environments where many users are trying to access the wireless network at the same time. When multiple users (e.g. smartphones, tablets, computers) begin accessing the AP simultaneously, or practically simultaneously, congestion can arise as the AP services the first user's request, while the second (and subsequent) users are forced to wait. MU-MIMO mitigates this situation by allowing multiple users to access AP functions, thus reducing congestion. MU-MIMO systems segment the available bandwidth into separate, discrete streams that share the connection equally. An MU-MIMO AP may have 2×2, 3×3, 4×4, or 8×8 variations, where the designation refers to the number of streams (two, three, four, or eight) that are created by the AP. To obtain the desired improvements, the AP must enable MU-MIMO and beamforming functionality. The streams are spatial in nature; while interference is minimized if two devices are in proximity to each other, they still share the same stream. Prior to 802.11ax, the MU-MIMO procedure only applies to DL connections; this may be fine for home users that need faster downloading speeds for 4K video content, but it is less useful for business users who need faster uploads for two-way high-quality video conferencing applications [14]. IEEE 802.11ax supports bidirectional MU-MIMO. In addition to allowing 8×8 arrays, the 802.11ax standard addresses the use of uplink (UL) MU-MIMO.

The PHY entity is based on OFDM or OFDMA. OFDM is a type of digital modulation that uses frequency-division multiplexing (FDM) principles. The method subdivides an RF channel into a large number of contiguous subchannels to provide reliable high-speed communications. All subcarrier signals within a subchannel are orthogonal to one another. The subcarrier frequency signals that are being modulated are selected such that the subcarriers are orthogonal to each other whereby cross-talk between the subchannels is minimized or eliminated; note that inter-carrier guard bands are not needed. OFDM transmitters and receivers are relatively simple; in particular, a separate filter for each of the subchannel is not required. Modulation is achieved by encoding signals on multiple carrier frequencies. In this scheme, multiple closely spaced orthogonal subcarrier signals with minimally overlapping spectra are transmitted such that they can carry information in parallel. Each individual subcarrier signal can be modulated with a traditional modulation scheme at a low symbol rate, for example, using Quadrature Amplitude Modulation (QAM). Demodulation utilizes Fast Fourier Transform (FFT) methods. While the total data rates in an OFDM scheme are generally similar to conventional single-carrier modulation in the same aggregate bandwidth, the key advantage of OFDM over single-carrier schemes is its ability to function in environments with challenging channel conditions, for example, with attenuation of high frequencies components in a cable, and with channel interference including fading due to multipath reflections. OFDM is broadly deployed for wideband digital communication, including digital television, Digital Subscriber Line (DSL) internet access, wireless networks, and 4G/5G mobile communications.

OFDM, in its basic form, is a digital modulation technique being employed for transferring a data stream from a single user over an aggregate communication channel, utilizing a sequence of OFDM symbols. Nonetheless, OFDM can be combined with multiple access techniques to support multiple users utilizing time, frequency, or coding separation of the various users. In OFDMA, Frequency-Division Multiple Access (FDMA) is achieved by assigning different OFDM subchannels to different users. IEEE 802.11ax WLANs utilize OFDMA for high-efficiency and simultaneous communication; OFDMA is also used in wide-area applications including but not limited to WiMAX, 3GPP LTE 4G mobile broadband standard DL, and the 3GPP 5G NR (New Radio) fifth-generation mobile network standard for the DL and for the UL.

In either OFDM or OFDMA PHY layers, a station is capable of transmitting and receiving PPDUs that are compliant with the mandatory PHY specifications. A PHY specification defines a set of MCSs and a maximum number of spatial streams. Some PHY entities define DL and UL Multi-User (MU) transmissions having a maximum number of STSs per user and employing up to a predetermined total number of STSs. A PHY entity may provide support for 20, 40, 80, and 160 MHz contiguous channel widths and support for an $80 + 80$ MHz noncontiguous channel width. Each channel includes a plurality of subcarriers, which may also be referred to as tones. A PHY entity may define fields denoted as Legacy Signal (L-SIG), Signal A (SIG-A), and Signal B (SIG-B) within which some necessary information about PHY Service Data Unit

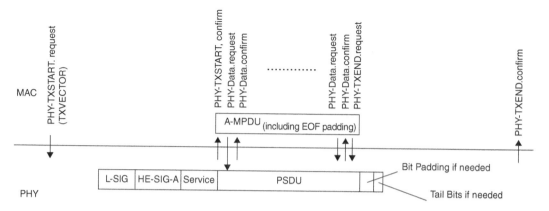

FIGURE 2.4 Example of PHY transmit procedure [2].

(PSDU) attributes are communicated. For example, a High-Efficiency (HE) PHY entity may define an L-SIG field, a HE Signal A (HE-SIG-A) field, and an HE Signal B (HE-SIG-B) field. In the IEEE Std 802.11ac, SIG-A and SIG-B fields are called VHT SIG-A and VHT SIG-B fields. In IEEE Std 802.11ax, SIG-A and SIG-B fields are, respectively referred to as HE-SIG-A and HE-SIG-B fields. See Figure 2.4.

In some of the 802.11 standards, such as 802.11ah and beyond, the identity of the BSS (e.g. as managed by an AP of the BSS) is indicated in a PPDU by a set of bits that described the "color" of the BSS. The color of a BSS corresponds to an identifier (ID) of the BSS that is shorter than the Basic Service Set Identifier (BSSID) defined by 802.11. The BSS color may be contained in the PHY Signal (SIG) field in a PHY header of a PPDU, whereas the BSSID is typically included in a MAC portion of PPDUs. A device (e.g. an AP or client) in a BSS can determine whether a PPDU is from the BSS to which the device belongs (the "same-BSS") or some other BSS (e.g. an overlapping BSS [OBSS]), by decoding the SIG field and interpreting BSS color bits included therein [6].

Machine To Machine (M2M) communication technology has been discussed as a next-generation communication technology. The technological standard for supporting M2M communication in an IEEE 802.11 WLAN system has been developed as IEEE 802.11ah (other standards or recommendations have been advanced by ETSI). Regarding M2M communication, a scenario of occasionally communicating a small amount of data at low speed in an environment in which numerous devices are present is considered. Communication in a WLAN system is performed in a medium shared by all devices. When the number of devices is increased like M2M communication, there is a need to enhance a channel access mechanism more effectively to reduce unnecessary power consumption and interference [15].

2.3.2 MAC Layer Operation

IEEE 802.11 defines a data frame exchange process that enables the stations and APs, to negotiate the timing of the exchange of data between devices over the various shared channels in the 2.4 and 5 GHz bands. In WLAN systems using the IEEE 802.11 standards, frames exchanged between stations (including APs) are classified into management frames, control frames, and data frames. The management frame is a frame used for exchanging management information that is not forwarded to higher layers of a communication protocol stack. The control frame is a frame used for controlling access to the transmission medium. The data frame is a frame used for transmitting data that will be forwarded to higher layers of the communication protocol stack [2]. Each frame's type and subtype are identified using a type field and a subtype field included in a control field of the frame, as described in the applicable standard.

Traditionally, at the frame level, Wi-Fi WNs use Request to Send/Clear to Send (RTS/CTS) to take advantage of the shared medium. The AP only issues a CTS packet to one WN at a time. When the WN receives the CTS, it sends its entire frame to the AP; the WN then waits for an acknowledgment (ACK) frame from the AP indicating that it received the packet correctly; if the WN does not get the ACK in a specified amount of time, it postulates that the packet in fact collided with some other WN transmission – at that point the WN transitions into a period of binary exponential backoff; it will try to access the medium and retransmit its packet after the backoff time expires.

Clearly, data are transmitted using MAC framing and channel management mechanisms along with PHY resources. As alluded to earlier, at the MAC layer, the following frames are utilized:

- A data frame is used for the transmission of data forwarded to a higher protocol layer. The WLAN device transmits the data frame after performing backoff if a Distributed Coordination Function (DCF) Inter-Frame Space (IFS) (known as DIFS) interval has elapsed, during which such DIFS interval, the medium has been idle.
- A management frame is used for exchanging management information that is not forwarded to a higher protocol layer. Subtype frames of the management frame include a beacon frame, an association request/response frame, a probe request/response frame, and an authentication request/response frame.
- A control frame is used for controlling access to the medium. Subtype frames of the control frame include a RTS frame, a CTS frame, and an ACK frame.

IFSs are waiting periods between transmission of frames operating in the MAC sublayer. These waiting periods are used to prevent collisions as defined in IEEE 802.11-based WLAN standards; they represent the time period between completion of the transmission of the last frame and starting transmission of the next frame, apart from the variable backoff period. These are techniques used to prevent collisions as defined in IEEE 802.11-based WLAN standard. Specifically, IFS is the time period between the completion of the transmission of the last frame and the start of transmission of the next frame, apart from the variable backoff period [16]. The list that follows enumerates the different types of IFSs starting from the shortest duration (highest priority) to the longest duration (lowest priority):

- Reduced Inter-frame Space (RIFS)
- Short Inter-frame Space (SIFS)
- Point Coordination Function (PCF) Inter-frame Space (PIFS)
- Distributed Coordination Function (DCF) Inter-frame Space (DIFS)
- Arbitration Inter-frame Space (AIFS)
- Extended Inter-frame Space (EIFS)

The DCF is a required technique utilized to prevent collisions in 802.11-based WLANs. When using the DCF, a station is required to sense the status of the wireless channel before it can place its request to transmit a frame. DIFS is the time interval that a station must wait before it sends its request frame. SIFS is the time interval required by a WLAN device between receiving a frame and responding to the frame. See Table 2.3.

Figure 2.5 illustrates IFS relationships; the figure illustrates a SIFS, a PIFS, a DIFS, and an AIFS corresponding to an Access Category (AC) "i" (AIFS[i]) [2].

Before making a transmission, a WLAN device assesses the availability of the wireless medium using a CCA procedure. If the medium is occupied, CCA establishes that it is busy, while if the medium is available, CCA determines that it is idle. A WLAN device performs a backoff

TABLE 2.3   Inter-frame Space Types

Inter-frame Space Type	Description
Inter-frame Space (IFS)	The time period between completion of the transmission of the last frame and starting transmission of the next frame. Various types of waiting times Distributed Coordination Function (DCF) scheme are listed below (these values typically depend on the standard, e.g. 802.11n, 802.11 ac).
Reduced Inter-frame Space (RIFS)	A very short-duration IFS spacing that has been used to send a burst of high-priority frames. When a station needs to send multiple frames, RIFS is introduced between the individual frames to ensure that no other station finds an opportunity to occupy the channel within the frame burst. The use of RIFS is obsoleted from 802.11ac amendment onward (for compatibility, it is now set to zero).
Short Inter-frame Space (SIFS)	The time interval required by a station that runs between receiving a frame and responding to the frame. It is used in the DCF scheme, which is the baseline technique used to prevent collisions. It is the IFS spacing maintained before and after the transmission of an acknowledgment frame and Clear To Send (CTS) frame. For example, in 802.11ac the SIFS is 16 μs.
Point Coordination Function (PCF) Inter-frame Space (PIFS)	An IFS spacing used in the DCF. In coordinating the communications centrally, the AP waits for PIFS duration to acquire the channel. Because PIFS is less than the DIFS duration, the AP always has the priority to access the channel over the other stations. For example, in 802.11ac, the slot time is 9 μs and the PIFS is 25 μs.
Distributed Coordination Function (DCF) Inter-frame Space (DIFS)	The time interval that a station should wait before it sends its request frame: with DCF, a station needs to sense the status of the wireless channel before it can place its request to transmit a frame. The following relationship holds: DIFS = SIFS + 2 x Slot Time. For example, in 802.11ac, the slot time is 9 μs and the DIFS is 34 μs; as noted above, SIFS was 16 μs, that the resulting DIFS value of 34 μs.
Arbitration Inter-frame Space (AIFS)	Timing to support stations that need to operate in a prioritized manner based upon the Access Category (e.g. for video or voice traffic). Here the waiting period (that is, the AISF) of a station is shortened or expanded before the station can transmit its frame: higher priority stations are assigned shorter AISF (a higher priority station has to wait for a shorter time interval before it can transmit its frame).
Extended Inter-frame Space (EIFS)	An additional waiting period used in case of corrupted frames: If a previously received frame contains an error, then a station must defer EIFS duration instead of DIFS before transmitting a new frame.

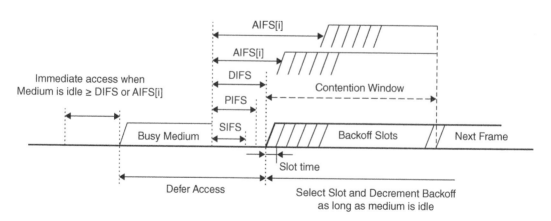

FIGURE 2.5   Inter-frame Space relationships [2].

procedure when the WLAN device that is ready to transfer a frame finds the medium busy. In addition, a WLAN device operating according to the IEEE 802.11n and 802.11ac standards performs the backoff procedure when the WLAN device infers that a transmission of a frame by the WLAN device has failed. The backoff procedure includes determining a random backoff time composed of N backoff slots, each backoff slot having a duration equal to a slot time and N being an integer number greater than or equal to zero. The backoff time may be determined according to a length of a Contention Window (CW). All backoff slots occur following a DIFS or EIFS period during which the medium is determined to be idle for the duration of the period. When the WLAN device detects no medium activity for the duration of a particular backoff slot, the backoff procedure decrements the backoff time by the slot time. When the WLAN determines that the medium is busy during a backoff slot, the backoff procedure is suspended until the medium is again determined to be idle for the duration of a DIFS or EIFS period. The WLAN device may perform transmission or retransmission of the frame when the backoff timer reaches zero. The backoff procedure operates so that when multiple WLAN devices are deferring and execute the backoff procedure, each WLAN device may select a backoff time using a random function, and the WLAN device selecting the shortest backoff time may win the contention, reducing the probability of a collision [2]. When the control frame is not a response frame of another frame, the WLAN device transmits the control frame after performing backoff if a DIFS has elapsed, during which DIFS, the medium has been idle. When the control frame is the response frame of another frame, the WLAN device transmits the control frame after a SIFS has elapsed without performing backoff or checking whether the medium is idle.

A WLAN device that supports a Quality of Service (QoS) functionality may transmit the frame after performing backoff if an AIFS for an associated Access Category (AC) (AIFS[AC]), has elapsed. When transmitted by the QoS station, any of the data frame, the management frame, and the control frame, which is not the response frame, may use the AIFS[AC] of the AC of the transmitted frame.

Figure 2.6 illustrates a CSMA/CA-based frame transmission procedure for avoiding collision between frames in a channel. The figure shows a first station, STA1 transmitting data, a second station, STA2 receiving the data, and a third station, STA3 that may be located in an area where a frame transmitted from the STA1, a frame transmitted from the second station STA2, or both can be received. STA1 may determine whether the channel is busy by carrier sensing. The STA1 may determine the channel occupation based on an energy level in the channel or autocorrelation of signals in the channel or may determine the channel occupation by using a Network Allocation Vector (NAV) timer. After determining that the channel is not used by other devices (that is, that the channel is IDLE) during a DIFS (and performing backoff if required), STA1 may transmit an RTS frame to STA2. Upon receiving the RTS frame, after a SIFS, STA2 may transmit a CTS frame as a response of the RTS frame. If Dual-CTS is enabled and the second station STA2 is an AP, the AP may send two CTS frames in response to the RTS frame: a first CTS frame in the legacy non-HT format and a second CTS frame in the HT format.

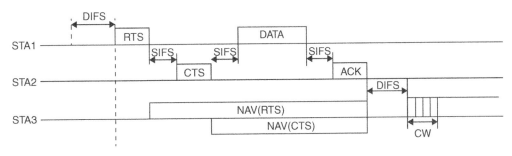

FIGURE 2.6   Carrier Sense Multiple Access/Collision Avoidance-based frame transmission procedure.

When the third station STA3 receives the RTS frame, it may set a Network Allocation Vector (NAV) timer of the third station, STA3 for a transmission duration of subsequently transmitted frames (for example, a duration of SIFS+CTS frame duration+SIFS+data frame duration+SIFS+ACK frame duration) using duration information included in the RTS frame. When the third station STA3 receives the CTS frame, it may set the NAV timer of the third station STA3 for a transmission duration of subsequently transmitted frames using duration information included in the CTS frame. Upon receiving a new frame before the NAV timer expires, the third station STA3 may update the NAV timer of the third station STA3 by using duration information included in the new frame. The third station STA3 does not attempt to access the channel until the NAV timer expires. When STA1 receives the CTS frame from the second station STA2, it may transmit a data frame to the second station STA2 after SIFS elapses from a time when the CTS frame has been completely received. Upon successfully receiving the data frame, the second station STA2 may transmit an ACK frame as a response of the data frame after SIFS elapses. When the NAV timer expires, the third station STA3 may determine whether the channel is busy using carrier sensing. Upon determining that the channel is not used by other devices during a DIFS after the NAV timer has expired, the third station STA3 may attempt to access the channel after a CW according to a backoff process elapses [2].

When Dual-CTS is enabled, a station that has obtained a transmission opportunity (TXOP) and that has no data to transmit may transmit a CF-End frame to cut short the TXOP. An AP receiving a CF-End frame having a BSSID of the AP as a destination address may respond by transmitting two more CF-End frames: a first CF-End frame using STBC and a second CF-End frame using non-STBC. A station receiving a CF-End frame resets its NAV timer to 0 at the end of the PPDU containing the CF-End frame.

Figure 2.6 also shows the second station STA2 transmitting an ACK frame to acknowledge the successful reception of a frame by the recipient.

2.4 HARDWARE ELEMENTS

Figure 2.2, called out above, further illustrates components of a wireless device configured to transmit data, including a Transmission (Tx) Signal Processing Unit (TxSP), an RF transceiver, an antenna unit, and four illustrative antennas. The TxSP, RF transmitter, and antenna unit may be components of the transmitting signal processing unit, RF transmitter, and antenna unit of the WLAN device. Each spatial stream needs its own dedicated transmit/receive chain; for example, 802.11ac 8×8 AP capable of supporting all eight spatial streams needs eight independent radio chains and antennas.

The RF transceiver includes an RF transmitter and an RF receiver. The RF transceiver is configured to transmit information received from the baseband processor to the WLAN, and provide information received from the WLAN to the baseband processor. The antenna unit includes one or more antennas; when MIMO or MU-MIMO is used, the antenna unit may include a plurality of antennas [2].

The TxSP includes a stream encoder; a stream parser; first and second interleavers; first and second mappers; a diversity encoder; a spatial mapper; in this example, a first to fourth inverse Fourier Transformers (iFTs), and in this example, a first to fourth Guard Interval (GI) inserters.

The stream encoder receives and encodes data. The stream encoder includes a Forward Error Correction (FEC) encoder. The FEC encoder may include a Binary Convolutional Code (BCC) encoder, a Low-Density Parity-Check (LDPC) encoder, or one or more combinations thereof. There are two types of coding, Convolutional Coding (CC) and Block Coding (BC). CC is state machine-based, while BC is an algebra-based approach. In WLAN, BCC has been the

TABLE 2.4   Basic Error Coding Schemes

Type	Approach	Method	Timeframe	Application	Feature
Convolutional	State Machine	BCC Viterbi	1960	WLAN (11a/g)	Simple and widely used
		Turbo	1993	3G, 4G	Iterative decoder (close to Shannon limit)
Block Code	Algebra	Hamming	1950	Computer memory	Simple Detect up to 2 simultaneous bit error and can correct 1 bit
		Reed-Solomon	1960	CD/MP3, Satellite, DVB	Widely used in digital storage and communication
		LDPC	1962/1996	WLAN(11n~) 5G NR	Low density and complexity
		Polar	2009	5G NR	

mandatory coding method. In terms of performance, LDPC is better than BCC, but complexity is much higher. LDPC is used from 11ax onward. See Table 2.4, loosely based on reference [17].

The stream parser is configured to divide outputs of the encoder into one or more spatial streams. The stream parser may allocate consecutive blocks of bits to the one or more spatial streams in a round robin fashion. The blocks of bits typically have a length according to number of bits on an axis of a constellation point of a modulation and coding scheme, such as the length being 2 bits for 16-QAM, 3 bits for 64-QAM, 4 bits for 256-QAM [2]. The respective bits of the first and second spatial streams are interleaved by first and second interleavers when BCC encoding is used.

The first and second mappers map the sequence of bits of the first and second spatial stream to first and second sequences of constellation points, respectively. A constellation point may include a (mathematical) complex number representing an amplitude and a phase. Within each of the first and second sequences of constellation points, the constellation points are divided into groups. Each group of constellation points corresponds to an OFDM symbol to be transmitted, and each constellation points in a group correspond to a different subcarrier in the corresponding OFDM symbol [2]. The diversity encoder is configured to spread the constellation points from the spatial streams into a plurality of space–time streams in order to provide diversity gain.

In Figure 2.2, the diversity encoder is shown mapping two spatial streams into four space–time streams (the Number of Spatial Streams N_{ss} is equal to 2 and the Number of Space–Time Streams [STS] N_{STS} is equal to 4). Each space–time-stream corresponds to a different transmitting antenna or a different beam of a beamformed antenna array. The diversity encoder spreads each input constellation point output by the mappers onto first and second output constellation points. The first output constellation point is included in a first space–time stream and the second output constellation point is included in a second space–time stream, different from the first space–time stream. The first output constellation point has a value corresponding to a value of the input constellation point, and the second output constellation point has a value corresponding to a complex conjugate of the value of the input constellation point or to a negative of the complex conjugate (i.e. a negative complex conjugate) [2]. The first output constellation point is at a different time slot (that is, in a different OFDM symbol period) than the second output constellation point when Space–Time Block Coding (STBC) is used. The first output constellation point is at a different frequency (that is, transmitted using a different subcarrier) than the second output constellation point when Space-Frequency Block Coding (SFBC) is used.

The spatial mapper maps the space–time streams to one or more transmit chains. The spatial mapper maps the space–time stream to the transmit chains using a one-to-one correspondence when direct mapping is used. The spatial mapper maps each constellation point in each space–time stream to a plurality of transmit chains when spatial expansion or beamforming is used. Mapping the space–time streams to the transmit chains may include multiplying constellation points of the space–time streams associated with an OFDM subcarrier by a spatial mapping matrix associated with the OFDM subcarrier [2].

The first to fourth iFTs convert blocks of constellation points output by the spatial mapper to a time domain block (i.e. a symbol) by applying an Inverse Discrete Fourier Transform (iDFT) or an Inverse Fast Fourier Transform (iFFT) to each block. The number of constellation points in each block corresponds to the number of subcarriers in each symbol. A temporal length of the symbol corresponds to an inverse of the subcarrier spacing. When MIMO or MU-MIMO transmission is used, the TxSP may insert cyclic shift diversities to prevent unintentional beamforming; the cyclic shift diversity may be specified per transmit chain or per space–time stream [2].

The first to fourth GI inserters prepends a guard interval to the symbol. The TxSP may optionally perform windowing to smooth the edges of each symbol after inserting the GI.

2.5 KEY IEEE 802.11AC MECHANISMS

New mechanisms were introduced with 802.11ac [18] to increase nominal speed and throughput. A DL channel refers to a communication channel from a transmit antenna of the AP to a receive antenna of a WN/STA, and an UL channel refers to a communication channel from a transmit antenna of a WN/STA to a receive antenna of the AP; DL and UL may be referred to as forward link and reverse link, respectively. New mechanisms 802.11ac included but were not limited to: (i) extended channel binding; (ii) optional 160 MHz and mandatory 80 MHz channel bandwidth for stations; (iii) (as noted), more MIMO spatial streams, also with Downlink Multi-User MIMO (DL-MU-MIMO) – this DL-MU-MIMO formulation allows up to four simultaneous clients; (iv) multiple STAs (WNs) each having one or more antennas, to transmit or receive independent data streams simultaneously; (v) 256-QAM, rate 3/4 and 5/6, added as optional modes (as compared with 64-QAM, rate 5/6 maximum in 802.11n); and (vi) beamforming with standardized sounding and feedback for compatibility between vendors. Some features (e.g. low-density parity-check code; 400 ns short guard interval; five to eight spatial streams; 160 MHz channel bandwidths – contiguous 80 + 80; and 80 + 80 MHz channel bonding including discontiguous sections [19]) are optional.

2.5.1 Downlink Multi-User MIMO (DL-MU-MIMO)

As discussed earlier, MIMO systems may use multiple transmit antennas to provide beamforming-based signal transmission. Typically, beamforming-based signals transmitted from different antennas are adjusted in-phase (and optionally amplitude) such that the resulting signal power is focused toward a receiver device. See Figure 2.7. A wireless MIMO system may support communication for a single user at a time or several users concurrently; transmissions to a single user (e.g. a single receiver device) are referred to as Single-User MIMO (SU-MIMO), while concurrent transmissions to multiple users are referred to as MU-MIMO. An AP (e.g. a base station [BS]) of an 802.11-based MIMO system employs multiple antennas for data transmission and reception; each user STA employs one or more antennas. MIMO channels corresponding to transmissions from a set of transmit antennas to a receive antenna are referred to as spatial streams since precoding (e.g. beamforming) is employed to direct the transmissions toward the receive antenna [20]. A MIMO-based system provides improved

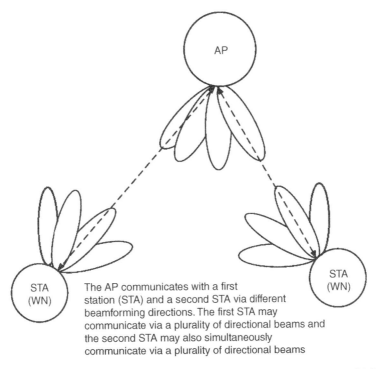

The AP communicates with a first station (STA) and a second STA via different beamforming directions. The first STA may communicate via a plurality of directional beams and the second STA may also simultaneously communicate via a plurality of directional beams

FIGURE 2.7 Distributed MIMO communication with beamforming [20].

performance (e.g. higher throughput and/or greater reliability) using the additional spatial streams.

The 802.11n standard introduced MIMO to the LAN environment, allowing a maximum of four MIMO streams to be transmitted to a WN at a time; 802.11ac increased the maximum (theoretical) number of single-user MIMO streams received by a WN to eight, effectively doubling the network throughput with 802.11ac compared to 802.11n (note that 802.11ac MU-MIMO specification defines radio configurations that support up to four simultaneous MIMO channels[5]).

Specifically, 802.11ac supports MU-MIMO affording a major improvement over SU-MIMO (also just called MIMO). See Figure 2.8. This capability is supported in the DL, and the process is known more specifically as DL-MU-MIMO. APs typically have four antennas (APs with eight antennas are also available), but most of the client devices are limited to 1–2 antennas; thus, in a SU-MIMO channel operation, and the full capacity is rarely achieved. For example, a 4×4 Wi-Fi 11ac AP supports a peak PHY rate of 1.7 Gbps. But a smartphone or tablet with one antenna can only support a peak rate of 433 Mbps, leaving 1.3 Gbps capacity of the AP unused – this difference is called the MIMO gap. 802.11ac addresses the MU-MIMO gap, allowing an AP to support up to four simultaneous full-rate Wi-Fi connections (say, 433 Mbps each) where each of these connections is assigned to a different client device such as smartphone, laptop, or tablet. The total bandwidth of 1.7 Gbps is utilized, representing the systems' bandwidth. In this manner, MU-MIMO improves performance by affording the AP more options to support the BSS

---

[5]"Stream" refers to spatial concepts: streams are used for per client Wi-Fi communication (multiple antenna communication); "channel" refers to the RF facility used to support simultaneous MU Wi-Fi client communications; there are two or three bands in recent AP hardware that uses 2.4 GHz and one or two 5 GHz – the so-called Dual-Band APs or the so-called Tri-Band APs.

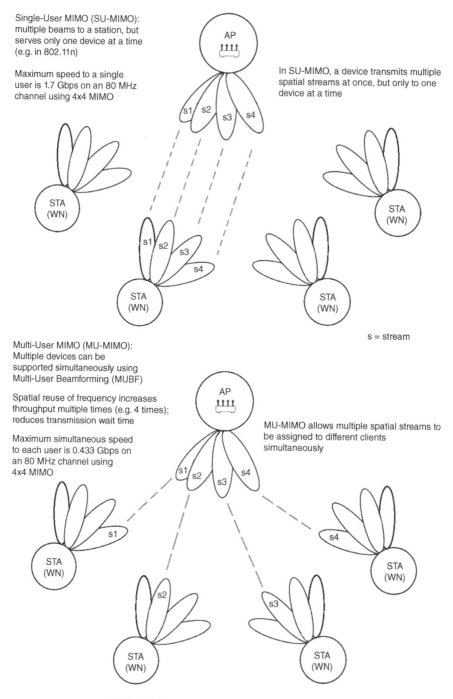

Single-User MIMO (SU-MIMO): multiple beams to a station, but serves only one device at a time (e.g. in 802.11n)

Maximum speed to a single user is 1.7 Gbps on an 80 MHz channel using 4x4 MIMO

In SU-MIMO, a device transmits multiple spatial streams at once, but only to one device at a time

s = stream

Multi-User MIMO (MU-MIMO): Multiple devices can be supported simultaneously using Multi-User Beamforming (MUBF)

Spatial reuse of frequency increases throughput multiple times (e.g. 4 times); reduces transmission wait time

Maximum simultaneous speed to each user is 0.433 Gbps on an 80 MHz channel using 4x4 MIMO

MU-MIMO allows multiple spatial streams to be assigned to different clients simultaneously

FIGURE 2.8   SU-MIMO versus MU-MIMO.

clients and enabling the AP to make full use of the total system throughput. In summary, MU-MIMO provides increased throughput and reduced latency: the efficient use of available spectrum increases the total capacity of a network by a factor of 2×–3×, and since client devices do not time-share connections with other clients on the network, each device incurs reduced wait time. To achieve the full MU-MIMO benefit with an 8×8 AP would require an 8×8 client configuration; unfortunately, this is not practical, especially with mobile devices and limited battery power (typical mobile clients support 1×1 or 2×2 configuration). The maximum throughputs in the 5 GHz band are:

- 4×4/4-stream: 1.733 Gbps max rate
- 3×3/3-stream: 1.300 Gbps max rate
- 2-stream 802.11ac: 0.867 Gbps max rate
- 1-stream 802.11ac: 0.433 Gbps max rate

As noted, in 802.11ac, only a single-user WN is allowed to transmit (in the UL direction) at a point in time; multiuser DL transmission from an AP to non-AP WNs is supported through DL-MU-MIMO beamforming. The more WNs active in the network, the longer the stations may need to wait before they are allowed to transmit UL a buffered frame. The issue is improved in the 802.11ax specification.

2.5.2 Beamforming

Beamforming is a methodology that focuses the AP's transmit energy of the spatial stream toward the targeted WN. Channel estimation is employed to introduce a small difference in the phase and amplitude in the transmitted signal (a process called precoding) to enable the AP to focus the signal in the direction of the receiving WN. 802.11n had previously defined a number of methods of beamforming, and consequently, chipset vendors implemented various non-interoperable techniques, keeping beamforming from general acceptance. To address the issue, the 802.11ac specification defined a single closed-loop SU/MU Transmit Beamforming (TxBF) method where the AP transmits a "special sounding signal" to all WNs – each WN estimates the channel and reports its channel feedback information back to the AP. In the sounding mechanisms, each WN provides channel feedback, which the AP uses to give its spatial streams the necessary mobility. Once channel probing request to the WN results in the WN providing the AP with a characterization of its environment, the AP uses MU-MIMO beam-shaping capabilities to maximize signal in the desired direction and squelch the signal in the undesired direction. MU-MIMO capitalizes on the transmit beamforming capabilities to establish up to four simultaneous directional RF links: this technique provides each of the four users with its own dedicated full-bandwidth channel. In practice, however, the beamforming process is imperfect, and some of the energy of a spatial stream appears in sidelobes for several degrees off-axis. Two adjacent MU-MIMO streams start to interfere with each other as soon as their sidelobes begin to overlap. The presence of this interference adds to the overall noise floor of the channel at the AP. Analysis shows that adding additional MU spatial stream adds intra-stream interference but increases the number of usable spatial streams; this requires a design tradeoff analysis for specific environments and applications [21].

2.5.3 Dynamic Frequency Selection

The 802.11ac system throughput is at, or greater than 1 Gbps and single-link throughput of at least 0.5 Gbps; 800 ns guard intervals are supported. Figure 2.9 depicts available frequencies for the 802.11ac LAN environment. Dynamic Frequency Selection (DFS) is a Wi-Fi function that enables WLANs to use 5 GHz frequencies that are generally reserved for radars; these are less-crowded Wi-Fi bands and can be utilized to increase the number of available Wi-Fi channels, especially in (residential) multi-dwelling units. When support for DFS is enabled, it will be necessary for the AP to verify that any radar in proximity is not using DFS frequencies; this is done by a process called Channel Availability Check, which is executed during the boot process of the AP and also as during its normal operations. See Table 2.5 [22].

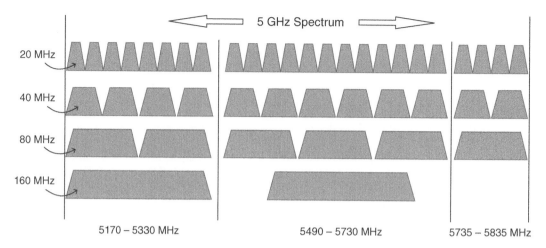

FIGURE 2.9 5GHz spectrum usability for IEEE 802.11ac LANs.

TABLE 2.5 5 GHz Wi-Fi Frequencies

Band	Channel	Frequency (MHz)
U-NII-1	36–48	5170–5250
U-NII-2A/DFS	52–64	5250–5330
U-NII-2C/DFS	100–140	5490–5710
U-NII-3	149–165	5735–5835

2.5.4 Space–Time Block Coding

In addition to the standard WLAN mechanisms at the MAC and PHY layers, IEEE 802.11ac incorporates STBC. Space–time Codes (STCs) involve the transmission of multiple redundant copies of the information to deal with fading and thermal noise with the expectation that some copies may arrive at the receiver in a better condition than other copies; this is known as diversity reception. In the particular case of STBC, the data stream to be transmitted is encoded in blocks, which are distributed among spaced antennas and across time [23–26]. While one must have multiple transmit antennas, it is not always necessary to have multiple receive antennas, although having multiple receive antennas improves performance.

STBC improves data transfer reliability in wireless systems by transmitting a data stream and variations of the data stream across multiple antennas. STBC is a method to transmit multiple copies of a data stream across a number of antennas and to utilize the various received versions of the data to endeavor to improve the quality and assurance of the information transfer. An STBC receiver combines all the copies of the received signal to extract as much usable information from each copy as possible. In general, scattering, absorption, reflection, multipath, refraction, and receive-point amplifier thermal noise typically result in (some) corruption of the signal, such that some of the received copies of the information may be more faithful to the original signal than other copies. The redundancy achieved by STBC implies that there is an opportunity to use one or more of the received copies to correctly decode the received signal. An STBC is usually represented by a matrix where each row represents a time slot, and each column represents an antenna's transmissions over time.

The environment of the WLAN often distorts both the transmitted data stream and the transmitted variations of the data stream. Typically, the distortion of the transmitted data stream

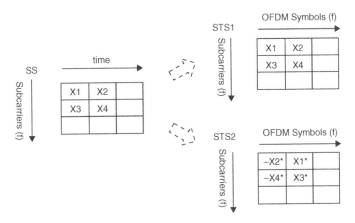

FIGURE 2.10 Space–time block coding [2].

is different from the distortions of the transmitted variations of the data stream. A receiver receives the distorted data stream and the distorted variations of the data stream. STBC combines the distorted data stream and the distorted variations of the data stream to extract as much information from each of them as possible [2].

In IEEE 802.11ac, STBC is used to expand the spatial streams into twice as many space–time streams; that is, 1, 2, 3, and 4 spatial streams may be expanded into 2, 4, 6, and 8 space–time streams, respectively. Alamouti's scheme is used to provide full transmit diversity gain with low complexity for a system with two antennas [20]. Each spatial stream is expanded separately using Alamouti's code as follows: for first and second symbols x_1 and x_2 (in a time domain), a first spatial stream transmits the symbols x_1 and x_2 in their original order, and a second spatial stream transmits symbols $-x_2^*$ and x_1^* having values corresponding to a negative complex conjugate of x_2 and a complex conjugate of x_1, respectively. Thus, as shown in Figure 2.10, the symbols x_1 and x_2 are transmitted using first and second transmitter outputs y_1 and y_2 at first and second times, respectively, as may be expressed by Eq. 2.1:

$$y_1 = \begin{bmatrix} x_1 \\ -x_2^* \end{bmatrix}, y_2 = \begin{bmatrix} x_2 \\ x_1^* \end{bmatrix}.$$

(2.1)

wherein for each transmitter output at each time, a top element is a symbol transmitted using a first antenna, and a bottom element is a symbol transmitted using a second antenna. Notably, the first symbol x_1 is transmitted at a different time than the complex conjugate of the first symbol x_1^*, and the second symbol x_2 is transmitted at a different time than the negative complex conjugate of the second symbol $-x_2$ [2].

First and second received symbols r_1 and r_2 at a receiver having two antennas may be expressed by Eq. 2.2:

$$r_1 = [h_{11}\ h_{12}] \begin{bmatrix} x_1 \\ -x_2^* \end{bmatrix} + n_1, r_2 = [h_{21}\ h_{22}] \begin{bmatrix} x_2 \\ x_1^* \end{bmatrix} + n_2.$$

(2.2)

where h_{ab} is a path gain for a path including an ath transmitting antenna and a bth receiving antenna, and n_1 and n_2 represent first and second additive white noise, respectively. The receiver can recover the transmitted symbols x_1 and x_2 using linear processing [2].

TABLE 2.6   802.11 Cheat Sheet

Topic	Description
Observation	Band, Channel, and Stream have their special definitions.
Band	There are two general public shared bands 2.4 and 5 GHz for Wi-Fi operation.
Channel	Channel is the divided small portions of frequency within each band. For example, there are 11 channels in 2.4 GHz as originally used in 802.11b which utilize 20 MHz per channel, with 15 MHz overlapping to cross over 100 MHz.
Stream	Stream is used since 802.11n (the first implementation of MIMO and known as Wi-Fi 4). One stream in a single 2.4 GHz band and 40 MHz channel (with 400 ns GI) can deliver a maximum of 150 Mbps.
	A four-stream Wi-Fi 802.11n AP can deliver up to 4×150 Mbps = 600 Mbps (one needs to equip with 4×4 antenna in such AP).
Practical/ commercial example	802.11ac (known as Wi-Fi 5) still maintains the same 802.11n maximum of 4 streams per band. It operates in 5 GHz band; thus, the throughput increases to 433 Gbps per stream (often called "450" – it is almost 3 times data rate than 802.11n).
	Most commercial 802.11ac AP in the market are dual-band. They only implement three streams in 5 GHz band (even though in the specification, it can support 4 streams), complemented by four streams in the 2.4 GHz band. Thus, such a Wi-Fi AP could support 1900 Mbps of system throughput capacity with the following configuration: 3×433 (= 1300 Gbps) + 4 x 150 (= 600 Mbps).

2.5.5   Product Waves

The Wi-Fi Alliance separated the introduction of 802.11ac wireless products into two phases ("wave"), named "Wave 1" and "Wave 2". Initially, (2013) products were based on the IEEE 802.11ac Draft 3.0; it supported three spatial streams (with three antennas). Wave 2 certification became available in 2016; Wave 2 products achieve higher bandwidth and system capacity than Wave 1 products. Wave 2 included newer features such as MU-MIMO, wider 160 MHz channel width support, additional 5 GHz channels, and four spatial streams (with four antennas; compared to three in Wave 1 and 802.11n) (IEEE's 802.11ax supports eight.) See Table 2.6 for a "Cheat Sheet" on 802.11ac concepts and system throughput.

2.6   BRIEF PREVIEW OF IEEE 802.11AX

As a quick initial comparison, note that an amendment to the IEEE Std 802.11 (the IEEE 802.11ax amendment) was being developed at press time by the IEEE 802.11ax Task Group. The amendment defines a high-efficiency WLAN for enhancing the system throughput in high-density scenarios. Unlike previous amendments that focused on improving aggregate throughput, the IEEE 802.11ax amendment is focused on improving metrics that reflect the user experience, such as average per station throughput, the fifth percentile of per station throughput of a group of stations, and area throughput. Improvements aimed at targeting environments such as wireless corporate offices, outdoor hotspots, dense residential apartments, and stadiums. The principal focus of the IEEE 802.11ax amendment is on indoor and outdoor operation of the WLAN. The target for increases in average throughput per station is in the range of 5–10 times, depending on a given technology and scenario of the WLAN. Outdoor operation is limited to stationary and pedestrian speeds [2]. This HEW system marketed as Wi-Fi 6 by Wi-Fi Alliance saw initial deployment in late 2019.

As is the case in other 802.11 standards, 802.11ax is designed to operate in the Industrial Scientific and Medical (ISM) bands located between 1 and 6 GHz, including the 2.4 and 5 GHz

TABLE 2.7   Modulation and Coding Schemes for Single Spatial Stream

Modulation and Coding Scheme (MCS)	Modulation Scheme	Coding Rate	20 MHz Channels 1600 ns GI Data Rate (Mbps)	40 MHz Channels 1600 ns GI Data Rate (Mbps)	80 MHz Channels 1600 ns GI Data Rate (Mbps)	160 MHz Channels 1600 ns GI Data Rate (Mbps)
0	BPSK	1/2	8	16	34	68
1 or 2	QPSK	1/2 or 3/4	16 or 24	33 or 49	68 or 102	136 or 204
3 or 4	16-QAM	1/2 or 3/4	33 or 49	65 or 98	136 or 204	272 or 408
5 or 6 or 7	64-QAM	2/3 or 3/4 or 5/6	65 or 73 or 81	130 or 146 or 163	272 or 306 or 340	544 or 613 or 681
8 or 9	256-QAM	3/4 or 5/6	98 or 108	195 or 217	408 or 453	817 or 907
10 or 11	1024-QAM	3/4 or 5/6	122 or 135	244 or 271	510 or 567	1021 or 1134

Note: a GI of 800 ns results in slightly higher data rates (up to about 10% higher).

bands traditionally utilized; additional bands between 1 and 6 GHz may be added as they become available. An aggregate theoretical data rate exceeding 10 Gbps is achievable. For dense deployments, throughput speeds can be four times higher than the throughput speed achieved with IEEE 802.11ac systems (the nominal data rate, however, is only around 37% faster under optimal circumstances). Another key goal of 802.11ax is to improve spectrum efficient utilization. To achieve this goal, better power-control methods are utilized to minimize or avoid interference with neighboring networks. Also, OFDMA, higher order modulation at 1024-QAM (see Table 2.7), and UL of MIMO and MU-MIMO combined with DL of MIMO and MU-MIMO are all utilized to further increase throughput. Additionally, dependability improvements of power consumption and enhanced security protocols, specifically, WPA3 were added.

The average throughput per station is directly proportional to both an aggregate BSS throughput and an area throughput. A fifth percentile measure of the per station throughput, that is, a measure of the throughput achieved by 95% of the stations, may be used to determine that a desired distribution of throughput among a number of stations in an area is satisfied. Since the values of the metrics will depend on the scenario, the IEEE 802.11ax amendment focused on a relative improvement of the metrics compared to previous IEEE 802.11 revisions (e.g. IEEE Std 802.11-2012 in a 2.4 GHz band and IEEE 802.11ac in a 5 GHz band). The amendment includes a capability to handle multiple simultaneous communications in both spatial and frequency domains, in both UL and DL directions [2]. Design goals include robustness in outdoor channels, higher indoor efficiency, and the use of OFDMA. Chapter 5 will expand on these concepts.

REFERENCES

1. Liu, G. and Jiang, D. (2016). 5G: vision and requirements for mobile communication system towards year 2020. Chinese Journal of Engineering 2016: 5974586. https://doi.org/10.1155/2016/5974586.
2. Noh, Y., Lee, D.W., Ahn, J.H. et al. (2019). Support Of Frequency Diversity Mode For Block Code Based Transmission in OFDMA. US Patent 10, 523, 483, 31 December 2019. Uncopyrighted.
3. Minoli, D. (2020). Positioning of blockchain mechanisms in IoT-powered smart home systems: a gateway-based approach. Elsevier IoT Journal, Special Issue on IoT Blockchains 10: 100147. https://doi.org/10.1016/j.iot.2019.100147. https://www.sciencedirect.com/science/article/pii/S2542660519302525.
4. Hedayat, A.R. and Kwon, Y.H. (2019). Mixed Fine/Coarse Sounding Methods For HE STAs for MIMO and OFDMA. US Patent 10, 505, 595, 10 December 2019. Uncopyrighted.
5. Kim, J., Ryu, K., Choi, J., and Cho, H. (2019). Method for Operating In Power Saving Mode In Wireless LAN System And Apparatus Therefor. US Patent 10, 524, 201, 31 December 2019. Uncopyrighted.

6. Chu, L., Wang, L., Zhang, H. et al. (2019). Method And Apparatus For Uplink Orthogonal Frequency Division Multiple Access Communication In A WLAN. US Patent 10, 524, 290, 31 December 2019. Uncopyrighted.

7. IEEE. (2012). Part 11: Wireless LAN Medium Access Control (MAC) and Physical Layer (PHY) Specifications. IEEE Standards 802.11.TM.-2012 (Revision of IEEE Standard 802.11–2007), 29 March 2012, 1–2695, IEEE (The Institute of Electrical and Electronic Engineers, Inc.), New York, NY, USA.

8. IEEE. (2013). Part 11: Wireless LAN Medium Access Control (MAC) and Physical Layer (PHY) Specifications, Amendment 4: Enhancements for Very High Throughput for Operation in Bands below 6 GHz. IEEE Standards 802.11ac.TM.-2013, 2013, 1–395, IEEE (The Institute of Electrical and Electronic Engineers, Inc.), New York, NY, USA.

9. Part 11: Wireless LAN Medium Access Control (MAC) and Physical Layer (PHY) Specifications, Amendment 2: Sub 1 GHz License Exempt Operation. IEEE http://P802.11ah.TM./D5.0 March 2015, 1–604, IEEE (The Institute of Electrical and Electronic Engineers, Inc.), New York, NY, USA.

10. https://standards.ieee.org/standard/802_11-2012.html (accessed 3 January 2020).

11. CWAP Staff. 802.11 PHY Layers. Certified Wireless Analysis Professional, https://www.cwnp.com. Available online on November 2, 2020 at http://media.techtarget.com/searchMobileComputing/downloads/CWAP_ch8.pdf.

12. Menon, M.P., Ketchum, J.W., Wallace, M. et al. (2005). Beam-steering And Beam-Forming For Wideband MIMO/MISO systems. US Patent 6, 940, 917, Qualcomm, 6 September 2005. Uncopyrighted.

13. Sakai, E., Yamaura, T., and Sakoda, K. (2019). Radio Communication Apparatus. US Patent 10, 512, 078, 12 December 2019. Uncopyrighted.

14. Shaw, K. (2018). What is MU-MIMO and why you need it in your wireless routers. *Networkworld*, 26 January 2018 https://www.networkworld.com/article/3250268/what-is-mu-mimo-and-why-you-need-it-in-your-wireless-routers.html

15. Choi, H., Ryu, K., Kim, J. et al. (2019). Method and Device For Transmitting And Receiving Frame Related To Multi-User Transmission In Wireless LAN System. US Patent 10, 524, 287, 31 December 2019. Uncopyrighted.

16. Tutorialspoint Staff. Inter – frame spaces (RIFS, SIFS, PIFS, DIFS, AIFS, EIFS). https://www.tutorialspoint.com/inter-frame-spaces-rifs-sifs-pifs-difs-aifs-eifs

17. Staff. FEC Coding. https://www.wlanpedia.org/tech/phy/coding-bcc-ldpc (accessed 6 March 2020.

18. IEEE. *IEEE 802.11ac-2013 - IEEE Standard for Information technology – Telecommunications and information exchange between systems – Local and metropolitan area networks – Specific requirements – Part 11: Wireless LAN Medium Access Control (MAC) and Physical Layer (PHY) Specifications – Amendment 4: Enhancements for Very High Throughput for Operation in Bands below 6 GHz.* https://standards.ieee.org/standard/802_11ac-2013.html.

19. Aruba Staff. (2014). 802.11ac In-Depth White paper. https://www.arubanetworks.com/assets/wp/WP_80211acInDepth.pdf (accessed 30 October 2020).

20. Zhou, Y. and Pramod, A. (2018). Distributed MIMO Communication Scheduling In An Access Point Cluster. US Patent 10, 820, 333; 27 October 2020; filed 8 March 2018. Uncopyrighted.

21. Qualcomm Staff. 802.11ac MU-MIMO: Bridging the MIMO Gap in Wi-Fi. White Paper; January 2015; Qualcomm Atheros, Inc., 1700 Technology Drive, San Jose, CA 95110. https://www.qualcomm.com/media/documents/files/802-11ac-mu-mimo-bridging-the-mimo-gap-in-wi-fi.pdf (accessed 11 November 2020).

22. Gridelli, S. (2018). How to Use DFS Channels in WiFi. 05 December 2018. https://netbeez.net/blog/dfs-channels-wifi (accessed 30 October 2020).

23. Tarokh, V., Jafarkhani, H., and Calderbank, A.R. (1999). Space–time block codes from orthogonal designs. IEEE Transactions on Information Theory 45 (5) https://doi.org/10.1109/18.771146.

24. Foschini, G. and Gans, M. (1998). On limits of wireless communications in a fading environment when using multiple antennas. Wireless Personal Communications 6: 311–335. https://doi.org/10.1023/A:1008889222784.

25. Tarokh, V., Seshadri, N.I., and Calderbank, A.R. (1998). Space–time codes for high data rate wireless communication: performance analysis and code construction. IEEE Transactions on Information Theory 44 (2) https://doi.org/10.1109/18.661517.

26. Alamouti, S.M. (1998). A simple transmit diversity technique for wireless communications. IEEE Journal on Selected Areas in Communications 16 (8) https://doi.org/10.1109/49.730453.

3 Traditional DAS Technologies

This chapter covers cellular technology at a very basic level, with particular emphasis on Distributed Antenna Systems (DASs). A DAS can be implemented to provide adequate cellular telephone and Internet access coverage within an area (typically an indoor area), where the propagation of a Radio Frequency (RF) signal is often impaired. DASs are typically used in large office complexes, hospitals, airports, stadiums, and other venues, to enhance and/or facilitate indoor reception due to RF attenuation caused by obstructions, furniture, walls, and various other signal-absorbing materials. The higher the cellular frequency in use (for example, with the newer 4G/5G system), the greater the potential for indoor attenuation and blind spots. While some fundamental concepts are covered herewith, it is not the goal of this chapter to go beyond a basic description of cellular communication.

A DAS can be configured to support a variety (or combination) of services, including cellular communications services, Wireless Local Area Networks (WLANs) services, Personal Area Networks (PANs) services, RF identification (RFID) and Bluetooth beacon tracking services, Location-Based Services (LBSs), and generic Internet of Things (IoT) sensor support services. The implicit focus of this chapter is on cellular communications services, delivered through traditional User Equipment (UE) such as a multimedia cellular smartphone handling voice, data, video, and telemetry; other UE could include devices that may have embedded cellular adapters such as laptops, tablets, wearable devices, and Machine-Type Communication (MTC) devices (for example, devices capable of Machine to Machine [M2M] communications), or other cellular-oriented IoT devices or sensors. The UE under discussion can have one or more antenna panels having vertical and horizontal RF elements.

3.1 OVERVIEW

Figure 3.1 illustrates a basic wireless (cellular) communication system comprised of Base Stations (BSs) and UEs. A typical BS consists of (i) a baseband unit (BBU – aka Base Transceiver Station [BTS][1]), (ii) an RF processing unit, which can be in immediate proximity or further away (hundreds or thousands of feet away), then known as Remote Radio Unit (RRU), and (iii) an adequate (sector) antenna. More specifically, a typical BBU/BTS comprises (i) a Transceiver (TRX) handling transmission and reception of signals to or from downstream network entities; (ii) the RF elements consisting of a Combiner for combining feeds from several TRXs so that they could be sent out through a single antenna, a Power Amplifier for providing signal amplification

---

[1]Note: the BBU is also often called Base Transceiver Station (BTS) as in [1, 2], among others. The RRU is also called a Remote Radio Head (RRH), as in [3, 4], among others; or, a Distributed Antenna Unit (DAU) as in [5], among others; or, a Digital Remote Unit (DRU) as in [1], among others. The term DAS Access Point (DASAP) can also be used. While various terms may have slightly different nuances, they all describe the same general network elements.

High-Density and De-Densified Smart Campus Communications: Technologies, Integration, Implementation, and Applications, First Edition. Daniel Minoli and Jo-Anne Dressendofer.
© 2022 John Wiley & Sons, Inc. Published 2022 by John Wiley & Sons, Inc.

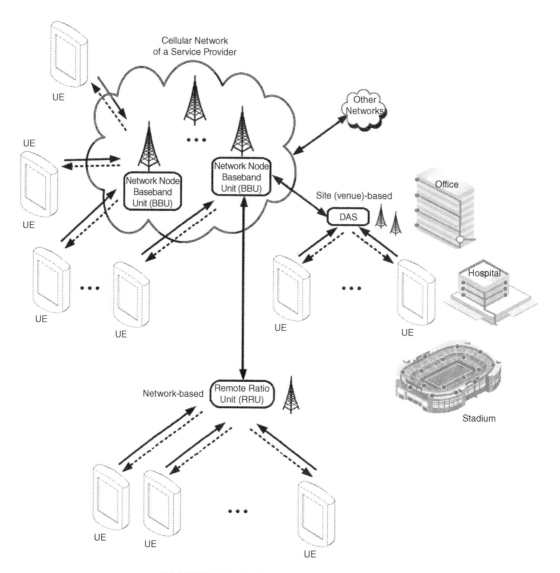

FIGURE 3.1 Basic cellular and DAS setup.

from TRX for transmission via the antenna, and a Duplexer for separating sending and receiving signals to or from the antenna; and (iii) an Antenna (or antennas). Baseband refers to the original frequency range of a transmission signal before it is modulated. A BS is often deployed in the proximity of the antenna tower, although distant placement is possible with fiber-optic links.

Within the BS, typically a BBU connects to both (i) the Mobile Switching Center (MSC),[2] which manages the handoff between towers as mobile users traverse the covered geography, and also the connectivity to the traditional Public Switched Telephone Network (PSTN); and (ii) to the RRU; the RRU, in turn, connects to RF transmitters/receivers antenna located on the tower structure – when using fiber-optic systems, the RRU can be miles away. The RRU devices can be primarily used for transmission and reception of radio signals from UEs (e.g. RF processing), while the BBU devices can be used as the processing unit of telecom systems.

[2]MSCs are also known as Mobile Telephone Switching Offices (MTSOs) or Mobile Switching Offices (MSOs).

The MSC acts as a control center in a cellular system; it connects calls between subscribers by switching voice packets between communicating endpoints. The MSC is positioned between the BS and the PTSN. All mobile communications are routed from the BS through the MSC. The MSC manages call setup between subscribers and is responsible for handling voice calls and Internet services; it also manages billing and account monitoring. Furthermore, the MSC is responsible for inter-Base Station Controllers (BSCs) handovers and for inter-MSC handover (between mobile switching centers). A BSC initiates an inter-BSC handover from the MSC when it notices a cellphone approaching the edge of its cell through a power thresholding mechanism. After the request is made by the BSC, the MSC scans through a list of adjacent BSCs to determine the best handoff and then proceeds to hand over the mobile device to the appropriate BSC. The MSC also operates with the Home Location Register (HLR) – which stores location information and other relevant information – to keep track of the mobility status of mobile devices: the MSC utilizes HLR's database to determine the location of each mobile device in order to provide proper routing of active calls.

Several basic BS architectures are in general use, including the following [6]:

- Traditional architecture, with all the requisite equipment located inside the remote cell hut, with a coax connection to the top of the tower and a fiber/copper connection to the MSC (illustrated in Figure 3.2 top).
- Split architecture design, with the BBU located indoors and an RRU located on the tower (illustrated in Figure 3.2 bottom).
- "Hoteling" approach that uses a single BTS hut but connects to multiple towers.
- All-outdoor, zero-footprint BTS, with all components located on the tower.

While some UEs in the system of Figure 3.1 communicate directly with network nodes, the figure also illustrates UEs communicating with a (quite-remote) RRU that is in turn connected to a BBU/BS device via a communications link. A DAS arrangement is also shown in Figure 3.1. The DAS offers a commonly used approach that wireless cellular service providers can use to improve the coverage provided by a given BS or group of BSs in institutional venues such as hospitals, large office buildings, stadiums, and the like. In Figure 3.1, the dashed lines from the network nodes to the UE represent a wireless downlink (DL), and the solid arrow lines from the UE to the network nodes represent wireless uplink (UL). The communication is bidirectional.

The wireless core network can span various types of technologies, principally traditional cellular networks,[3] including 5G New Radio (NR) systems, femtocell networks, picocell networks,

---

[3]For example, a mobile system can be of any variety, and operate in accordance with standards, protocols (also referred to as schemes), and network architectures, including but not limited to (as described in [8]): Global System For Mobile Communications (GSM), 3GSM, GSM Enhanced Data Rates for Global Evolution (GSM EDGE) radio access network (GERAN), Universal Mobile Telecommunications Service (UMTS), General Packet Radio Service (GPRS), Evolution-Data Optimized (EV-DO), Digital Enhanced Cordless Telecommunications (DECT), Digital AMPS (IS-136/TDMA), Integrated Digital Enhanced Network (iDEN), Long-Term Evolution (LTE), LTE Frequency Division Duplexing (LTE FUD), LTE time division duplexing (LTE TDD), Time Division LTE (TD-LTE), LTE Advanced (LTE-A), Time Division LTE Advanced (TD-LTE-A), Advanced eXtended Global Platform (AXGP), High-Speed Packet Access (HSPA), Code-Division Multiple Access (CDMA), Wideband CDMA (WCMDA), CDMA2000, Time-Division Multiple Access (TDMA), Frequency-Division Multiple Access (FDMA), Multi-carrier Code Division Multiple Access (MC-CDMA), Single-carrier Code Division Multiple Access (SC-CDMA), Single-carrier FDMA (SC-FDMA), Orthogonal Frequency Division Multiplexing (OFDM), Discrete Fourier Transform Spread OFDM (DFT-spread OFDM), Single-Carrier FDMA (SC-FDMA), Filter Bank-Based Multi-carrier (FBMC), zero tail DFT-spread-OFDM (ZT DFT-s-OFDM), Unique Word OFDM (UW-OFDM), Unique Word DFT-spread OFDM (UW DFT-Spread-OFDM), Cyclic Prefix OFDM (CP-OFDM), resource-block-filtered OFDM, Generalized Frequency Division Multiplexing (GFDM), Fixed-mobile Convergence (FMC), Universal Fixed-mobile Convergence (UFMC), Multi Radio Bearers (RAB), Wi-Fi, and Worldwide Interoperability for Microwave Access (WiMax).

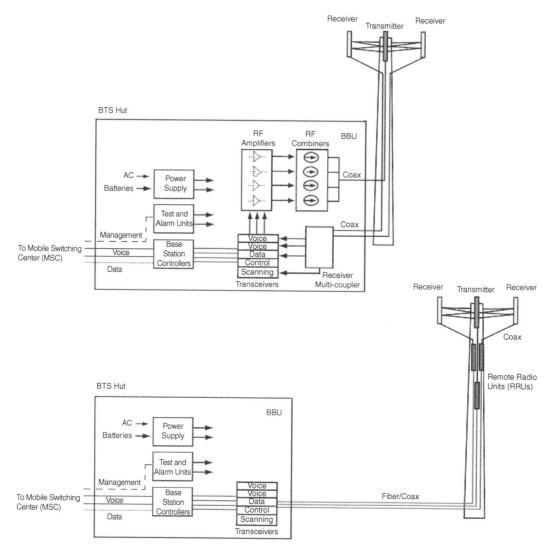

FIGURE 3.2 Examples of base station architectures. Top: traditional. Bottom: use of remote RRUs.

and microcell networks; it can also encompass Internet Protocol (IP) networks and Wi-Fi hot-spot networks. Network nodes typically include transmission electronics and multiple antennas to handle receive/transmit functions for one or more cells or geographic sectors. Network nodes can comprise NodeB and/or Evolved NodeB (eNB or eNodeB) devices; Radio Access Network (RAN) devices; Radio Network Controllers (RNCs); BSCs; or nodes supporting a DAS – the eNB is the BS component of the Long-Term Evolution (LTE) network providing coverage for mobile broadband services.[4,5]

[4]An Evolved Universal Mobile Telecommunications System (E-UMTS) is an advanced version of a traditional Universal Mobile Telecommunications System (UMTS); E-UMTS is generally referred to as a "LTE system"; technical details related to E-UMTS are provided in Release 7 and Release 8 of the "3rd Generation Partnership Project; Technical Specification Group Radio Access Network," among other specifications.
[5]In the context of this discussion, make note of the Voice over Long-Term Evolution (Voice over LTE or VoLTE) concept. VoLTE, term advocated by the GSM Association, refers to the use of Voice over IP (VoIP) mechanisms where both the signaling and media components are transported over a 4G LTE packet switched (PS) data path. This contrasts

In modern networks, the radio equipment, the RRU, is often remote to the BTS/NodeB/eNodeB, and distributed BSs have emerged as cost-effective solutions to the build out of large regional cellular networks. RRUs are used to extend the coverage of a BTS/NodeB/eNodeB in rural areas, highways, tunnels, venues, or other challenging environments. The RRUs contain the BS's RF circuitry, A/D-D/A converters, and up/down frequency converters. RRUs are typically connected to the BTS/NodeB/eNodeB with a fiber-optic using a standardized interface.

To meet increasing wireless service demand, without necessarily having to expand in newer RF bands, small cell deployments are being implemented with cloud RAN systems,[6] where a portion of a BS device – for example, the BBU device of a network node – may support multiple RRU devices. "Small cells" refer to relatively low-powered radio communications equipment and ancillary antennas and/or towers that provide mobile, Internet, and IoT services within localized areas. Small cells typically have a range up to 1–2 km but can also be smaller – small cells are generally categorized as follows: microcell: about 2 km in radius; picocell: less than 0.2 km; femtocell approximately 10 m (Personal Area Network [PAN] ranges). The terms femtocells, picocells, microcells, urban microcell (UMi), and metrocells are effectively synonymous with the "small cells" concept. On the other hand, a typical mobile macrocell (such as urban macro-cellular [UMa] or rural macrocell [RMa]) has a range of several kilometers up to 10–20 km. Small(er) cells have been used for years to increase area spectral efficiency – the reduced number of users per cell provides more usable spectrum to each user. However, the smaller cells in 5G at higher frequencies are also dictated by the propagation characteristics. In fact, cell size typically depends on the underlying frequency of operation; for example, in a CDMA2000 network, systems operating at 950 MHz can typically support cell sizes with radii of about 27 km; at 1800 MHz, 14.0 km; and, at 2100 MHz about 12.0 km. In practical terms, lower frequencies allow networks to provide coverage over a larger area, while higher frequencies allow networks to provide service to more customers in a smaller area due to a wider available spectrum. In the *5G context,* a UMi typically has a radius of 5–120 m for Line-of-Sight (LOS) and 20–270 m in Near LOS (NLOS); a UMa typically has a radius of 60–1000 m for LOS and 50–1500 m for NLOS [7].

In the small cell environment, BBU devices can be of a smaller, modular design allowing for higher integration, lower power consumption, and easier deployment. In other arrangements, a BBU device can be placed in the equipment room and connected with the multiple RRUs via communications links (e.g. optical fiber); this is the "hoteling" arrangement cited earlier; see Figure 3.3. In some implementations, including DAS environments, the RRUs can be physically located at some distance from the BBUs; thus, instead of deploying more network nodes having the full capabilities performed by a collocated (or replicated) RRU and BBU, the network can have many RRUs coupled to a single BBU [8]. Another alternative approach to supporting

---

with handling voice using Circuit Switch (CS) methods, which requires 4G phones to employ a secondary 3G/4G radio channel through (what might be considered by some as) an inefficient access infrastructure and licensed spectrum. GSM Association's *Permanent Reference Document (PRD)* IR.92 describes the application of the IP Multimedia Subsystem (IMS) and the Evolved Packet Core (EPC) for the transportation of voice (as well as short message service traffic over IP); the *Reference* lists a minimum subset of mandatory 3GPP standard features, including 19 mandatory supplementary services that UEs are required to implement. Certain functionality and signaling capabilities need to be supported. IR.92 also specifies the process by which calls fallback to circuit switched voice if the UE were to roam out of a 4G LTE coverage area. VoLTE uses the Session Initiation Protocol (SIP) for registration, authentication addressing, call establishment, and call termination; as is typical in SIP environments, the Session Description Protocol (SDP) is employed for media and bandwidth negotiation as related to the Real-time Transport Protocol (RTP) end-to-end session and peering. VoLTE makes use of the LTE QoS Class Identifier (QCI) codepoint "1," which provides a Guaranteed Bit Rate (GBR) with sub-100 ms packet delay and an error rate not higher than 10^{-2}. The *Reference* also describes a Voice over Wi-Fi (VoWi-Fi), which is covered in Chapter 5.

[6] Cloud RANs are also referred to as Cloud-RANs, C-RANs, CRANs, centralized-RANs.

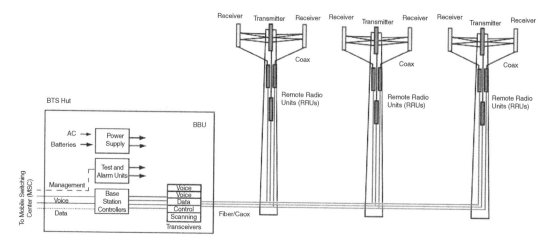

FIGURE 3.3 Hoteling concept for RRUs.

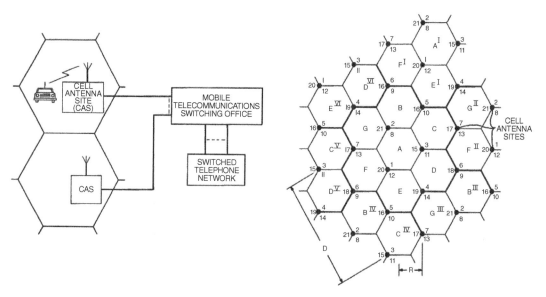

FIGURE 3.4 Frequency reuse [11].

increasing wireless service demand (besides using more efficient modulation schemes) is to operate in newer RF bands that either incrementally add entirely new spectrum or increase the usable bandwidth within a given spectrum, as discussed next.

3.2 FREQUENCY BANDS OF CELLULAR OPERATION

3.2.1 Traditional RF Spectrum

Cellular systems make extensive use of "frequency reuse" (the published literature on this topic includes tens of thousands of articles, but also including [9, 10]). Figure 3.4 [11] provides the basic view of this concept. The outdoor geography is divided into a number of cells distributed in a tessellation pattern to preclude co-channel interferences and provide coverage of mobile and fixed subscriber units operating within the service area of the system. Each cell includes a

BS that employs RF transceiver equipment, antennas, and wireline communication equipment. Mobile/fixed subscriber units within the geographic area of the cell site use RF transceiver equipment to communicate with RF transceivers within the BS. The BS relays voice and data traffic to/from the subscriber mobile units or devices (e.g. a smartphone) and to/from a MSC or Access Service Gateway, which in turn are connected to a central network such as the PSTN or packet-switched networks such as the Internet [5]. Cells are typically divided into sectors; or they are subdivided into smaller cells where the traditional BS is replaced with lower cost (but reduced capability) micro or picocells.

Different generations of cellular systems have evolved over the years (1G thru 5G), which span different technologies, different systems, and even different world geographies. Systems/technologies have included Frequency-Division Multiple Access (FDMA), Time-Division Multiple Access (TDMA, utilized in the Global System for Mobile Communications [GSM]), and Code-Division Multiple Access (CDMA, the basis of 3G); others were listed in footnote 3.

Cellular transmission towers usually have sets of three sector antennas aimed in three different directions with 120° for each cell to cover a general all-encompassing geography and receiving/transmitting with three different frequencies. In the United States, the Federal Communications Commission (FCC) limits sector transmission signals to a maximum of 100W of Effective Radiated Power (ERP). If the tower uses directional antennas, then the FCC allows the carrier to broadcast up to 500W of ERP (carriers use directional signals to improve reception along highways, high-traffic areas, and inside buildings and large venues). The RF spectrum used for cellular networks differ in ITU Regions, which are as follows:

- ITU Region 1 (Europe, Middle East, Russia, and Africa): encompasses Europe, Africa, the Former Soviet Union (FSU), Mongolia, and the Middle East west of the Persian Gulf.
- ITU Region 2 (The Americas): encompasses the Americas, Greenland, and some eastern Pacific Islands.
- ITU Region 3 (Asia, Australia, and Oceania): encompasses most of eastern Asia (non-FSU localities) and most of Oceania.

The first commercial cellular system in the United States was Advanced Mobile Phone System (AMPS), which operated in the 800 MHz frequency band; in Scandinavia, the first system operated in the 450 MHz band. The worldwide GSM standard initially used the 900 MHz band, but as demand grew, carriers acquired licenses in the 1800 MHz band. GSM phones typically supported three bands (900 MHz/1800 MHz/1900 MHz, sometimes 850 MHz/1800 MHz/1900 MHz – tri-band phones) or four even bands (850 MHz 900 MHz/1800 MHz/1900 MHz – quad-band phones).

LTE is the upgrade trajectory for carriers with both GSM/UMTS and CDMA2000 networks – LTE was the mechanism used in transitioning beyond the cellular 3G services, e.g. GSM to UMTS to HSPA to LTE, or CDMA to LTE. LTE is a standard developed by the 3GPP (3rd Generation Partnership Project) for wireless broadband communication based on and evolved from the GSM/EDGE and UMTS/HSPA systems. It was specified in Release 8 of their document series (minor enhancements published in Release 9). Although 3GPP had previously developed standards for the GSM/UMTS protocol suite, the LTE standards were completely new. LTE has been marketed both as "4G LTE" or "4G," but it falls somewhat short of the technical criteria of a 4G wireless service, as specified in the 3GPP Release 8 and 9 document series (however, in 2010, the ITU-R decided that LTE together with some enhancements can be called 4G technology). LTE aimed at (e.g. [12–21]):

- Efficient spectrum utilization: Improved spectral efficiency: 5 bps/Hz for DL and 2.5 bps/Hz for UL.
- Higher data rate: Increased peak data rate: 100 Mbps for DL with 20 MHz (2 Rx Antenna at UE), 50 Mbps for UL with 20 MHz.

- Flexible spectrum allocation.
- Improved system capacity.
- Improved coverage.
- Reduced latency.

The LTE physical layer transmission is based on Orthogonal Frequency Division Multiplexing (OFDM) scheme in the DL and on Single-Carrier FDMA (SC-FDMA) scheme in the UL. The modulation schemes supported in the DL and UL are QPSK, 16-QAM, and 64-QAM (more recently, 256-QAM). Multiple-Input and Multiple-Output (MIMO) technology with two or four antennas is supported; multiuser MIMO is supported in the DL and UL.

Figure 3.5 provides a diagram highlighting the functional and protocol architecture of a typical LTE/Evolved Packet Core (EPC) environment [22]. An eNodeB provides endpoints of a user plane and a control plane to the UE. The Serving Gateway (S-GW), along with the Mobility Management Entity (MME), provides session and mobility management functions for UE. The eNodeB is generally a fixed station that communicates with a UE and may also be referred to as a BS or an Access Point (AP). One eNodeB may be deployed per cell; a facility for transmitting user traffic or control traffic may be used between geographically dispersed eNodeBs. The eNodeB typically performs functions of selection for gateway; routing toward the gateway

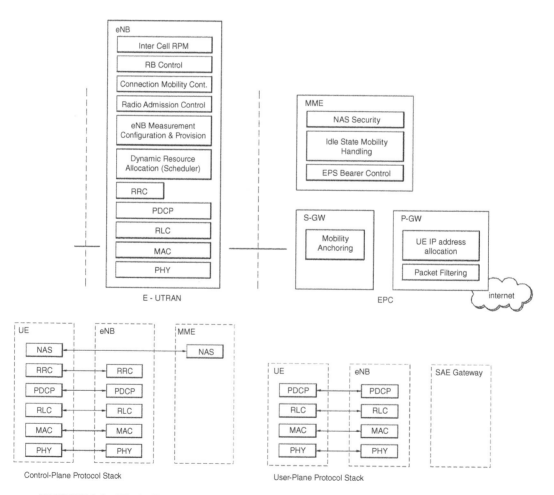

FIGURE 3.5 Block diagram depicting architecture of a typical LTE/EPC environment [22].

during a Radio Resource Control (RRC) activation; scheduling and transmitting of paging messages; scheduling and transmitting of Broadcast Channel (BCCH) information; dynamic allocation of resources to UEs in both UL and DL; configuration and provisioning of eNodeB measurements; radio bearer control; Radio Admission Control (RAC); and connection mobility control. The Non-Access Stratum (NAS) is a set of protocols used to convey non-radio signaling between the UE and MME. The MME provides various functions, including NAS signaling to eNodeBs; NAS signaling security; AS Security control; Inter Core Network (CN) node signaling for mobility between 3GPP access networks; Tracking Area list management; Packet Data Network (PDN) GW and S-GW selection; Roaming; Authentication; Bearer management functions. The S-GW gateway host provides several functions, including Per-user-based packet filtering (by, e.g. deep packet inspection); Lawful Interception; UE IP address allocation; Transport level packet marking in the DL; UL and DL service level charging; gating and rate enforcement; and DL rate enforcement.

Table 3.1 depicts LTE bands usable in North America based on 3GPP's *Technical Specification TS 36.101: Evolved Universal Terrestrial Radio Access (E-UTRA); User Equipment (UE) radio transmission and reception, Release 16*, that lists the specified frequency bands of LTE and the channel bandwidths each band supports [23]. Typical (sub)channel bandwidths in MHz are 1.4,

TABLE 3.1   Bandwidths for LTE Systems in North America

3GPP/LTE Bands for North America	Duplex Mode	UL (MHz)	DL (MHz)	Max Channel Bandwidth (MHz)	Notes
02	FDD	1850–1910	1930–1990	20	PCS
04	FDD	1710–1755	2110–2155	20	
05	FDD	824–849	869–894	10	
07	FDD	2500–2570	2620–2690	20	Canada (specific operators)
12	FDD	698–716	728–746	10	
13	FDD	777–787	746–756	10	
14	FDD	788–798	758–768	10	U.S. (specific operators)
17	FDD	704–716	734–746	10	
25	FDD	1850–1915	1930–1995	20	PCS; U.S. and Canada (specific operators)
26	FDD	814–849	859–894	15	U.S. & Canada (specific operators)
29	SDL	N/A	717–728	10	U.S. & Canada (specific operators)
30	FDD	2305–2315	2350–2360	10	U.S. & Canada (specific operators)
41	TDD	2496–2690		20	
42	TDD	3400–3600		20	C-band
46	TDD	5150–5925		20	
48	TDD	3550–3700		20	CBRS
66	FDD	1710–1780	2110–2200	20	
70	FDD	1695–1710	1995–2020	15	
71	FDD	663–698	617–652	20	U.S. (specific operators)

Notes:
- LTE bands 2 and 4 are only usable for roaming in ITU Region 2; LTE band 5 usable for roaming in ITU Regions 2 and 3.
- Other 3GPP LTE bands are omitted from this table; in particular, LTE bands 7, 28, 38, and 40 (not shown) are usable for global roaming in ITU Regions 1, 2, and 3.

3, 5, 10, 15, 20 (e.g. Band 2, 4, and others) or 5, 10, 15, 20 (e.g. Band 71). The smaller the (sub) channel bandwidths, the higher the number of simultaneous customers that can be supported in one cell with that given LTE Band/bandwidth allocation (e.g. 12 customers on a 20 MHz channel with 1.4 MHz subchannels), but then the maximum bandwidth available to the user will be reduced (e.g. with a simple but high-end 64-QAM modulation – without MIMO – one gets 8 bits-per-Hz or around 10 Mbps on the 1.4 MHz subchannel; with 16-QAM one would get only about 5 Mbps on that subchannel). Naturally, one usually can get a better Signal to Noise Ratio (SNR) on the DL side; thus, one can use a more efficient modulation scheme and get higher throughput on the DL than on the UL.

Frequency-division Duplexing (FDD) (aka duplex mode or offset mode) is designed so that the transmitter and receiver operate using different carrier frequencies. The UL and DL sub-bands are separated by the frequency offset. With FDD, different BSs do not "hear" each other (since they transmit and receive in different sub-bands) and, thus, will typically not interfere with each other. This is also advantageous because BSs must be able to send and receive a transmission simultaneously, and the use of different frequencies avoids antenna and transmission issues. As the name implies, Time-division Duplexing (TDD) uses time-division multiplexing to separate DL and UL signals within a given frequency band; various transmissions can take place over allocated time slots. Obviously, system-wide synchronization is needed between the BS and the UE. Furthermore, consideration must be given to keep guard times between neighboring BSs (or otherwise, also synchronize the BSs, but this coordination will increase complexity and overall cost). As seen in the table, TDD is used in the more "modern" schemes such as Citizen Broadband Radio Service (CBRS) and the (satellite) C-Band reallocation (but only a portion thereof).

Advances in waveforms, modulation, higher-order coding schemes (e.g. 64-QAM/256-QAM), and MIMO antennas are theoretically enabling multi-Gbps speeds over cellular bands. For example, LTE Advanced Pro (also known as 4.5G or Pre-5G, 5G Project) is a brand name for 3GPP Release 13 and 14, which was introduced in the late 2010s as the next-generation LTE version that can support data rates at the 3 Gbps range utilizing 32-carrier aggregation. LTE Advanced Pro supports 256-QAM, Massive MIMO (4×4 or better), LTE IoT, LTE-Unlicensed, and paring of licensed and unlicensed spectrum (all intended to facilitate the evolution of existing networks to 5G). Actual networks support lower aggregation, such as two to four and up to seven carriers (depending on the provider). LTE-A-Pro also uses License Assisted Access (LAA) as a methodology for carrier aggregation across the unlicensed spectrum in the sub-6 GHz arena. However, in some environments, devices operate in very noisy conditions that impact the efficacy of the newer technical improvements, thus limiting or reducing the data rate and the maximum system capacity. In addition, achieving a higher MIMO implementation is still a challenge for device manufacturers endeavoring to place multiple antennas in a limited space (both at the UE and the BS); densifying the networks and intensifying the distribution of small cells can help, but this deployment process presents economic challenges [24]. Therefore, an approach for mitigating these challenges is to make use of new spectrum, possibly with larger (sub)channel bandwidth, as discussed next.

3.2.2 Citizens Broadband Radio Service (CBRS)

Recently the U.S. FCC expanded the use of the 3.5 GHz band to broadband operators and service providers, enterprises, venue managers, municipalities, and public agencies by modifying the use of the 3550–3700 MHz band, previously the exclusive domain of legacy users, including the US Military (3550–3700 MHz), costal communication operators (3650–3700 MHz), satellite operators (3400–3700 MHz), and Wireless Internet Service Providers (WISPs – 3650–3700 MHz) (noticing that the CBRS spectrum falls into the C-Band: the satellite C-band covered no less than 3600–3700 MHz, but portions of the C-band have been vacated, as described Section 3.2.3).

Spectrum sharing will enhance the utilization of this band and free-up valuable spectrum for fixed and mobile users. The CBRS band maps to the 3GPPP Band 48 cited in Table 3.1.

CBRS facilitated new approaches to use spectrum in the 3.5 GHz band in the United States by introducing a new regulatory framework for spectrum sharing. Currently, the use of this band is limited because it is restricted to legacy users. With CBRS, the FCC expanded the use of the 3.5 GHz band to other users, for example, Wireless Service Providers (WSPs), who are now able to operate over a CBRS licensed channel with a Priority Access License (PAL); users with less traffic-intensive needs can also utilize the band under the General Authorized Access (GAA) provision. The CBRS band will be utilized both for mobile and fixed access (this implies that in some locations, fixed and mobile users may contend for the same spectrum). This use of the CBRS spectrum is different from the use in the unlicensed Industrial, Scientific, and Medical (ISM) bands, where multiple users (such as Wi-Fi users) share the spectrum opportunistically, using listen-before-talk mechanisms to determine who should transmit at any given time, without the intermediation of an external administrative entity [25]. Under the new regulation, the CBRS spectrum will be shared among users and made available on a dynamic basis, based on priority tiers. Legacy users retain the right to utilize the spectrum whenever they need it. PAL holders collectively retain access to the 3550–3650 portion of the spectrum (up to seven 10 MHz channels for a total of 70 MHz) in a license area, with up to 40 MHz of spectrum per PAL holder; they must protect legacy users from harmful interference, but they receive protection from interference by GAA users. GAA users collectively have access to spectrum not being used by legacy users and PAL holders in a given area (this is up to 150 MHz of bandwidth); GAA users do not receive interference protection from legacy users or PAL holders. See Figure 3.6. Spectrum Access Systems (SASs) entities have been created to dynamically monitor and authorize the use of specific spectrum resources for PAL and GAA users based on this priority order, using geolocation databases and policy management servers; the SASs authorize users to use the spectrum and ensure that sharing among users is fair.

On 8 May 2018, the CBRS Alliance launched the OnGo™ brand. OnGo provides wireless connectivity with LTE using spectrum sharing in the 3.5 GHz band. OnGo enables multiple users to share the CBRS spectrum, and while doing so, each has use of its assigned channel based on the priority of the tier the user is in. Although unprotected, GAA access is similar to Wi-Fi in unlicensed bands, combined with the ability to obtain higher-tier protected operation as a PAL holder if desired. The 3.5 GHz band (3GPP 48 band) is available in many countries, and it is emerging as a key 5G licensed band. Although the regulatory framework is different (shared spectrum in the United States and licensed band elsewhere), equipment vendors will be able to develop equipment for worldwide deployability [25]. In the United States, the CBRS environment is technology-neutral, but LTE-based systems are an initial technology candidate; as 5G technology becomes available, OnGo users will be able to upgrade their CBRS systems

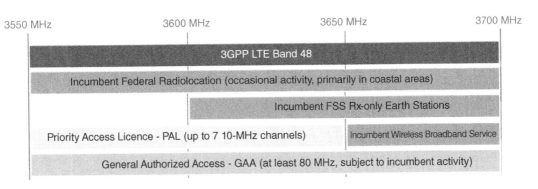

FIGURE 3.6   CBRS spectrum.

with that new technology. Some large enterprises or government entities may opt to deploy private LTE networks using the CBRS spectrum, but this remains to be established as a cost-effective way to deploy usable wireless technology (as compared with using carrier-provided services).

3.2.3 Freed-up Satellite C-Band

There is an expectation for *a dramatic growth of mobile broadband (MBB) data traffic that will likely happen over the next 10–15 years* [24]. Additional spectrum is needed to support these evolving requirements. Traditionally, the satellite operators have had access to the C-Band, the Ku-band, and the Ka band. The C-band is at the lower portion of the spectrum and is a robust band that is used not only for content distribution (for example, commercial headend video) but also for Telemetry, Tracking, and Control (TT&C) to manage the orbit and operation of the satellites [26, 27]. The C-band includes 500 MHz of spectrum between 3.7 and 4.2 GHz, is used today by satellite providers to deliver video programming to cable providers (on the DL side). There is also an "Extended C-band" allocation that covers the 3.4–3.7 GHz range. See Table 3.2. The FCC plans to auction 280 MHz of the block of spectrum while preserving about 200 MHz that will continue to be used for TV programming. Specifically, within the 3.7–4.2 GHz band, the FCC is allocating the 3.7–4.0 GHz portion of the band for mobile use, and 280 MHz from 3.7 to 3.98 GHz band will be auctioned by the FCC for wireless services in the contiguous US Satellite providers were provided with monetary grants to enable them to develop new satellites and possibly new antennas to deal with the changes: operators with C-band licenses will receive compensation for (i) the basic cost to relocate the spectrum (e.g. having to build and deploy new satellites) and for (ii) incentivization to undertake accelerated relocation. Up to $9.7 B of compensation will be shared between Intelsat, SES, Telesat, Eutelsat, and Star One) [28]. Satellite operators must clear a certain amount of spectrum in Phase 1 by 5 December 2021, and additional spectrum by 5 December 2023 (this entails an acceleration since the initial FCC draft order had set those dates in September of 2021 and 2023, respectively); the entire 3.7–3.98 GHz spectrum must be cleared by 5 December 2025.

TABLE 3.2 Satellite Bands, Generalized View per IEEE Standard 521-1984

Band Designator	Frequency (GHz)
L band	1–2
S band	2–4
C band	4–8 bands. More specifically, from 3.7 to 4.2 GHz (space-to-Earth, DL) and 5.925 to 6.425 GHz (Earth-to-space, UL) are called as "Standard" C-band and the bands from 3.4 to 3.7 GHz (space-to-Earth) and 6.425 to 6.725 GHz (Earth-to-space) are known as "Extended" C-band
X band	8–12
Ku band	1–18
K band	18–27
Ka band	27–40
V band	40–75
W band	75–110

Note: The transmit/receive throughput increases as one moves from the L band to the W band (due to increased spectrum allocation); the antenna size becomes smaller as one moves from the L band to the W band; but there is increased susceptibility to rain fading as the frequencies increase (generally, for cellular applications the UL has more of a challenge since the total transmit power is limited at the mobile device).

TABLE 3.3   Performance Measures, eMBB/5G

Performance Measure	Value		
Peak data rate	The minimum requirements for peak data rate are as follows: • Downlink peak data rate is 20 Gbps • Uplink peak data rate is 10 Gbps		
User-experienced data rate	Dense urban • Downlink user-experienced data rate is 100 Mbps • Uplink user-experienced data rate is 50 Mbps		
5th percentile user spectral efficiency	Test environment Indoor hotspot – eMBB Dense urban – eMBB Rural – eMBB	Downlink (bps/Hz) 0.3 0.225 0.12	Uplink (bps/Hz) 0.21 0.15 0.045
Average spectral efficiency	Test environment Indoor hotspot – eMBB Dense urban – eMBB Rural – eMBB	Downlink (bps/Hz/TRxP) 9 7.8 3.3	Uplink (bps/Hz/TRxP) 6.75 5.4 1.6
Latency	• User plane latency: 4 ms for eMBB • Control plane latency: The minimum requirement for control plane latency is 20 ms. Proponents are encouraged to consider lower control plane latency, e.g. 10 ms		
Connection density	The minimum requirement for connection density is 1 000 000 devices per km^2		
Bandwidth	Bandwidth is the maximum aggregated system bandwidth. The bandwidth may be supported by single or multiple radio frequency (RF) carriers. The requirement for bandwidth is at least 100 MHz. Bandwidths up to 1 GHz for operation in higher frequency bands (e.g. above 6 GHz)		

New spectrum is important for the mobile industry to support the Enhanced Mobile Broadband (eMBB) services, which are contemplated for 5G and may be targeted with LTE-A-Pro. The peak rate and user-experienced data rates require significant spectrum bandwidth, as implied in Table 3.3 ([29] and also [30]). 5G transmission bands are broadly grouped as sub-6 GHz and supra-6 GHz. To start with, the supra-6 GHz spectrum of interest corresponds to the millimeter wave (mmWave) at 24–29.5 GHz (Table 3.5), which tends to align with the ka band being used in a number of High Throughput Satellite (HTS) systems, thus possibly sharing some of the technological advancements, perhaps at the chipset level. Some of the performance metrics can be supported with the sub-6 GHz bands, but still, additional spectrum is needed (but other metrics need the supra-6 GHz spectrum).

The new sub-6 GHz spectrum offers a compromise between the broad coverage of lower frequencies and the higher capacity of mmWave [24]; in particular,

- Economical pragmatics: overlay the C-band on top of existing macro-cellular or small-cell grids without needing new cell sites, unlike what mmWave would require.
- Technical pragmatics: access to a range of the spectrum with fewer challenging propagation conditions than mmWave. Due to its wavelength, propagation at these high frequencies is very complex and often requires LOS conditions between the BS and the device. These conditions require highly directional beams and massive MIMO antennas that track users in real time. The sub-6 GHz approach can operate with transmission in a NLOS environment and facilitates indoor penetration on a scale like lower-frequency bands.

- Technical pragmatics compared with traditional lower frequencies: as seen in the earlier table, C-Band can be used as a time-division-duplex technology (TDD-LTE), allowing transmission and reception on the same channel, thus eliminating the use of a dedicated diplexer to isolate transmission and receptions, thereby, reducing costs (the FDD-LTE implementations require a paired spectrum with different frequencies and affiliated guard bands).

3.2.4 5G Bands

5G systems plan to use all the above-cited frequencies as well as mmWave spectrum. The latter is important for truly high-throughput applications, although these frequencies have transmission challenges such as shorter range, weather-related issues, and LOS requirements. In the *3GPP Technical Specification TS 38.101–1: NR; User Equipment (UE) radio transmission and reception, Release 15* [31], frequency bands for 5G NR are defined into two frequency ranges as follows:

- Frequency Range 1 (FR1): encompasses sub-6 GHz frequency bands; some bands have been utilized by previous 3G/4G standards; other bands include potential new spectrum ranging from 410 to 7125 MHz (some of these frequency regions may or may not be available in various parts of the world). Various systems have a variety of channel bandwidths, with many following the traditional repertoire such as 5, 10, 15, 20 MHz (e.g. PCS), or newer repertoires such as 5, 10, 15, 20, 25, 30, 40, 50 MHz (e.g. IMT-E) or 5, 10, 15, 20, 40, 50, 60, 80, 90, 100 MHz (CBRS), to list just a few. It should be noted that in some cases, 5G will allow larger (sub) channels in the same band that was allowed in LTE. See Table 3.4 for an example.
- Frequency Range 2 (FR2): encompass the supra-6 MHz frequency bands from 24.25 to 52.6 GHz. The channel bandwidth spans the ranges of 50, 100, 200, or 400 MHz, thus allowing very high user-experienced data rates when the technology is finally (reliably) deployed (e.g. at a high-end of 10 bits-per-Hertz, the 400 MHz channel would allow an endpoint's throughput of 4 Gbps). See Table 3.5.

5G networks were being deployed at press time in many parts of the industrialized world.

TABLE 3.4   Frequency Range 1 (FR1)

Band	Duplex Mode	Common Name	Frequency of Operation (MHz)	Channel Bandwidths (MHz)
n48/5G	TDD	CBRS (US)	3550–3700	5, 10, 15, 20, 40, 50, 60, 80, 90, 100
n77/5G	TDD	C-Band	3300–4200	10, 15, 20, 25, 30, 40, 50, 60, 70, 80, 90, 100
n78/5G	TDD	C-Band	3300–3800	10, 15, 20, 25, 30, 40, 50, 60, 70, 80, 90, 100
n79/5G	TDD	C-Band	4400–5000	40, 50, 60, 80, 100
48/LTE	TDD	CBRS (US)	3550–3700	5, 10, 15, 20
49/LTE	TDD	C-Band	3550–3700	10, 20

TABLE 3.5   Frequency Range 2

NR Band	Uplink/Downlink (GHz)	Traditional Nomenclature
n257	26.50–29.50	LMDS
n258	24.25–27.50	K-band
n259	39.50–43.50	V-band
n260	37.00–40.00	Ka-band
n261	27.50–28.35	Ka-band

3.2.5 Motivations for Additional Spectrum

The use of higher (sub)channel bandwidth, new spectrum, and mmWave is driven by the requirement to meet the demand for wireless video-oriented applications now becoming ubiquitous, also as highlighted in Table 3.3. While IoT is also an important 5G area of support, these applications tend to currently require lower bandwidth but extensive geographic coverage. The ITU-R has assessed usage scenarios in three classes: Ultra-Reliable and Low-Latency Communications (URLLC), Massive Machine-Type Communications (mMTC), and eMBB. eMBB is probably the earliest class of services being broadly supported and implemented. Some examples of eMBB use cases include smartphones, home/enterprise/venues applications, UHD (4 K and 8 K) broadcast, and virtual reality/augmented reality. mMTC use cases include smart buildings, logistics, tracking and fleet management, and smart meters. URLLC cases include traffic safety and control, remote surgery, and industrial control. 5G design goals that drive bandwidth requirements include [30]:

- 10 to 100× higher user data rate than current systems, especially for peak rates (e.g. 1–20 Gbps – compared with a 1 Gbps maximum on even the best 4G networks).
- 1000× higher mobile data volume per area than current systems.
- 10 to 100× higher number of devices than current systems (i.e. dense coverage).

Other 5G design goals include:

- 10× longer battery life for low-power IoT devices than current systems (up to 10-year battery life for machine type communications).
- 5× reduced end-to-end latency than current systems: tight latency, availability, and reliability requirements to facilitate applications related to video delivery, healthcare, surveillance, and physical security, logistics, automotive locomotion, and mission-critical control, among others, particularly in an IoT context.
- High system spectral efficiency – for example, about 3.5 times that of spectral efficiency of LTE systems.
- A panoply of data rates, up to multiple Gbps, and tens of Mbps to facilitate existing and evolving applications.
- Lower infrastructure development costs, and higher reliability of the communications: pragmatic deployment cost metrics, along with acceptable service price points across the gamut of applications and data rates.
- Network scalability and cost-effectiveness to support both clustered users with very high data rate requirements as well a large number of distributed devices with low complexity and limited power resources, particularly in an IoT context, where, as noted, a rapid increase in the number of connected devices is anticipated.

To aid in increasing capacity, the upcoming 5G access network will, at some point soon, utilize the higher frequencies (e.g. supra-6 MHz). Currently, much of the mmWave spectrum, the band of spectrum between 30 gigahertz (GHz) and 300 GHz is underutilized, offering the availability of large swaths of unused spectrum. The millimeter waves have shorter wavelengths that range from 10 to 1 mm. While these new spectrum bands in higher frequency do hold the promise of more spectrum, and therefore the ability to meet the increasing demands of the mobile industry as stated in Section 3.2.1, the use of these higher frequency bands also comes with some significant challenges and hurdles. One of the key issues is the poorer propagation that radio waves experience in these high-frequency bands, as these mmWave signals can experience severe path loss, penetration loss, and fading [8]. Satellite services have already used mmWaves at the ka band for a couple of decades but using millimeter waves to connect mobile users with a nearby BS is new [26].

Because of the propagation performance at high frequency, the use of mmWave will require small cells [30]. Small cells are miniature BS that can be easily deployed, for example, on poles or multistory buildings, have relatively small antennas, and require low power to operate. To prevent signals from being dropped, carriers could install thousands of these stations in a city placed every 1000 ft. or so (rural environments may require different arrangements). The use of small cells facilitates frequency reuse, which also facilitates the provision of high bandwidth to many users.

3.2.6   Private LTE/Private CBRS

Wi-Fi remains the most prevalent wireless technology in enterprise networks, but private 4G LTE cellular technology and 5G might play a role for certain use cases [32]. Private LTE is a cellular network that is run specifically for the benefit of an organization, such as a utility, factory, or police department. Only authorized users of that organization have access to the network. The organization decides where there will be coverage, how the network will perform, who has access and priority. This is in contrast to a public LTE network, which is operated for the benefit of anyone willing to pay the monthly fee, such as Verizon, AT&T, and T-Mobile [33]. Private LTE and 5G networks are referred to as "non-public networks" by 3GPP. Private LTE and 5G networks are networks that use licensed, shared, or unlicensed wireless spectrum and LTE or 5G cellular networking BSs, small cells, and other RAN infrastructure to transmit voice and data to edge devices, including smartphones, embedded modules in "IoT things," routers, and gateways. Technologically, private LTE and 5G networks work the same as public LTE and 5G networks operated by national WSPs such as AT&T Mobility, T-Mobile, Verizon Wireless, and other Mobile Network Operators (MNOs).

The difference between public and private LTE and 5G networks relates to who has a license or priority access to the wireless spectrum and who owns and operates the network's BSs and infrastructure. With public LTE and 5G networks, the MNO owns and operates the spectrum and the network infrastructure; all of the MNO's customers (with the exception of first responders or similar public safety organizations) have the same access rights to the network. With private LTE and 5G networks, private organizations own, operate, or have some level of priority access to the network's infrastructure or spectrum. The amount of network infrastructure and spectrum owned and operated can vary: (i) With Full Private LTE and 5G networks, the organization owns the wireless spectrum it uses for the network, as well as the network BSs and other infrastructure; (ii) With Private Shared and/or Hybrid Private LTE and 5G networks, parts of the network are either owned, shared, or operated by the MNO or another organization. Full Private LTE and 5G networks require a higher initial capital investment than Wi-Fi and other networks [34].

In the United States, recent ruling on rebanding by the FCC, particularly the CBRS band, as well as availability at the 900 MHz spectrum, allows large organizations, particularly utilities, to consider this option. Private LTE offers the data speeds, technical flexibility, signal prioritization, and security necessary for the multitude of endpoints and smarter utility applications. Many utilities use 900 MHz currently for voice and narrowband data for land mobile radio long-range data communications and CBRS for ultrafast coverage of smaller areas such as substations, storage yards and office spaces. With new spectrum becoming available for utilities, the opportunity for more robust data communication can now include video, sensors, and analytics. In the United States, in 2020, the FCC approved the Notice of Proposed Rulemaking (NPRM) Realignment Rules to repurpose the 900 MHz spectrum band (spectrum owned by Anterix) for uses by utility and enterprise private networks. This broadband spectrum is ideal not only for protection of critical infrastructure, but also for smart grid systems and other Smart Community applications (note parenthetically that the FCC was requiring Part 90 Subpart Z users – in the 3650–3700 MHz band – to transition to Part 96 [CBRS], starting in October 2020 [33]).

Past hurricanes have exposed flaws in public LTE networks, with widespread and lengthy interruptions to service impeding recovery efforts; natural disasters are growing in intensity and frequency; utilities are thus looking to complement the public service with other solutions [33]. Other applications are directed at Industrial IoT (IIoT). In this context, until recently, if an organization wanted to deploy a private wireless network at a factory, office building, transit hub, another facility, or over a utility service, its options were limited to Wi-Fi or proprietary network technologies such as LoRa or Sigfox. These types of private networks were adequate for connecting laptops to the Internet and for basic IIoT applications; however, the coverage and security limitations of these networks, their incompatibility with public cellular networks, and their ongoing management costs have limited the overall deployment. Some see private LTE systems as alleviating these issues, but the concepts are relatively new as of press time. Note that with Full Private LTE and 5G networks, the organization needs to deploy its own infrastructure; furthermore, they require edge devices that have been certified for the wireless spectrum being used. On the other hand, the idea of providing a private network over a public infrastructure goes back to at least the mid-1980s, when the concept of Private Virtual Networks (VPNs) was first applied to obtaining a closed network over the PSTN before the idea was "highjacked" by the data world in the form of VPNs over the Internet.

Any organization can set up and operate their own private LTE or 5G network: to do so, they need spectrum, network infrastructure equipment, and edge devices that can connect to this equipment. Three basic groups include [34]:

- **MNOs:** an increasing number of MNOs are considering deploying private networks to supplement their existing wireless services in areas where there is high demand or they have limited licensed spectrum.
- **Neutral Hosts:** neutral hosts are private LTE and 5G networks that supplement existing public wireless networks in a particular location. For example, a neutral host might set up a private cellular network in an airport, office building, stadium, or hotel. The neutral host network can provide faster and better connectivity to the travelers, office workers, sports fans, or hotel guests at the location. The facility owner will pay the neutral host network provider for improved connectivity in their facility. MNOs may also compensate the neutral host provider for offering connectivity to their public LTE or 5G networks in facilities where the MNO's own coverage is limited.
- **Private Enterprises:** Any organization that is able or willing to make the time, technical, and infrastructure investments to deploy such a network. To do this, the organization needs:
  - Wireless spectrum purchased from the government or provided to them by an MNO or third-party spectrum provider. They can also use unlicensed spectrum or spectrum that is "shared," like CBRS spectrum in the United States. *In particular, a private LTE CBRS network is* a private LTE network using CBRS wireless spectrum. As noted earlier, although the FCC auctioned some CBRS licenses, PALs, companies can still use GAA CBRS spectrum without obtaining a license, sharing this spectrum with PAL license owners (who have priority access to the spectrum) and other GAA users; this allows both PAL license owners and GAA users to build and operate private LTE networks in the United States using the CBRS 3.5 GHz band of wireless spectrum. Some governments in Europe and elsewhere are making available (for purchase) wireless spectrum for private LTE or private 5G networks, e.g. Germany has allocated spectrum in the 3.4–3.8 GHz band and France has allocated spectrum in the 2.57–2.62 GHz band. In the future, mmWave spectrum may also be used.

○ LTE or 5G infrastructure – BSs, mini-towers, small cells, and other equipment – purchased from network infrastructure equipment providers.

○ Smartphones, embedded modules, routers, gateways, and other edge devices with SIM cards and modems to connect to their private LTE or 5G network. This will need a unique Subscriber Identity Module (SIM) card that identifies it and allows it to securely communicate over the private cellular network. If the private network is operated by an MNO or by a neutral host or other organization with an operating agreement with the MNO, the device typically can use the MNO's SIM to connect to the network. Organizations with Full Private LTE or 5G networks that do not connect their private LTE network to an MNO will need a unique SIM card that connects to their own private LTE network. Some edge devices – including smartphones, embedded modules, routers – that have been certified to use the CBRS band by the FCC can connect to these private CBRS LTE networks (e.g. Apple's new iPhones support CBRS and Sierra Wireless's AirPrime EM7511 module).

In the United States, several industry groups are promoting private LTE standards, including the CBRS Alliance, the Wireless Innovation Forum, and the MulteFire Alliance.

As already implied in Section 3.2.6., there are various pros and cons to consider when comparing Wi-Fi to private LTE/5G networks. Once the cost assessments are undertaken and accepted, private LTE and private 5G networks can offer a number of advantages over Wi-Fi networks. Notably, these networks deliver improved wireless coverage than Wi-Fi over large geographic areas; furthermore, there will be less spectrum contention than at the typical Wi-Fi bands. Venues, where private LTE/5G networks are being considered include: warehouses, particularly for robotic product picking, product tracking, and other IoT/IIoT warehouse applications; airports, stadiums, and campuses, which often need reliable coverage both inside their facility and outdoors to support the connectivity needs of their employees, IoT devices and the large crowds of visitors – and utilities, as noted earlier. Refer to Section 5.3.7 for additional discussion on this topic.

3.2.7   5G Network Slicing

5G network slicing is an intrinsic network capability that enables the overlaying of independent virtualized logical networks on the same physical 5G network infrastructure, realizing a service-oriented perception of the network. 5G network slicing is the use of network virtualization to segment a single network into multiple distinct virtual domains that provide different amounts of resources to different types of traffic[s]. Each network slice is isolated from other slices and/or users and is created by Software-Defined Networking (SDN) methods to meet specific requirements needed by a particular organization or application. In addition to SDN mechanisms, slicing also relies on Network Function Virtualization (NFV). Network slicing can be traced back to the late 1980s with the introduction of the concept of VPN in the (voice) network environment alluded to earlier.

Proponents and developers see network slicing as the mechanism to adapt and provide for the specific needs of various vertical industries. One example could be the delivery of services over the common infrastructure with different service-level requirements: each use case can receive a unique set of optimized resources and network topology – covering certain Service Level Agreements (SLA)-specified factors such as connectivity, speed, and capacity – that suit the needs of that application [35]. This topic is further discussed in Section 10.

3.2.8   Supportive Technologies

mmWave, small cells, massive MIMO, full duplex, and beamforming are critical elements of 5G solutions to meet the goals highlighted in Section 3.2.5.

Performance, as measured by throughput, can be improved if both the transmitter and the receiver are equipped with multiple antennas: the multi-antenna technique can improve the spectral efficiency of transmissions, thereby significantly increasing the overall data-carrying capacity of a wireless system. Chapter 2 discussed MIMO in the context of Local Area Networks (LAN). The same concepts clearly apply to cellular environments. The use of MIMO techniques in cellular environments was introduced by 3GPP, and the technique has been in use in LTE (and other 3G/4G systems). MIMO is planned for use in 5G systems both in the sub-6 MHz and in the supra 6 MHz (including mmWave systems). Systems using Massive MIMO incorporate dozens of antennas on a single array. Massive MIMO has been recently tested in a number of field trials, setting new records for spectrum efficiency.

As implied in the discussion in Section 3.2.1, the shorter wavelengths at mmWave frequencies allow for more antennas to be packed at the same physical location, which allows for large-scale spatial multiplexing and highly directional beamforming. Multi-antenna techniques can significantly increase the data rates and reliability of a wireless communication system. The MIMO technique uses a commonly known notation $(M \times N)$ to represent MIMO configuration in terms of number of transmit (M) and receive antennas (N) on one end of the transmission system. The common MIMO configurations used for various technologies are: (2×1), (1×2), (2×2), (4×2), (8×2) and (2×4), (4×4), (8×4). The configurations represented by (2×1) and (1×2) are special cases of MIMO known as transmit diversity (or spatial diversity) and receive diversity. In addition to transmit diversity (or spatial diversity) and receive diversity, other techniques such as spatial multiplexing (comprising both open-loop and closed-loop), beamforming, and codebook-based precoding can also be used to address issues such as efficiency, interference, and range [8]. Systems incorporating a very large number of antennas (degrees of freedom) at the network node, for example, greater than 8×8 (8 transmit and 8 receive antennas), can be referred to as "massive MIMO" systems (also known as Large-Scale Antenna Systems, Very Large MIMO), which are expected to be a differentiator between currently deployed LTE (4G) mobile networks and 5G mobile networks of the future. Figure 3.7 illustrates an example of a multi-antenna transmission system having multiple antenna ports.

The challenge is that installing a large number of antennas to handle cellular traffic at a single location also causes added interference if those signals cross; therefore, a key challenge for massive MIMO is to reduce interference while transmitting more information from many more antennas at once. To address the issue, 5G stations must incorporate beamforming. Beamforming is a traffic-signaling system for cellular BSs that identifies the most efficient information-delivery route to a particular user, and it reduces interference for nearby users in the process. Beamforming can help massive MIMO arrays make more efficient use of the spectrum around them. At massive MIMO BSs, signal-processing algorithms plot the best transmission route through the air to each user; then, they can send individual data packets in many different directions, bouncing them off buildings and other objects in a precisely coordinated pattern.

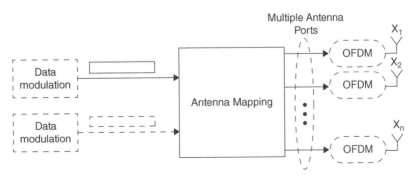

FIGURE 3.7 Example of a multi-antenna transmission embodiment having multiple antenna ports [8].

By managing the packets' movements and arrival time, beamforming allows many users and antennas on a massive MIMO array to exchange much more information at once [36]. For mmWave environments where LOS is needed and signals attenuate rather quickly, beamforming is used to focus a signal in a concentrated beam that points only in the direction of a user, rather than transmitting the signal over a wide angle (e.g. 60°), which is typically the case in traditional cellular systems.

Current BSs and cellphones rely on transceivers that alternate the communication process when they are transmitting and receiving information over the same frequency (systems would need to utilize different frequencies if a user wishes to transmit and receive information simultaneously). On the other hand, 5G will utilize full duplex techniques so that a transceiver will be able to transmit and receive data at the same time, on the same frequency.

3.3 DISTRIBUTED ANTENNA SYSTEMS (DASs)

3.3.1 Technology Scope

Employees and consumers are increasingly demanding reliable and ubiquitous wireless communications services, such as cellular communications services and Wi-Fi connectivity. However, many large indoor environments, for example, high-rise office towers, convention centers, hospitals, airports, and stadiums are often poorly serviced by conventional cellular networks. A traditional solution to enhancing indoor cellular access has been the use of DASs[7]: DASs can extend coverage to areas within the cell coverage area in cases where RF propagation may be limited by obstructions, such as in buildings and tunnels, or to areas where the amount of traffic (revenue) does not justify the investment required for a complete BS. DASs allow the RF coverage to be adapted to the specific environment[8]; they enable the distribution and rebroadcasting of various cellular, LTE, 5G and other RF frequencies within a building or confined/defined structural environment [30]. DASs can be particularly useful when deployed inside buildings or other indoor environments where the wireless communication devices may not otherwise be able to effectively receive RF signals from a distant external tower. DASs are typically used in connection with the distribution of wireless communications that employ licensed cellular RF spectrum, also including the newer bands discussed earlier (e.g. CBRS, C-Band, mmWave). Besides the goal of enhancing connectivity in commercial buildings and venues, typical current business drivers for DAS deployments include the predicament that during anticipated peak times – e.g. in a building or in some public venue as a stadium – users may experience coverage deficiencies, congestion, blocked connections, reduced data speeds, and other service degradations. Safety is another driver, the goal being to eliminate blind spots for E911 support or support of various first responders. While DAS is often used in large urban office buildings, DAS can also be used in open spaces such as campuses, conference centers, stadiums, hospitals, airports, train stations, tunnels, hotels, and cruise ships.

A DAS is configured to handle one or more cellular bidirectional RF bands; this is especially the case when the DAS is "carrier-neutral"; namely, where smartphones typically serviced natively by various wireless providers, are in fact, supported by the DAS natively without requiring a carrier-to-carrier handoff. Current commercially available systems support CDMA, EVDO, GSM, HSPA, UMTS, and LTE, among other cellular services; future systems will support 5G, and they become even more prevalent. In some cases, a DAS can also be used with

---

[7]The technology is also is sometimes called "in-building cellular."

[8]Even in open spaces, DASs may also be employed to optimize the RF distribution in larger cells, in order to increase capacity at an even lower cost than with pico and/or micro cells; this approach allows the reuse of a limited number of expensive RF channels without incurring the costs of one or more complete base stations [5].

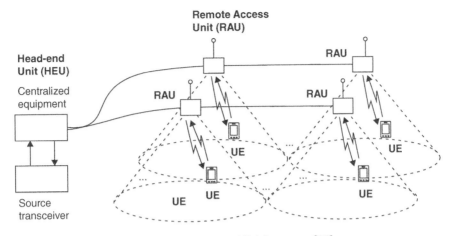

FIGURE 3.8 General DAS concept [37].

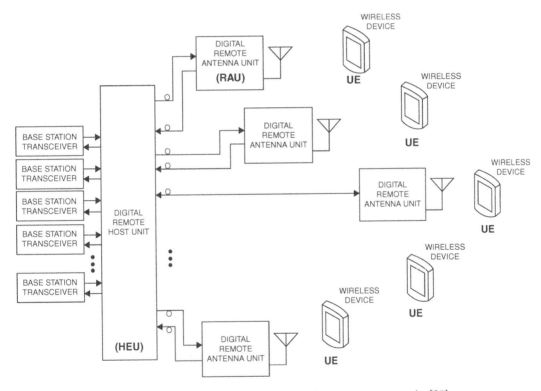

FIGURE 3.9 A DAS system using digital remote antenna units [38].

wireless communications that make use of unlicensed RF spectrum. Additionally, DASs can also support cellular-based IoT (e.g. LTE-M, NB-IoT, and 5G IoT.)

Figures 3.8 and 3.9 depict the general DAS concept [37, 38]. A DAS is a network of a (large) number of (small) (indoor or on-location) antennas connected to a common cellular source via an appropriate communication channel, providing cellular/wireless service within a given physical structure. A DAS usually includes a Head-End Unit (HEU)[9] connected down-chain to

---

[9]HEUs are also known as host units, as in [38].

remote equipment in specific zones of interest, and up-chain to BSs of one or more cellular providers, thereby creating coverage areas that facilitate communications with wireless UEs located in these areas; the remote equipment is generally known as Remote Access Units (RAUs).[10] As an example, the HUE could be located in the basement of a large office building and the RAUs are distributed throughout the various floors. RF signals are communicated between the HUE and one or more RAUs: RAUs are configured to receive and transmit signals to client UEs within the antenna range of the RAUs. Many DASs use Radio-over-Fiber (RoF) connectivity methods, where RF signals are sent back to the HEU over optical fibers. Traditionally, DASs have supported voice-oriented services for access to a cellular provider; however, more sophisticated deployments may rely entirely on a Wi-Fi infrastructure allowing VoWiFi solutions with direct handoff of the traffic to a VoIP/Internet provider.

The HEU receives downstream signals corresponding to a number of downstream frequency bands, each band associated with a respective RF channel, possibly licensed by various service providers. Each of the bidirectional RF bands distributed by the DAS includes a separate RF band for each direction of communication. One direction of communication, the DL direction, goes from the BS's transceivers to the wireless device via the RAU; the other direction of communication, the UL direction, goes from the wireless device to the BS's transceivers via the RAU. Each of the distributed bidirectional RF bands includes a DL band in which downstream RF channels are communicated for that bidirectional RF band and a UL band in which upstream RF channels are communicated for that bidirectional RF band. The host unit/HEU can be connected to one or more BSs directly, for example, utilizing coaxial cabling. The HEU can also be connected to one or more BSs wirelessly, for example, using a donor antenna and a bidirectional amplifier. RF signals transmitted from the BS are received at the host unit/HEU. The HEU uses the DL RF signals to generate a DL transport signal that is distributed to one or more of the RAUs. Each such RAU receives the DL transport signal and reconstructs the DL RF signals based on the DL transport signal and causes the reconstructed DL RF signals to be radiated from at least one antenna in the RAU. A similar process is performed in the UL direction: RF signals transmitted from the UEs are received at one (or more) RAUs; each RAU uses the UL RF signals to generate a UL transport signal that is transmitted from the RAU to the host unit; the HEU reconstructs the UL RF signals received at the RAUs and communicates the reconstructed UL RF signals to the BS [38].

Indoor Usage As discussed in Section 3.3.1, to address coverage problems, a DAS includes HEU components that receive an input RF signal and convert it, for example, to an optical signal to be distributed by fiber-optic cables to areas where RF signals are blocked, e.g. inside the buildings. The DAS antennas can be placed close to the possible locations of mobile or portable terminals, originated from a utility or service room, and then arranged to form a star-like topology. The DAS also entails components that reconvert the wired signals back to the RF signals. As implied in Figures 3.8 and 3.9, elements of a building- or venue-based DAS include:

- (Small) Broadband antennas and amplifiers in the indoor space (typically one or more per floor) that shape the coverage. An indoor DAS has a confined antenna coverage area that is anchored about the antenna of the RAU. These antennas typically cover the entire spectrum of the cellular service (for/from multiple service providers).
- Coax or fiber-optic cabling to connect the structure antennas to a local BS.
- Remote Radio Head, the host unit/HEU, effectively a local BS ("small cell"), typically in the basement.

---

[10]The terms Remote *Antenna* Unit (as in [38]), Remote Units (RUs), and Remote Hub Units (RHUs) are also used.

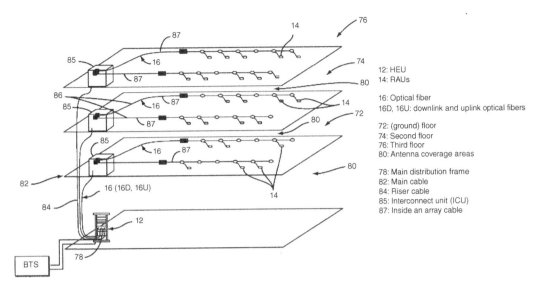

12: HEU
14: RAUs

16: Optical fiber
16D, 16U: downlink and uplink optical fibers

72: (ground) floor
74: Second floor
76: Third floor
80: Antenna coverage areas

78: Main distribution frame
82: Main cable
84: Riser cable
85: Interconnect unit (ICU)
87: Inside an array cable

FIGURE 3.10   In-building installation of a DAS [39].

- Fiber-optic connection to an aggregation point (typically in a carrier colocation space when the indoor coverage service is offered by a "building local exchange carrier" [BLEC]) (or the use of an outdoor donor antenna to a specific cellular provider). The former supports carrier-neutral applications; the latter typically supports only one carrier. Physical connectivity from the colocation space to each of the wireless providers is needed, typically in the form of fiber connectivity or other telecom service.
- Business relationships with the wireless providers are needed to deliver the traffic to the core network, including the PSTN or Internet.

Figure 3.10 provides a topological pictorial diagram of a building infrastructure employing an optical fiber-based DAS. The DAS incorporates the HEU to provide various types of communication services to coverage areas within the physical building infrastructure. As described in reference [39], the DAS is configured to receive wireless RF communications signals and convert the signals into RoF signals to be communicated over the optical fiber (shown as item "16" in the figure) to multiple RAUs to provide wireless services inside the building infrastructure. The floors are serviced by the HEU (shown as item "12" in the figure) through a main distribution frame to provide antenna coverage areas in the building zones and subzones. The main cable has several different sections that facilitate the placement of many RAUs in the building infrastructure. Each RAU, in turn, services its own coverage area in the antenna coverage areas. The main cable can include, for example, a riser cable that carries all of the DL and UL optical fibers (shown as "16D" and "16U" in the figure) to and from the HEU. The riser cable may be routed through an interconnect unit. The interconnect unit may also provide power to the RAUs via the electrical power and provided inside an array cable.

Indoor DASs' antenna coverage areas can have a radius in the range from a few feet up to 60 ft., as an example. Deploying a number of DAS Access Point (DAP) devices (the RAUs) creates an array of coverage areas. The RAUs can be distributed throughout locations inside a building to extend wireless communication coverage throughout the building. Because, individually, the antenna coverage area covers a small zone, there are typically only a few users per antenna coverage area; this allows for minimizing the amount of RF bandwidth shared among the wireless system users. While extending the RAUs to locations in the building can provide seamless wireless coverage to wireless clients, other services may be negatively affected or not

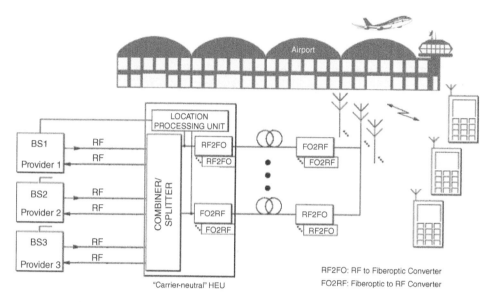

FIGURE 3.11   BS-to-HEU-RAU connectivity (partially based on [40]).

possible due to the indoor environment; for example, it may be desired or required to determine the location of client devices or provide localization services for client devices, such as Emergency 911 (E911) services. If the client device is located indoors, techniques such as the Global Positioning System (GPS) may not be used to determine the location of the client's device. Further, triangulation techniques may not be able to determine the location of the client device due to the remote antenna units typically being arranged to avoid overlapping regions between antenna coverage areas [40]. Aspects of this topic are discussed in Chapters 6–8.

Campus Usage   Many venues and campuses, especially venues where there are clusters of tightly connected multistory buildings with modern heavy-duty construction, need mechanisms to provide or enhance cellular connectivity. At the practical level, a single, carrier-neutral, consolidated system is often sought in a large building or large venue setting: a carrier-neutral system avoids multiplicity of antenna distribution and sharing of resources allows more coverage and higher capacity. See Figure 3.11. A carrier-neutral DAS supports an end-user system, for example, a smartphone, regardless of which service provider the user is subscribed to. With carrier-neutral DAS arrangements, the ownership of the system is shifted from the building/venue owner, or a specific cellular carrier, to a third-party venue system provider or a DAS integrator. Obtaining wireless carrier permission and coordinating between different wireless carriers is a key planning undertaking of any successful DAS rollout.

In order to reduce the costs associated with the development of their communication systems, multiple service providers often locate their BSs at the same geographical point; the providers can then share such elements as antennas, antenna towers, primary power drops, land costs, and regulatory costs. These service providers may employ multiple RF bands, multiple channels within the same RF band, and multiple air interface standards (e.g. CDMA, UMTS, LTE). However, the cost for each service provider to extend coverage to increase capacity by deploying their own micro/pico cells and/or their own DASs could high; furthermore, in some areas where RF propagation is poor, such as sporting venues or shopping malls, the owners of such facilities may not permit the installation of such equipment by multiple service providers for aesthetic reasons or because of space limitations [5]. A "carrier-neutral" DAS is a DAS that can be used by multiple WSPs to increase the capacity and the coverage area of multiple communication systems in a zone of interest, for example, an airport, without the need for each provider to incur the cost of deploying one or more micro/pico cells or DAS. Such a system needs to be capable of simultaneously distributing signals between collocated BSs, operated by multiple service providers, and remote or fixed subscriber units; the signals may encompass multiple RF bands, multiple channels within those bands, and multiple air interface standards [5]. The goal of a "carrier-neutral" DAS is to provide a solution for WSPs that allows the service providers to cover certain specific environments at a lower cost when compared to microcells or picocells. An administrator of the "carrier-neutral" DAS is needed, perhaps a BLEC.

The use of a "carrier-neutral" DAS is then a practical solution. Three scenarios can be considered:

- Scenario/Approach S1: A DAS integrator/provider wires up a remote building or space and drops a fiber link into an existing colo rack at an existing carrier-neutral provider, thus sharing all the Base Station Hotel (BSH) colo equipment and interfaces to the various wireless providers.
- Scenario/Approach S2: A DAS integrator/provider must build out the requisite BS equipment in the colo (the colo provider only provides power, rack space, Heating, Ventilation, and Air Conditioning (HVAC), and so on). The DAS integrator/provider must also build interfaces to the wireless providers and secure business arrangement with them. The DAS integrator/provider builds out the remote buildings or venues.
- Scenario/Approach S3: A DAS integrator/provider must build out the requisite BS equipment in the colo, but the DAS integrator/provider can make use of existing interfaces and equipment to the various wireless providers. The DAS integrator/provider builds out the remote buildings or venues.

A less desirable approach is to use "donor antennas," cited earlier. These antennas are installed on the roof of a building and are pointed at "donor" cell towers; typically, a single cellular vendor is supported. The in-building arrangement is like that of a carrier-neutral arrangement, except that there typically will not be a remote BS: a combination of fiber-optic cable, coaxial cable, and in-building antennas are used to amplify and distribute those signals within a given space; coordination with the given carrier is still needed to make sure that the concentrated traffic is accepted by the provider.

To deploy a DAS in a specific public venue (preferably a "carrier-neutral" DAS) or an in-building, a detailed RF design is needed to study the signal propagation/reception at various locations within the zone to validate acceptable reception; the system installer will then attempt to construct and/or commission the DAS system to closely match the RF design. After the installer connects the remote equipment at the zone locations indicated in the design, the service commissioner needs to calibrate all the settings for all the DAS system components. The deployment process is typical of other technology deployments with some "tweaks": (i) establish requirements, especially with an RF study of the signal presents throughout the zone of interest; (ii) establish project management goals (e.g. budgets, due dates); (iii) select the technology; (iv) prepare an engineering design; (v) establish an installation plan (how, which protocols, how much cost, etc.) with the WSP; (vi) install the equipment per the design; (vii) install a communications link with the WSP; (viii) connect the equipment to distribute live signals with the WSP's head-end components; and (ix) undertake a validation/optimization study of the zone of interest to ascertain that coverage is adequate and the output power of the remote equipment is correct.

Given the mmWave transmission issues mentioned in Section 3.2.5 (the small cells, the directionality, the free-space loss, and other attenuation factors), DASs will likely play a big role in 5G, both for regular voice and data services and for IoT; this applies to both indoors/venue environments as well as for carrier-based outdoor applications. S&P Global Market Intelligence

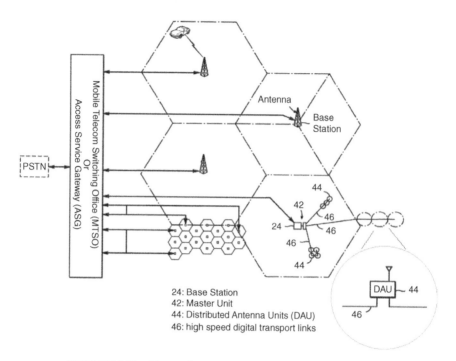

FIGURE 3.12 Illustrative example of carrier-specific DAS [5].

estimates that small-cell deployments reach approximately 850 000 in the United States by 2025 (with approximately 700 000 already deployed in 2019), with about 30% of small cell installations being outdoors; the same projection forecasts a total of 8.4 million small cells worldwide, with some regions of the world experiencing much higher deployments rates that in the United States, e.g. doubling the 2019 numbers by the year 2025. These data show that placement within buildings is a common alternative (there will be more in-building systems than outdoor systems) [41]. The large number of "small cells" forecasted (with about 70% of these considered to be indoors) supports the thesis that DASs will play a pivotal role in the future. They will be a key element of smart city IoT support, not only for traditional UEs but also for in-building sensors. In some cases, for example, in smaller/older buildings and/or in suburbia and/or for buildings very close to a 5G cell tower, a direct 5G IoT connection may suffice; but for high-density urban and smart building applications, the use of DASs seems inevitable.

Outdoor Carrier Usage DASs can also be used by a specific service provider to extend their own coverage, as depicted graphically in Figures 3.12 and 3.13; this may become more prevalent with 5G, especially at the supra-6 GHz frequencies. Figure 3.14 depicts a possible DAS arrangement for outdoor coverage in 5G environments using pole-deployed neighborhood "small cell" systems [42].

3.3.2 More Detailed Exemplary Arrangement

Figure 3.15 provides a more detailed view of a DAS system, some aspects of which are discussed next. The remote coverage areas are created by and centered on RAUs that are strategically deployed in the business zone of interest, based on the physical/RF topology of the zone. The client device can be any device that can receive (cellular) RF communication signals. The client device includes an embedded antenna (e.g. a wireless card) adapted to receive and/or send RF signals; typically, this is a smartphone. Each of the bidirectional RF bands distributed by the

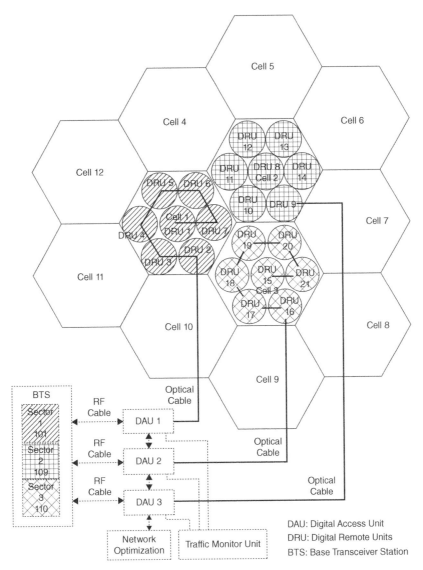

FIGURE 3.13 Illustrative example of carrier-specific DAS used to better support an entire cell [1].

DAS typically includes a distinct RF band for each of two directions of communications, UL and DL. Each of the distributed bidirectional RF bands includes a "downstream" band in which downstream RF channels are communicated for that bidirectional RF band and an "upstream" band, in which upstream RF channels are communicated for that bidirectional RF band [43].

The RAUs are connected to centralized equipment at the "headend" via a direct wireline/wireless (e.g. dedicated fiber or dedicated microwave) transmission link, or via a "donor antenna" directly to a cell tower – in the discussion in Section 3.2.5, the direct connection is assumed. Clearly, the RAUs are configured to receive the DL communications signals from the centralized equipment over this communications link. The RAUs are configured with filters and other signal processing circuits that are configured to support all or a subset of the specific communications services, that is, frequency spectrum and (sub)channel bandwidth, supported by the centralized equipment. Each of the RAUs includes an RF transmitter/receiver and a respective antenna connected to the RF transmitter/receiver to wirelessly distribute the communications services to UE within the respective remote coverage areas. The RAUs are

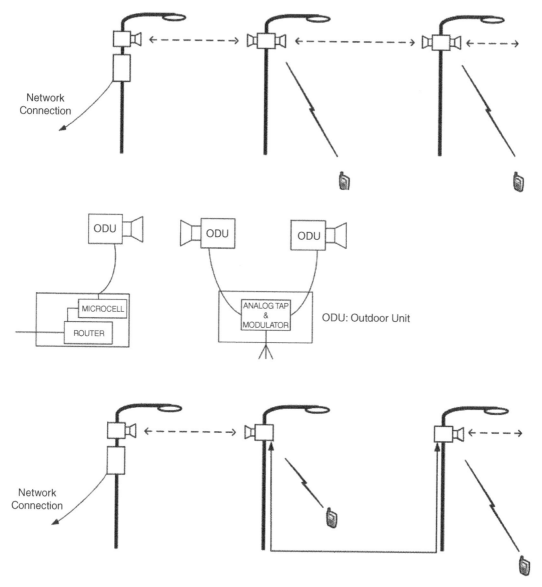

FIGURE 3.14   Possible DAS arrangement for outdoor coverage in 5G environments [42].

also configured to receive UL communications signals from the UE in the respective remote coverage areas to be distributed to the source transceiver [37].

The centralized equipment is typically coupled to a source transceiver, such as a BTS or a BBU. The host unit, the DAS headend, the HEU, is connected to the one or more BSs, BTSs, either directly (e.g. with a dedicated line, a point-to-point microwave link, or an intermediary network) or indirectly (e.g. via one or more donor antennas and one or more bidirectional amplifiers). The centralized equipment receives DL communications signals from the source transceiver, for example, a cellular carrier/Wireless Operator, which is then to be distributed to the RAUs; the DL carries voice and/or data (including video and multimedia) and/or signaling information. The centralized equipment has filtering elements that are configured to support a specific number of communications services in a particular frequency spectrum and (sub) channel bandwidth.

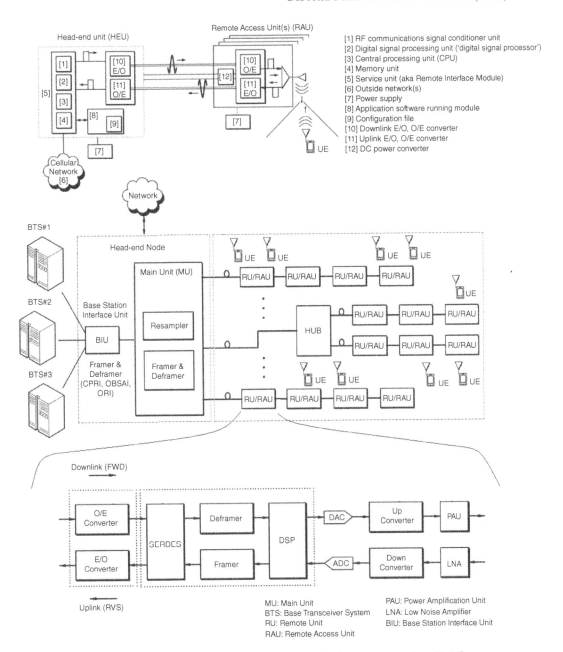

FIGURE 3.15 More detailed view of the DAS: two perspectives [2, 39].

A BTS is any station or source that provides an input signal to the HEU and can receive a return signal from the HEU. An RF source, such as a BTS, may be provided by a cellular service provider. In a typical cellular system, for example, a plurality of BTSs are deployed at a plurality of remote locations to provide wireless telephone coverage. Each BTS serves a corresponding cell, and when a mobile station enters the cell, the BTS communicates with the mobile station. The DAS provides point-to-point communications between the HEU and the RAU. Each RAU communicates with the HEU over a distinct DL and UL optical fiber pair to provide point-to-point communications. Multiple DL and UL optical fiber pairs can be provided in a fiber-optic cable to service multiple RAUs from a common fiber-optic cable. The DAS can support a wide variety of radio sources, such as LTE, GSM, CDMA, TDMA, and/or others. These radio sources can range from 400 to 2700 MHz, as an example [39].

The host unit (the DAS headend, the HEU) is connected to each remote antenna unit over a transport communication medium or media. The transport communication media can be implemented in various ways: the HEU can use any number of connectivity options to the RAUs, which may include point-to-point microwave links, fiber-optic links, free-space optical communications (which some call "wireless fiber," effectively point-to-point mmWave systems), or IP/MPLS tunnels or links.

When using fiber-optic links, an electrical-to-optical (E/O) converter is used to convert the DL electrical RF communications signals to DL optical RF communications signals to be transmitted over the DL optical fiber; in that case, the RAU(s) also include an O/E converter to convert received DL optical RF communications signals back to electrical RF communications signals to be communicated wirelessly through an antenna of the RAU to client devices located in the antenna coverage area. Similarly, in that connectivity case, an O/E converter provided in the HEU converts the UL optical RF communications signals into UL electrical RF communications signals, which can then be communicated as UL electrical RF communications signals back to a network or other source [37]. The RAU antenna is configured to receive wireless RF communications from the client devices in the antenna coverage area; the antenna receives wireless RF communications from the client devices and forwards electrical RF communications signals representing the wireless RF communications to communication link interface, that, when the link is a fiber-optic link, consists of an E/O converter in the RAU; the E/O converter converts the electrical RF communications signals into UL optical RF communications signals to be transmitted over the UL optical fiber.

The HEU includes an interface service unit that provides electrical RF service signals by passing (or conditioning and then passing) such signals from one or more outside networks via a network link. The service unit/radio interface module in the HEU can include an RF communications signal conditioner unit for conditioning the DL electrical RF communications signals and the UL electrical RF communications signals, respectively. The service unit can include a digital signal processing unit for providing an electrical signal to the RF communications signal conditioner unit that is modulated onto an RF carrier to generate a desired DL electrical RF communications signal. The digital signal processor is also configured to process a demodulation signal provided by the demodulation of the UL electrical RF communications signal by the RF communications signal conditioner unit.

In general, downstream RF signals transmitted by the BS are received at the host unit, HEU. The downstream RF signals include both downstream frequency bands distributed by the DAS. The downstream RF signals for each downstream frequency band are received on a respective downstream port of the host unit. The host unit then generates a digital representation of the downstream RF signals for each downstream frequency band. The host unit is typically configured to down-convert the downstream RF signals for each downstream frequency band to a respective lower frequency band (also called an "intermediate frequency" (IF) band. The host unit then digitizes the resulting downstream IF signals for each downstream band, thus producing digital samples of the downstream IF signals. The host unit then frames the downstream digital IF data for the downstream frequency bands together (along with appropriate overhead data) and communicates the frames to each of the remote antenna units over the respective optical fibers. The downstream signal that is communicated to each remote antenna unit is also referred to as a "downstream transport signal." The downstream transport signal that the host unit generates for each remote antenna unit is an optical signal that is produced by optically modulating a downstream optical carrier with the downstream framed data (which contains the downstream digital IF data for the downstream frequency bands). Each remote antenna unit receives the downstream transport signal that is communicated to that remote antenna unit over a respective optical fiber. In general, each remote antenna unit demodulates the optical downstream transport signal (or otherwise performs an O/E process) in order to recover the downstream framed data transmitted by the host unit. The remote antenna unit then extracts the downstream digital IF data for each of the downstream frequency bands. Each remote antenna unit, for each downstream frequency band, uses digital filtering techniques and/or digital signal processing on the downstream digital IF data for that downstream frequency band in order to apply one or more of the following: (i) pre-distortion to compensate for any nonlinearities in the downstream signal path; and (ii) phase and/or amplitude changes for beamforming or antenna steering. Then, for each downstream frequency band, the resulting digital IF data are applied to a digital-to-analog converter to produce a downstream analog IF signal for that downstream frequency band. The analog IF signal for each downstream frequency band is then up-converted to the appropriate RF frequency band and band-pass filtered to remove any unwanted harmonics and any other unwanted signal components. Thereafter, the resulting analog RF signal for each downstream frequency band is power amplified and is ready to be radiated from at least one antenna associated with the remote antenna unit [43]. The reverse process for the upstream direction is symmetric to the description in Section 3.2.5: the upstream RF signals for each upstream frequency band distributed by the DAS are received on at least one antenna at each remote antenna unit. Each remote antenna unit then generates a digital representation of the upstream RF signals for each upstream frequency band. And so on.

3.3.3 Traffic-aware DAS

One of the most difficult challenges faced by wireless network operators is the physical movement of subscribers from one location to another, particularly when wireless subscribers congregate in large numbers at one location. An example is a business enterprise facility during lunchtime when a large number of wireless subscribers visit a lunchroom or cafeteria location in the building. At that time, subscribers have moved away from their offices and usual work areas, and, likely, there are many locations throughout the facility with very few subscribers. Similar examples can be envisioned. A system for dynamically routing signals in a DAS would be useful.

An example of a traffic-aware DAS is discussed in [1]. This system uses a plurality of Digital Access Units (DAUs) that are interconnected and can route signals between them; the system also includes a plurality of Digital Remote Units (DRUs) which connect to the various DAUs transmitting UE signals to them. The DAUs receive DL signals from the BTS(s) (and, conversely); the DAUs translate the RF signals to optical signals for the purpose of sending (and conversely receiving) session information – specifically media – to the DRUs. The traffic monitoring capability and network optimization functionality provide dynamic traffic management. The traffic monitoring unit tracks the traffic load at each DAU in the network; the traffic load associated with each DAU is collected and stored in the network optimization unit. The optimization unit calculates the overall DAS network performance and determines the optimum reconfigured network to improve or maximize performance. Utilizing the traffic monitoring unit in conjunction with the DAUs provides optimization of the DAS network through a traffic monitoring unit external to the DAUs. The system of Figure 3.13 discussed earlier depicts such a system (although the figure was depicting a metropolitan setting rather than an office complex setting). Using these and comparable mechanisms, wireless network operators can address the challenge of building networks that effectively manage high data-traffic growth rates, especially in the context of high mobility and an increased level of multimedia content, such as video-enabled websites, YouTube content, and entertainment-quality streaming to mobile devices, not in a Wi-Fi footprint.

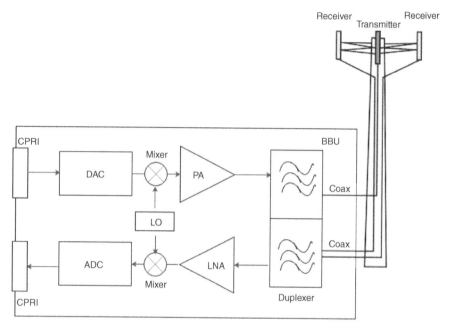

FIGURE 3.16 RRH sections.

3.3.4 BBU and DAS/RRU Connectivity

There is a practical preference for using fiber-optic systems to connect a BBU and RRU elements, rather than other connectivity arrangements (although other arrangements are not precluded). It was noted earlier that RRUs/RRHs are connected to the BTS/NodeB/eNodeB with a fiber-optic link using a standardized interface. As noted in footnote 1, a RRH is another term RRU used in some contexts: it is the RF circuitry of a BS enclosed in a small outdoor module, performing all RF functionality that includes transmit and receive functions, filtering, and amplification. It also contains analog-to-digital or digital-to-analog converters and up/down converters; additionally, it can also provide advanced monitoring and control features. The RRH/RRU is typically mounted near the antenna to reduce transmission line losses and is connected to the main, digital portion of the BS (specifically, the BBU) with an optical fiber. The Common Public Radio Interface (CPRI) is the standardized protocol for communication between the BBU and RRH. As seen in Figure 3.16, the RRH has two sections, a transmit part and a receive part. The transmit section usually consists of a DAC, Mixer, Power Amplifier and Filters; a digital signal is received via a CPRI interface, converted to analog, upconverted to an RF frequency, amplified, filtered, and then sent out via an antenna. The receive section consists of a filter, Low Noise Amplifier, Mixer, and an ADC. It receives a signal from the antenna, filters it, amplifies it, down-converts it to an IF frequency, and then converts it to a digital signal before sending it out via the CPRI to fiber for further processing, as discussed in Section 3.2.5.

Backhaul mechanisms connect the wireless network to the wired network by backhauling traffic from dispersed cell sites to MSCs. These links are typically either traditional transmission systems (such as SONET or point-to-point microwave at various operating bands) or Ethernet-over-Fiber links (e.g. 1 GbE or 10 GbE). A UMa site has BBU that processes user and control data, which is in turn connected to an RRU/RRH to generate radio signals transmitted over the air via the tower-mounted antennas.

Fronthaul is related to a new type of RAN architecture that is comprised of centralized baseband controllers and stand-alone radio heads installed at remote UMa or UMi sites, possibly many miles away. In the fronthaul model, the BBU and RU equipment is located further away

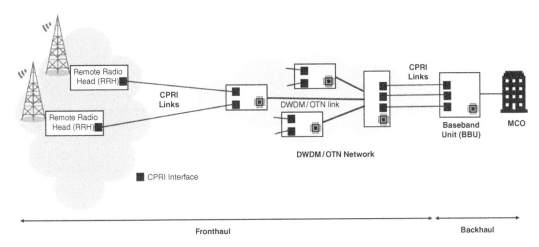

FIGURE 3.17   Fronthaul and backhaul.

from each other than is the case in the backhaul model. The RRU equipment (the RRH) is still located at the cell site, but the BBU is relocated to a centralized location where it supports multiple RRHs. See Figure 3.17. The optical links that interconnect the newly centralized BBU and the multiple RRHs are referred to as fronthaul. The use of fronthaul-based C-RAN architectures typically improves cell edge performance.

A few years ago, a consortium of wireless equipment vendors standardized the CPRI protocol that runs over these fronthaul links. CPRI is now an industry standard that supports connectivity between the BBU and RRU/RRH.[11] More recently, a newer eCPRI 1.0 interface has been defined; additionally, work is underway to define a more detailed interface. The tight performance requirements of CPRI/eCPRI – capacity, distance, and latency – drive toward fiber connectivity such as DWDM (or more specifically OTN [Optical Transport Network]) systems between centralized BBUs and the RRHs. Ethernet-based solutions have existed for several years using mmWave spectrum. Work is underway in 3GPP to define backhauling solutions using the same spectrum as access. A DAS can also interface with a BS through a digital interface standard, e.g. with CPRI.

According to the CPRI standards, the BS consists of two parts: a BBU (also called in this context a Radio Equipment Control unit) and an RRU (also called, in this context, Radio Access Equipment). The main feature of CPRI interfaces is the separation between the base frequency band and the RF band. Small size, ease of installation, and a variety of functions with low power consumption allow the use of the BBU on existing physical objects and support the extension in the cascade connection mode: due to its small size and weight, the RRU is easily mounted on a pole or wall near the antenna, maximizing its radio coverage. CPRI standardizes interfaces between the BBU and RRU, ensuring compatibility of equipment from different manufacturers. CPRI bit rates are referred to as "option numbers." There are 10 options, according to CPRI Specification V7.0 [44], as seen in Table 3.6.

---

[11] In 2002, Nokia, NEC, LG. and Samsung launched the Open Base Station Architecture Initiative (OBSAI) with the goal of standardizing the architecture of wireless base stations, internal interfaces, control modules, transmission modes, baseband band and radio frequencies. However, the usability of OBSAI was limited to a handful of manufacturers. In June 2003, Ericsson, Huawei, NEC, Nortel Networks, and Siemens created an alternative organization for the development of the CPRI, it also began developing universal standards for key internal interfaces with an emphasis on interfaces between the base frequency band and radio frequencies (NEC moved to the CPRI camp in the mid-2000s); currently, more than 100 manufacturers have joined the organization supporting CPRI [44].

TABLE 3.6   CPRI Rates

Option	Bit Rate (Gbps)	Line Coding
Option 1	0.6144	8B/10B
Option 2	1.2288	8B/10B
Option 3	2.4576	8B/10B
Option 4	3.0720	8B/10B
Option 5	4.9152	8B/10B
Option 6	6.144	8B/10B
Option 7	9.8304	8B/10B
Option 7A	8.11008	64B/66B
Option 8	10.13760	64B/66B
Option 9	12.16512	64B/66B
Option 10	24.33024	64B/66B

In a DAS environment, the HEU receives signals according to a digital interface standard through the corresponding BTS; from the BS, the HEU receives digital data according to the digital interface standard, e.g. CPRI. As described in [2], the HEU deframes CPRI data as raw data according to the digital interface standard, received from the BS via a BS-side deframer; the deframed digital data are converted into digital data with a frame rate according to a frame standard used in the DAS, through a resampling process. The digital data converted suitable for the DAS frame rate are again deframed through an RU-side framer to be transmitted to the RU as a lower node unit through a transport medium. The reason why the HEU resamples the digital data according to the digital interface standard (to be suitable for the DAS frame rate) is as follows: in the digital interface standard such as the CPRI, the sampling rate with respect to bandwidth is relatively high; stuffing bits are added to a payload portion corresponding to user data, and so, the number of bits in data transmission increases: the transmission efficiency with respect to the bandwidth is lowered due to line coding such as 8B/10B. In a DAS network, a plurality of digital interfaces is often employed to support a plurality of bands, providers, sectors, MIMO, etc., the transmission efficiency of data is lowered; therefore, the transmission efficiency of data is low in the actual payload transmission, therefore, the transmission capacity of data increases when the data are deframed.

3.3.5   Ethernet/IP Transport Connectivity of DAS

Legacy DASs are analog systems, where the DAS traffic is distributed between the master units/HEU and the RRUs in analog form; another type of DAS is a digital DAS where the traffic is distributed between the master units and the remote units in digital form. Existing IP/Ethernet infrastructure has typically not been used to distribute digital DAS traffic among nodes of a digital DAS along with other non-DAS traffic (for example, IT traffic or Ethernet-based RAN traffic). Some proposals have been made, as discussed in [45].

REFERENCES

1. Stapleton, S.P. and Hejazi, S.A. (2019). Optimization of traffic load in a distributed antenna system. US Patent 10,506,454; 10 December 2019; filed 24 August 2013. Uncopyrighted.
2. Kim, H. and Kim, D. (2019). Digital data transmission in distributed antenna system. US Patent 10,383,171; 13 August 2019; filed 20 December 2017. Uncopyrighted.
3. Ranson, C.G., Phillips, F.W., and Kummetz, T. (2020). Transport data reduction for DAS systems. US Patent 10,742,348; 11 August 2020; filed 33 October 2017. Uncopyrighted.

4. Rahman, I. and Kazmi, M. (2020). Indication to the master e-node B of successful primary secondary cell activation In dual connectivity. US Patent 10,728,944; 28 July 2020; filed 29 September 2015. Uncopyrighted.

5. Kummetz, T., McAllister, D.R., Pagano, C., et al. (2019). Distributed antenna system for wireless network systems. US Patent 10, 499, 253; 3 December 2019; filed 13 December 2012. Uncopyrighted.

6. Littelfuse Staff. Application Note: Distributed Base Stations. www.littelfuse.com (accessed 29 September 2020).

7. Sun, S., Rappaport, T., Rangan, S. et al. (2016). Propagation Path Loss Models for 5G Urban Micro and Macro-Cellular Scenarios. *2016 IEEE 83rd Vehicular Technology Conference* (VTC2016-Spring).

8. Ghosh, A. and Nammi, S. (2019). Adapting demodulation reference signal configuration in networks using massive MIMO. US Patent 10, 397, 052; 2 February 2019; filed 10 August 2017. Uncopyrighted.

9. Minoli, D. and Schneider, K. (1977). A Technique For Establishing The Minimum Number Of Frequencies Required For Urban Mobile Radio Communication. IEEE Trans. on Comm., pp. 1054–1056.

10. Minoli, D. (1975). Use of matrices in the four color problem. PME Journal 5, Spring (10): 503–511.

11. Frenkiel, R. (1979). Cellular Radiotelephone System Structured For Flexible Use Of Different Cell Sizes. US Patent 4, 144, 441, 3 March 1979; filed 22 September 1976. Uncopyrighted.

12. 3GPP TS 36.300: E-UTRA and E-UTRAN Overall Description; Stage 2.

13. 3GPP TR 25.913: "Requirements for Evolved UTRA (E-UTRA) and Evolved UTRAN (E-UTRAN)".

14. 3GPP TS 36.201: "Evolved Universal Terrestrial Radio Access (E-UTRA); Physical layer; General description".

15. 3GPP TS 36.211: "Evolved Universal Terrestrial Radio Access (E-UTRA); Physical Channels and Modulation".

16. 3GPP TS 36.212: "Evolved Universal Terrestrial Radio Access (E-UTRA); Multiplexing and Channel Coding".

17. 3GPP TS 36.213: "Evolved Universal Terrestrial Radio Access (E-UTRA); Physical Layer Procedures".

18. ETSI, "ETSI TS 136 331 V8.8.0 (2010–02) LTE; Evolved Universal Terrestrial Radio Access (E-UTRA); Radio Resource Control (RRC); Protocol specification (3GPP TS 36.331 version 8.8.0 Release 8)"; February 2010. ETSI, 650 Route des Lucioles, F-06921 Sophia Antipolis Cedex, France.

19. Parkvall, S., Dahlman, E., Skold, J. et al. (2008). 3G Evolution: HSPA and LTE for Mobile Broadband, 2e. Academic Press.

20. 3GPP TS36.300. "Evolved Universal Terrestrial Radio Access (E-UTRA) and Evolved Universal Terrestrial Radio Access Network (E-UTRAN): Overall Description".

21. Astély, D., Dahlman, E., Furuskär, A. et al. (2009). LTE: the evolution of mobile broadband. IEEE Communications Magazine 47 (4): 44–51.

22. Yi, S. (2020). Method for Transmitting Information for LTE-WLAN Aggregation System And A Device Therefor. US Patent 10, 785, 162; 22 September 2020; filed 22 February 2017. Uncopyrighted.

23. 3GPP. Technical Specification TS 36.101: Evolved Universal Terrestrial Radio Access (E-UTRA); User Equipment (UE) Radio Transmission and Reception, Release 16. a 2000-page document first published in 2007 and revised many numerous times, the latest at press time being V16.6.0 (2020–06).

24. Contento, M. (2019). C-Band Spectrum: The Next Step Toward Bringing 5G to Life. https://www.telit.com/blog/c-band-spectrum-the-next-step-towards-bringing-5g-to-life (accessed 18 September 2020).

25. Paolini, M. (2018). The total cost of ownership (TCO) for fixed OnGo in the 3.5 GHz CBRS band. *White Paper* www.senzafiliconsulting.com.

26. Minoli, D. (2015). Advances in Satellite Communication – The Industry Implications of DVB-S2x, HTS, UltraHD, M2M, and IP. Wiley.

27. Minoli, D. (2009). Satellite Systems Engineering in an IPv6 Environment. Francis and Taylor.

28. Federal Communications Commission. Public Notice: Wireless Telecommunications Bureau Announces Accelerated Clearing In The 3.7–4.2 Ghz Band. GN Docket No. 18–122, Released: 1 June 2020. FCC, 445 12th St., S.W. Washington, D.C. 20554.

29. ITU. Report ITU-R M.2410–0, Minimum Requirements Related To Technical Performance For IMT-2020 Radio Interface(s). M Series: Mobile, Radiodetermination, Amateur And Related Satellite Services; 2017.

30. Minoli, D. and Occhiogrosso, B. (August 2019). Practical aspects for the integration of 5G networks and IoT applications in smart cities environments", Special Issue titled. Integration of 5G Networks

and Internet of Things for Future Smart City, Wireless Communications and Mobile Computing 2019: 5710834, 30 pages. https://doi.org/10.1155/2019/5710834.

31. 3GPP. *Technical Specification TS 38.101–1: NR; User Equipment (UE) Radio Transmission And Reception; Part 1: Range 1 Standalone*, Release 15. Initially published 27 June 2017; updated 02 January 2018.

32. Doyle, L. (2019). When Private 4G LTE is Better Than Wi-Fi. Network World, 19 August 2019.

33. Motorola Staff. What Is Private LTE? www.motorolasolutions.com (accessed 10 December 2020).

34. Westrup, W. (2020). Should You Build a Private 5G or LTE Network? 23 October 2020, https://www.sierrawireless.com/iot-blog/what-are-private-lte-networks (accessed 10 December 2020).

35. Staff of SDxCentral Studios. What Is 5G Network Slicing? 2 January 2018. Available online on 10 December at https://www.sdxcentral.com/5g/definitions/5g-network-slicing.

36. Nordrum, A., Clark, K., and IEEE Spectrum Staff. (2017). Everything You Need to Know About 5G Millimeter waves, massive MIMO, full duplex, beamforming, and small cells are just a few of the technologies that could enable ultrafast 5G networks. IEEE Spectrum 27 January 2017.

37. Kruh, L.M. and Roark, B.R. Implementing a live distributed antenna system (DAS) configuration from a virtual DAS design using an original equipment manufacturer (OEM) specific software system in a DAS. US Patent 10, 390, 234; 20 August; filed 21 May 2018. Uncopyrighted.

38. Zavadsky, D. and Fischer, L.G. (2019). Distributed antenna system with dynamic capacity allocation and power adjustment. US Patent 10, 374, 665, 6 August 2019; filed 7 May 2018. Uncopyrighted.

39. Cune, W.P., Deutsch, B.A.M. Jason Elliott Greene, Thomas Knuth (2019). Distributed antenna system architectures. US Patent 10, 349, 156, 9 July 2019; filed 19 October 2018. Uncopyrighted.

40. Sauer, M. (2019). Apparatuses, systems, and methods for determining location of a mobile device(s) in a distributed antenna system(s). US Patent 10, 448, 205; 15 October 2019; filed 22 January 2018. Uncopyrighted.

41. Small Cell Forum, Small Cells Market Status Report. 3 December 2018, Figure 3–2. 19 February 2018. Published online at http://www.scf.io/en/documents/050_Small_cells_market_status_report_February_2018.php?utm_source=Email%20campaign&utm_medium=eshots&utm_campaign=member%20eshot (accessed 19 February 2018).

42. Barzegar, F., Barnickel, D.J., Gerszberg, I. et al. (2019). Remote distributed antenna system. US Patent 10, 484, 993; 19 November 2019; filed 29 August 2018. Uncopyrighted.

43. Stewart, K.A. and Fischer, L.G. (2019). Distributed antenna system architectures. US Patent 10, 389, 313; 20 August 2019; filed 28 August 2017. Uncopyrighted.

44. Vitols, A. CPRI vs OBSAI. https://edgeoptic.com/kb_article/cpri-vs-obsai (accessed 22 September 2020).

45. Lange, K.K. (2019). Hybrid RAN/digital DAS repeater system with Ethernet transport. US Patent 10, 355, 753; 16 July 2019; filed 29 January 2018. Uncopyrighted.

4 Traditional Sensor Networks/IoT Services

This chapter provides a high-level overview of the Internet of Things (IoT) field.[1,2] The basic concept of the IoT is to provide intelligent capabilities to monitor, collect, transmit, and process information generated by a large number of dispersed devices in order to support automated (machine-to-machine or people-to-machine) passive and/or active control of these devices or some aspects of their surroundings. The concept started to emerge in the early 1990s but has blossomed into a full-fledged discipline since the mid-2000s.

4.1 OVERVIEW AND ENVIRONMENT

Numerous definitions and descriptions of the IoT exist. One fundamental/descriptive quote from the author's previous work [1, 25] is as follows:

> The basic concept of the IoT is to enable objects of all kinds to have sensing, actuating, and communication capabilities, so that locally-intrinsic or extrinsic data can be collected, processed, transmitted, concentrated, and analyzed for either cyber-physical goals at the collection point (or perhaps along the way), or for process/environment/systems analytics (of a predictive or historical nature) at a processing center, often "in the cloud". Applications range from infrastructure and critical-infrastructure support (for example, smart grid, smart city, smart building, and transportation), to end-user applications such as e-health, crowdsensing, and further along, to a multitude of other applications where only the imagination is the limit.

In broad terms, IoT refers to an evolving generation of systems with integrated computational and physical capabilities that can interact with analytics and/or with humans in numerous ways: typical functionality of IoT includes embedded intelligence; device connectivity and authentication; status, parameter, or process sensing; data capturing; preliminary or optional data analysis at the network edge; data transfer over a backbone; and analytics at the core or

---

[1] Portions of this chapter are based on the 26 recent papers and textbooks authored (and/or co-authored) by the author on the IoT topic, [1–26].
[2] The term Cyber-Physical Systems (CPSs) has also been used to describe IoT systems. This term was originally introduced in 2006 at a National Science Foundation (NSF) workshop in Austin, TX, where it was defined as *a system composed of collaborative entities, equipped with calculation capabilities and actors of an intensive connection with the surrounding physical world and phenomena, using and providing all together services of treatment and communication of data available on the network* [27]. While some researchers consider CPSs as distinct from IoT systems, others broadly equate the two concepts, e.g. [28–33], as a short list of references – in fact, the National Institute for Standards and Technology (NIST) observes that "*CPS and related systems (including the Internet of Things, Industrial Internet, and more) are widely recognized as having great potential to enable innovative applications and impact multiple economic sectors in the world-wide economy,*" thus drawing a relationship between the two concepts [34].

High-Density and De-Densified Smart Campus Communications: Technologies, Integration, Implementation, and Applications, First Edition. Daniel Minoli and Jo-Anne Dressendofer.
© 2022 John Wiley & Sons, Inc. Published 2022 by John Wiley & Sons, Inc.

cloud (e.g. Big Data analysis, Artificial Intelligence [AI]) [35–37]. The IoT ecosystem consists of computer networks and a variety of devices with built-in processors that can control physical processes by means of feedback [38–40]; the IoT infrastructure allows IoT-enabled devices to operate remotely across a communications infrastructure to integrate the physical world into computerized systems utilizing a variety of networking technologies; in this environment, distributed entities, such as instrumented things, exchange and process information often without human intervention. IoT devices can be stationary or mobile, simple or complex [25, 34, 41, 42]. Some define the term "Internet of Everything (IoE)" as being the technological combination of the IoT technology and the Big Data processing technology through connections with cloud servers [43] (also including the Vehicle-to-Everything [V2X] environment). Figures 4.1–4.3 pictorially depict some aspects of the IoT ecosystem.

The value of IoT derives from the environmental and contextual information one can capture and from the actionable decision-making that comes from such aggregated data. Thus, IoTs are systems of computation-enabled entities that are integrally and intensively coupled with the surrounding physical world and its ongoing processes, simultaneously providing and using data-accessing and data-processing services available on a public or private cloud and/or on the Internet. Some perceive the IoT as an orchestration of computers and physical systems, where embedded computers monitor and control physical processes, usually making use of feedback loops, and where the physical processes affect computations, and vice versa [44–49]. There is a large body of recent and ongoing research concerning IoT technologies, including Wireless Sensor Networks (WSNs), Machine-to-Machine (M2M) networks, and Machine Type Communication (MTC) [26]. In terms of deployment, in the immediate or near future, "*it is expected that IoT will be used to automate nearly every field of human endeavor*" [50].

Traditionally, the IoT has been envisioned as supporting a large population of relatively low-bandwidth parameter-sensing devices, particularly in M2M environments and generally in stationary locations (e.g. data-collecting meteorological weather stations, electric meters, industrial control, and the like). However, the IoT has been taxonomized more inclusively as a computing paradigm comprising (i) the just-cited M2M communications, (ii) Human-to-Machine (H2M) communications, (iii) Machine in (or on) Humans (MiH) communications, and even (iv) Human-to-Human (H2H) communications. MiH devices include Global Positioning

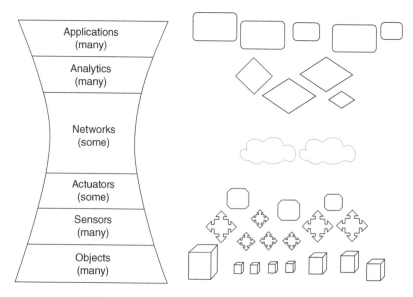

FIGURE 4.1 A logical view of an IoT ecosystem.

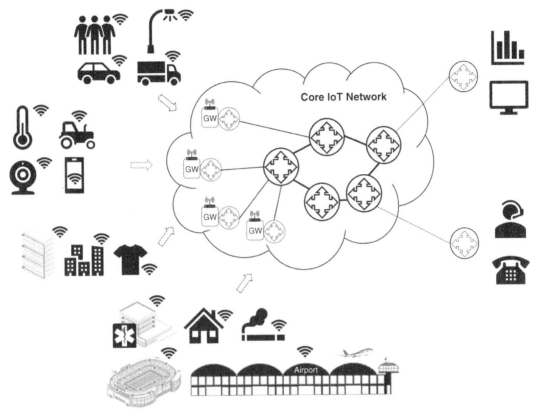

FIGURE 4.2 Applications scope of IoT/CPS (examples).

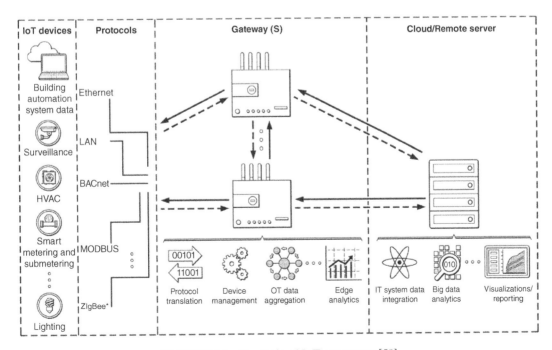

FIGURE 4.3 Example of IoT ecosystem [50].

System (GPS) tracking devices, embedded chips, and medical monitoring probes, among other devices [16, 25]. Mobility of the IoT devices is also becoming more prevalent.

The underlying technology of the IoT spans (i) sensors; (ii) access networks – especially wireless systems for Personal Area Networks (PANs), or, more generally, WSNs; (iii) edge networks (also called fogs by some); (iv) core networks – such as metro Ethernet systems, metro IP networks, encrypted Internet tunnels, or cellular links; and (v) cloud and/or Internet services and data analytics/Big Data systems. Sensors include devices that can address the following among others: ambient light, position, presence, proximity, motion, acceleration, tilt, force, torque, pressure, velocity, displacement, sound, vibration, temperature, humidity, gasses, ambient conditions, magnetic fields, optical surveillance video, and frame grabs. The data can be end-system data (e.g. device status, patient monitoring); proximity-environment sampled data (e.g. site temperature, vehicle or device location and/or proximity, traffic patterns, site parameters); or interactive data (e.g. one- or two-way multimedia streams, video surveillance). The commands can control (actuation) information, for example, resetting a device or system parameter or performing some action (e.g. for a road signal, a dam door, a grid transfer switch, a drone function, or a remote robot action – industrial process control may also entail object, people, or robot occupancy and positioning).[3]

As hinted in Figure 4.2, IoT applications span many realms and range from supporting mission-critical environments (e.g. Intelligent Transportation Systems [ITSs], smart grid, industrial process control, video surveillance, and e-health) to facilitating business-oriented transactions (e.g. contracts, insurance, banking, and various logistics). Documented applications include but are certainly not limited to: process control, physical security systems (access control and monitoring), asset management, smart buildings and campuses, smart grid and renewable energy systems, ITSs, automotive systems, air traffic control, and safety systems, biomedical and healthcare systems, manufacturing process control, military systems, aircraft instrumentation, water management systems, and distributed robotics and drones [35, 38, 44, 45, 52]. Smart-home devices include but are not limited to appliances, refrigerators, air conditioners, washers, TVs, lights, ovens, or outdoor grills, all of which are, or soon will be, capable of wireless communication, directly or via a gateway; smart-building (and/or smart venues) sensors include but are not

---

[3]Control of telecommunication Network Elements (NEs), in a broad sense, has existed for many years, as implied by the list that follows; however, in IoT, much lower-end entities are instrumented (for monitoring and control). Looking retrospectively, examples of real-time Network Elements control include: (i) control of routers detecting failure of a link transmission by resetting the internal routing table; (ii) control of routers detecting overload of a link transmission by resetting the internal routing table; (ii) control of routers detecting Quality-of-Service issues in overload buffers managing outgoing transmission by resetting internal transmission pointers and priorities in internal buffer-management and transmission processes; (iv) control of equipment handling Intranet handoff to Internet (or vice versa) when detecting congestion by resetting the load balancers setting for routing of flow control, content switchover, or failover; (v) Tracking, Telemetry and Control (TT&C) (also called tracking, telemetry and command by some) to establish precise location of satellites and issue commands to properly position them in their "box(es)" – e.g. adjusting the satellite's orbit, controlling orbital maneuvers, realigning the solar panels, performing system back-up, performing equipment tests, addressing anomalies and failures (geosynchronous satellite must be maintained within the 150 km assigned box and in cases of orbital slot colocation the multiple slots must kept at a suitable distance); (vi) implementation of traffic management policies in large carrier networks by real-time control of access routers, aggregation routers, core routers, and switches, more recently also in the context of Network Function Virtualization, where NEs are instantiated as needed; (vii) implementation of broadband fair use and traffic management policies in broadband networks, for example at edge devices: fair use (or fair usage) policies make up part of an Internet provider's contract – they are conditions placed on broadband customers designed to limit their data usage, especially at peak times. Some specific control functions in IoT environments include (a) controlling HVACs in smart building; (b) controlling lights based on occupancy or daylight; (c) site-based Demand/Response interactive power usage control to optimize power consumption and deal with peak times; and (d) controlling a connected grill ([51]). The sensors range from software probes, to integrated daemon modules, triangularization antennas, thermostats, optical sensors, consumption meters, and electronic sensors in the grill (or smart oven).

limited to smoke detectors, thermometers and thermostats supporting Heating, Ventilation, and Air Conditioning (HVAC) systems, lighting sensors, hygrometers, video cameras, video monitors, and/or various electronic nodes [1, 2, 25, 26]. Among the multitude of applications, the IoT can be used for active and passive asset monitoring; and among the multitude of active and passive asset monitoring applications, some deal with managing assets in a large commercial airport (as discussed in Chapter 9).

In retail and hospitality, IoT may be used in hotels, amusement parks, restaurants, sports and entertainment sites, vending machines, and so on. In healthcare, IoT may be used in hospitals, home care, healthcare plans, medical laboratories, biotechnology, and medical appliances and equipment. At the beginning of the 2020s decade, as countries across the world deal with the novel coronavirus, COVID-19, there has been a significant effort to integrate inputs from a panoply of sources, from surveillance cameras to point-of-sale systems and connected health devices; this trend has highlighted the power of IoT deployments for public data collection. South Korea, for example, has aggregated data from IoT devices and smartphones to create detailed logs of its citizens' behavior; the nation has documented where residents went, how long they stayed, who accompanied them, and whether they were wearing a mask [53]. Some other documented IoT use cases include [54]: (i) detecting symptoms with simple processors such as Raspberry Pi; (ii) using smartphones to report sickness; (iii) mapping the pandemic outbreak; (iv) caring for patients using wearables; and (v) measuring air quality with IoT sensors.

IoT systems can also provide or support services ranging among the following: predictive maintenance; process or quality control; fleet management; supply chain and inventory management; smart infrastructure; building optimization; remote asset management; secure access control; point-of-sale support; energy management; compliance; and safety, among many others [50].

Human interactions often benefit from visual or multimedia communications; for example, applications in the H2H and H2M environments typically require visual or multimedia communications, including voice, video, data, and sensory information; consequently, evolving IoT applications can also entail video/multimedia transmissions. It should be further noted that while legacy multimedia applications deal with the transmission of point-to-point video, point-to-multipoint video, a few-multipoint-to-point video (e.g. security monitoring), or at most, a-few-multipoint-to-a-few-multipoint video (e.g. a small multi-party video conference), multimedia-based IoT often also entails massive-multipoint-to-point applications (e.g. a large smart-city application to monitor an entire environment), or also entails massive-multipoint-to-massive-point applications (e.g. an advanced social media application, seen from an aggregate perspective). In some of these applications (e.g. mass surveillance), unlike the traditional multimedia use cases where humans are involved in processing and decision-making, the pertinent decisions are likely made by some AI system as a result of still image-, video-, and audio-processing through cloud computing and/or distributed fog processing.

To support these applications, various types of sensors are dispersed in the environment; these sensors can be stationary or mobile, depending on the application. Often, but not always, sensors rely on (limited) onboard power, in the form of small batteries intended to last up to 10 years. The sensors can forward the data locally to an aggregation point over some Local Area Network (LAN) system or to a remote aggregation point using specific Wide Area Network (WAN) technology. Very often (but not always, or exclusively), sensors utilize *wireless* technologies to forward their data to an aggregation point or, possibly, to a final destination point, such as an analytics engine; typically, Low Power WAN (LPWAN) technologies are utilized (these are further discussed below). Edge aggregation in the form of a routing point or node (or, also in the form of computing and processing – for example, data correlation, reduction, or accumulation), is often employed. This hierarchical model not only results in fewer nodes accessing the cloud-resident analytics directly, thus reducing network traffic, but also supports lower-power transmission (and, in turn, battery life) by requiring just enough transmission power to reach a

TABLE 4.1 IoT Deployment Synthesis

Area	Subdiscipline
IoT/M2M technologies	Sensors, including electric and magnetic field sensors; radio wave sensors; optical, electro-optic, and infrared sensors; radars; lasers; location/navigation sensors; seismic sensors; environmental parameter sensors (e.g. wind, humidity, heat); pressure-wave/presence sensors, biochemical and/or radiological sensors, gunshot detection/location sensors, and vital sign sensors for e-health applications
	Networking (especially wireless technologies for personal area networks (PANs), fogs, and cores, such as 5G cellular)
	Analytics
System architectures	Proposed IoT Architectures, e.g. Arrowhead Framework, Internet of Things Architecture (IoT-A), the ISO/IEC WD 30141 Internet of Things Reference Architecture (IoT RA), and Reference Architecture Model Industrie 4.0 (RAMI 4.0)
	M2M Architecture (ETSI High-level architecture for M2M)
	Architectures particularly suited for SCADA-based legacy systems
IoT/M2M standards	Layer 1, Wireless (ISM, PAN, LPWAN)
	Layer 2/3, IP, IPv6, MIPv6
	Upper Layers, e.g. TCP, UPnP
	Vertical-specific
Cybersecurity	Confidentiality
	Integrity
	Availability

neighborhood (or neighboring) node [3]. A snapshot of the requisite synthesis to achieve broad-scale deployment of the IoT is shown in Table 4.1 [10].

As implied above, IoT systems and methods utilized to support the connected ecosystem include the following functions: (i) acquisition, (ii) transport, (iii) aggregation, (iv) analysis, and (v) control, as listed in Table 4.2 [50]. Consider, as an example, the use case of environmental monitoring of an office building. Likely, there would be sensors installed throughout the building to monitor temperature and carbon monoxide levels; this is part of the data aggregation component of IoT. As part of the transport component, there are several lower-layer protocols that may be used to transport the collected data, including BACnet (Building and Automation Control) networks, Wi-Fi networks, BLE (Bluetooth Low Energy) networks, ZigBee networks, and others, that may use IoT gateway(s) to translate the sensors' data or communications to a cloud or remote backend server. Gateways provide the bidirectional wireless connectivity supporting the secure data transmission necessary to connect the IoT to the IT systems that ultimately extract and deliver valuable information. Historically, endsystem devices have been responsible for negotiating associations with nearby gateways to support their communications [55] (this being the case for the current Wi-Fi and cellular networks). From the gateway(s), data are communicated through an aggregation layer. The aggregated data reach a cloud-based analysis engine that enables the system to respond to changes in the monitored environment (e.g. such a system may be used to enforce a rule such that, if a room temperature gets too hot, a command is issued to the HVAC/chiller to lower the temperature). The cloud-based system can be used to administer multiple dispersed sites, for example, when a commercial real estate landlord owns and manages systems or HVACs in multiple buildings.

There are proposals for the development of IoT networks that support the integration of a plethora of sensing technologies, such as sound, light, electronic traffic, facial and pattern recognition, smell, vibration, into devices that are capable of autonomous organization; this

TABLE 4.2   IoT Systems and Methods to Support the Connected Ecosystem

Function	Description
Acquisition	Systems and methods that encompass the hardware – the IoT devices – that capture data from interactions with the environment, with other machines, with humans or other living things and make it available for transmission over a network. Sensors are the endpoints of IoT, and a growing assortment of devices are collecting data. There are cases where historical data may be analyzed to provide business practice improvements (not all data need to originate from sensors).
Transport	Systems and methods to take the acquired data from various sensors and move it over a network for aggregation and analysis. As the name implies, the final stages of data transport generally flow over communication protocols (e.g. TCP/IP and others referred to below) and may be transformed into Internet-friendly data formats like JSON or XML.
Aggregation	Systems and methods responsible for the collection and distribution of output data to designated consumers. Consumers can include databases, on-site services, analytics services, enterprise service buses, third-party cloud services and similar repositories.
Analysis	Systems and methods to take the aggregated data and turn it into operational insights by applying context-specific algorithms, rules, and predictive models. In most deployments, IoT analytics may include feedback capabilities so that the predictive models get better over time. In some cases, analytics may be run in the cloud or on the edge such as on an IoT gateway device.
Control	Systems and methods to provide the ability to act based upon insights gleaned from the analysis of IoT data. Different types of control actions available may range from graphical representation to humans, who can then take manual actions, to fully autonomous systems that can take orchestrated actions to improve operations, recognize and mitigate failures and threats, and prevent hazards.

environment is sometimes referred to as the Social Internet of Things (SIoT) [5].[4] The integration of sensory systems may allow systematic and autonomous communication and coordination of service delivery against contractual service objectives, orchestration and Quality of Service (QoS)-based swarming and fusion of resources [64]. Clusters of IoT devices may be able to communicate with other IoT devices as well as with a cloud network. This allows the IoT devices to form an ad hoc network between the devices, allowing them to function as a single device, which may be termed a fog device. The fog may be considered a massively interconnected network wherein many IoT devices are in communications with each other, for example, by radio links.

4.2   ARCHITECTURAL CONCEPTS

As already noted, an (IoT) ecosystem is comprised of many functional elements. Preferably, these elements can be obtained from a variety of vendors, with the assurance that they will interwork. Reference Architectures (RAs) are important in this context: architectures simplify the characterization of the system's constituent functional blocks and the manner in which these

[4]SIoT is an emerging paradigm where IoT-endowed devices collaborate with each other to achieve a specified goal by establishing durable logical and physical relationships; thus, an SIoT paradigm allows objects to have their own social networks. In this view, connected devices are given social meanings that render them unique and distinguishable from other things, entities, or devices. More specifically, objects establish social relationship by forfeiting their individuality in favor of common interest for federated service(s) to the larger community of objects or entities. Furthermore, in this view SIoT makes use of a Service-Oriented Architecture (SOA) model where a cadre of IoT devices can offer or request atomic services from each other as well as collaborate either as a whole or individually [56–63].

functional blocks interrelate to each other. A RA facilitates the orderly partition of functions, typically in a hierarchical fashion. Such partition not only reduces functional redundancy and promotes standardization with the possible definition of well-established layer-to-layer interfaces (also possibly including Application Programming Interfaces [APIs]), but also allows the intermingling of products from an open set of vendors' products, with the goal of layer-function cost optimization and/or usage of Best-In-Class technology for each layer. RAs typically embody a set of environment views: a *functional view* adopts a functional-decomposition perspective by specifying interfaces, interactions, and functionality; an *information view* adopts an information-structure perspective by defining semantics and information flows. In aggregate, RAs, frameworks and ensuing standards foster seamless connectivity and plug-and-play operations.

One of the early examples of a RA was the Open Systems Interconnection Reference Model that has proven instrumental in the development of seamless computer communications of all types. In recent years, the Information Technology (IT) field also developed a number of RAs, known as Enterprise Architecture Frameworks (EAFs), but due to the multidimensional nature of computing, a single architecture has not emerged as the canonical architecture. Some of the architectures that did emerge include, but are not limited to: the Open Group Architecture Framework (TOGAF), the Zachman International Model, the U.S. Departure of Defense Architecture Framework (DoDAF), the Federal Enterprise Architecture Framework (FEAF), the NIST Enterprise Architecture Model, the EA Cubed Framework, the DND/CF Architecture Framework (DNDAF), the NATO Architecture Framework (NAF), the Unified Modeling Language (UML), and the Decision Model and Notation (DMN), along with vendor or universities such as those promulgated by IBM or Gartner. Some of these RAs are also possibly useful in an organization-wide IoT environment [65].

An IoT architecture can help define fundamental connectivity and data management building blocks, which, in turn, determine information flows (and traffic). In the context of assessing the traffic flows, one should make note that the IoT benefits from architectural formulation and standardization. As noted, architecture deals with how things are assembled, including hierarchy. Standards relate to the ability to deploy the technology in a commodity fashion, assuring simple and reliable, end-to-end interoperability.

Several IoT-specific RAs and architectural frameworks have emerged of late, each concentrating on some specific abstraction IoT ecosystem (e.g. see [27,66–79] among others). Some IoT RAs have been developed by standardization bodies, while other architectures have been advanced by vendors. See Table 4.2 for a partial list of RAs from standards organizations. Each of the IoT architectural models that have been proposed addresses or focuses on some specific functionality, formulation, or abstraction of the concept. Open architectures include but are not limited to the following: the architecture formulated by the International Society of Automation (ISA), ISA-95; the 5C architecture; the IoT Reference Architecture (IoT RA – ISO/IEC WD 30141); the IoT Architecture (IoT-A); the Industrial Internet Reference Architecture (IIRA); the IEEE P2413 WG's Standard for an Architectural Framework for the IoT; and the ETSI High-Level Architecture for M2M. Some of the open IoT architectures are shown in Table 4.3. Each of these architectures has strengths and weaknesses [79].

At this juncture, there is not yet a broadly accepted IoT RA, although some convergence is anticipated. As is the case in general, several distinct perspectives have been assumed for IoT RAs, for example, a networking/internetworking connectivity perspective, a data-level/data-exchange-level semantics perspective, and a physical-level perspective, namely the hardware and/or software required to acquire and collect the endpoint information and transfer it upstream. Given the large number of applications in play (also known as Use Cases), a single RA may not suffice, and the stakeholder may need to work with one or more architectures.

Recently, the author defined a basic seven-layer Open Systems IoT Reference Model (OSiRM) that recognizes the fog/edge functionality in an IoT RA environment [15]. The layers of the OSiRM are as follows (also see Figure 4.4):

TABLE 4.3 Partial Listing of Key IoT RAs

Model/RA	Description
Industrial Internet Reference Architecture (IIRA)	General RA concentrating on functionality domains
Internet of Things Architecture (IoT-A)	RA with a functional and data perspective emphasis
Reference Architecture Model Industrie 4.0 (RAMI 4.0)	RA applicable to smart factories
Standard for an Architectural Framework for the Internet of Things (IoT)	RA advanced by the IEEE P2413 WG with a focus on security and safety
Arrowhead Framework	RA advanced by European Union for networked automation of embedded devices
ETSI High-level architecture for M2M	Well-known RA, but principally focused on M2M, concentrating on networking and communication
Internet of Things Reference Architecture (IoT RA)	RA developed by ISO/IEC WD 30141 seeking to define IoT domains to facilitate interoperability among IoT entities
Open Systems IoT Reference Model (OSiRM)	RA that has a physical hierarchy and allows layer-by-layer security

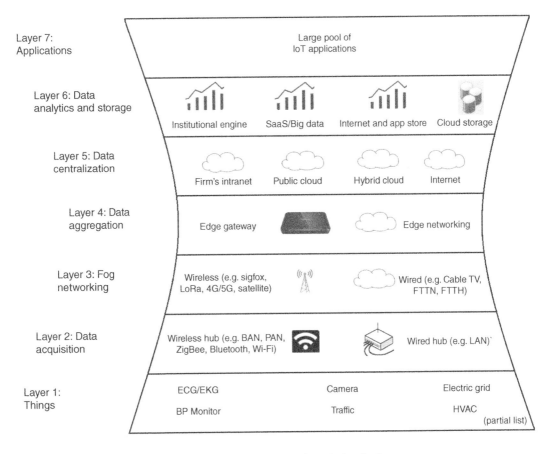

Layer 7: Applications: a vast array of horizontal and/or vertical applications

Layer 6: Data analytics and storage: data analytics and storage functions

Layer 5: Data centralization: "data centralization" function (traditional core networking)

Layer 4: Data aggregation: "data aggregation," data summarization or protocol converstion

Layer 3: Fog/Edge networking: "fog networking," site - or edge-localized networking

Layer 2: Data acquisition: "data acquisition" capabilities

Layer 1: Things: a vast universe of "things"

FIGURE 4.4 OSiRM: open systems IoT reference model (transaction stack).

- **Layer 1**: The "nearly-unlimited" universe of "things" that can partake of the automation provided by the IoT. Examples include but are not limited to smartphones, wearables, medical monitoring devices, home appliances, home security devices; homes and buildings mechanical systems (e.g. lighting, HVAC, occupancy); cameras, vehicles, and utility grid elements.
- **Layer 2**: The "data acquisition" capabilities supported by sensors, embedded microprocessors, embedded electronics, and the corresponding sensor hubs based on various short-range (on location) transmission technologies, whether wireless or wired. The acquired information includes ambiance data, video, and positioning data. Layers 1 and 2 are in a state of symbiosis, being that things endowed with sensors become the IoT clients or endpoints of the hubs (and ultimately on the entire system).
- **Layer 3**: The "fog networking" and/or "device cloud" mechanism, specifically the Neighborhood Area Network (NAN), Campus Area Network (CAN), or local site network that is the first link of the IoT device end-to-end connectivity. Fog networking is preferably optimized to the IoT operating environment and may utilize specialized lightweight protocols. It is often a wireless network (say at the 900 MHz, 2.5 GHz, 5 GHz, mmWave, or infrared/Li-Fi bands), but can also be a wire-based link (e.g. on a factory LAN supporting a process control or robotics application).
- **Layer 4**: This layer supports an optional "data aggregation" function. It may entail in-network processing, for example, some type of protocol conversion (for example, converting from a light, low-complexity protocol to a more traditional networking protocol), data summarization, or other edge-networking functionality related to the outer/access tier of a traditional network, utilizing well-known communication protocols. This function is typically handled in a "gateway" device.
- **Layer 5**: This is a "data centralization" function, corresponding to the classical core networking functions. It encompasses organization-owned (core) networks, public, private, hybrid cloud-type connectivity, and Internet paths (tunnels). These networks typically consist of carrier-provided telecommunication services and may be wired and/or wireless.
- **Layer 6**: This layer spans the various "data analytics and storage functions." The data analytics functions are typically application-specific.
- **Layer 7**: The "applications" layer spans a large swath of horizontal and/or vertical application domains (e.g. Use Cases). In practice, the list of applications is effectively "unlimited" in scope.

In conclusion, at this juncture, there is still no universally accepted IoT RA, standardized protocol suite, or canonical profiles. Possibly due to the sizable use case set, it may not be possible to converge to one canonical architecture or protocol suite, but rather several nonoverlapping RAs and protocol suites may emerge. However, the existence of some RAs highlights a step in the right direction. Broader-context standards, such as but not limited to IEEE 802.11, TCP/IP, HTTP, Moving Pictures Expert Group (MPEG), have clearly demonstrated the value of standardization both in terms of ubiquitous access/interoperability as well as product cost.

4.3   WIRELESS TECHNOLOGIES FOR THE IoT

Many IoT applications and use cases rely on the use of wireless technologies. As noted, these span the WAN, LAN, or PAN environment. Table 4.4, [80], highlights three types of wireless communications technologies that are typically used in IoT systems; also, see Figures 4.5 and 4.6. Wireless technologies are fundamental to a large number of IoT applications, and as it is tautological, these technologies are intrinsically part of WSNs [26, 81, 82]. Wireless access is

TABLE 4.4   Wireless Communications Technologies that are Typically Used in IoT Systems

Wireless Communications Technologies	Description
Long-range	Typically include those with a distance range on the order of miles, for example, those using radio communications standards such as LoRa (Long Range) spread spectrum modulation techniques using a LoRaWAN protocol, the cellular LTE-M (Long-Term Evolution Machine Type Communication) standard, and the NB-IoT (Narrowband Internet of Things) standard.
Medium-range	Typically include those with a distance range on the order of hundreds of feet, for example, ten feet to several hundred feet. Examples include BLE (Bluetooth Low Energy), Wi-Fi technologies, and RFID (Radio Frequency Identification) UHF (ultra-high frequency) technologies. In some cases, medium-range wireless communication technologies can be implemented as zero-power passive technologies by using the received radio waves to power radio communications and data processing operations.
Short-range	Typically include those with a distance range on the order of feet or less, for example about one inch to several feet. Examples of short-range wireless communication technologies include RFID HF (high frequency) and NFC (Near-Field Communication) protocols. In some cases, short-range wireless communication technologies can be implemented as a zero-power passive technology by using the received radio waves to power radio communications and data processing operations.

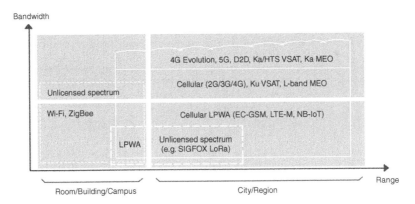

FIGURE 4.5   Typical wireless technologies usable in the IoT context.

applicable both in a building or campus environments, as well as in city or regional environments (applications also exist for worldwide IoT applications using satellite services of various kinds). Short-distance LAN-oriented technologies, such as Wi-Fi, have been used in some implementations; city-wide and regional WAN-oriented technologies such as 2G, 3G, Long-Term Evolution (LTE)/LTE-A (4G) or 5G cellular systems (along with other LPWAN services discussed below) are also used. Table 4.5 provides a comparison of metro-level WAN and LAN wireless technologies applicable to the IoT.

In practical terms, there are three classes of IoT applications:

- Localized, in a (smart) home, smart building, smart campus, smart venues.
- Geographically dispersed applications, as entailing a smart city, a smart region, a smart geography.

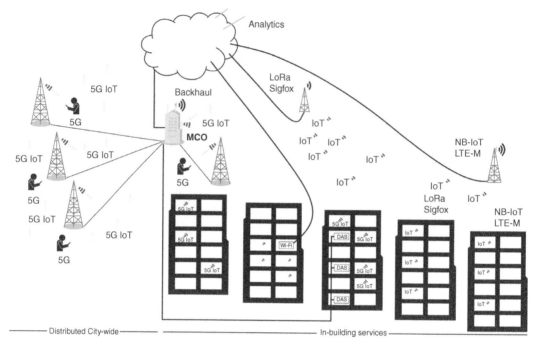

FIGURE 4.6 The pre-5G and the 5G IoT connectivity ecosystem.

TABLE 4.5 Comparison of Key Fog/Edge and Core Wireless Technologies Applicable to the IoT

IoT Technology	Basic Features
NB-IoT (Narrowband IoT)	• Mostly edge but can also be a small-geography core • Outdoor usability: up to about 20 miles • Several bands, licensed spectrum • Falls under the 5G "umbrella"; LTE-based • 0.1–0.2 Mbps data rates, battery ~10+ years • Low cost, low modem complexity, low power, energy-saving mechanisms (high battery life) • Does not require a gateway: sensor data are sent directly to the destination server (other IoT systems typically have gateways that aggregate sensor data, which then communicate with the destination server) • Reasonable building penetration (improved indoor coverage) • High number of low throughput devices (up to 150 000 devices per cell)
LTE-M (Long-Term Evolution Machine Type Communications) Rel 13 (Cat M1/Cat M)	• Mostly edge but can also be a small-geography core • Outdoor usability: up to about 20 miles • Cellular network architecture, LTE compatible, easy to deploy, new cellular antennas not required • Falls under the 5G "umbrella"; uses 4G-LTE bands below 1 GHz, licensed spectrum • Considered the second generation of LTE chips aimed at IoT applications • Caps maximum system bandwidth at 1.4 MHz • Cost-effective for LPWAN applications where only small amount of data are transferred, e.g. smart metering • 1 Mbps upload/download, battery ~10 years • Relatively low complexity and low power modem • Can be used for tracking moving objects (location services provided through cell tower mechanisms)

TABLE 4.5   (Continued)

IoT Technology	Basic Features
5G	• Evolving, not yet widely deployed • Several bands, low latency, high sensor density • Cellular network architecture • Licensed spectrum, 0.01 Mbps in some implementations, battery ~10 years • Broadband features available for surveillance/multimedia • Cost-effective • Expected to be available worldwide • Building penetration may need Distributed Antenna Systems (DASs)
LoRa	• End-to-end (edge and core) • Does not fall under the 5G "umbrella" • Outdoor usability: 6–30 miles with LOS (the higher number in rural environments) • Band below 1 GHz • IoT-focused from the get-go • Proprietary • Low power
Sigfox	• End-to-end (edge and core) • Does not fall under the 5G "umbrella" • Outdoor usability: 30 miles in rural environments; 1–6 miles in city environments • Band below 1 GHz • Narrowband • Low power • Proprietary • Star topology
Wi-Fi	• Several bands • In 2018 the FCC allowed the expansion of the 6 GHz band to next-generation Wi-Fi devices with 1.2 GHz of additional spectrum spanning 5.925–7.125 GHz (current Wi-Fi networks operate at 2.4 and 5 GHz with a few vendors offering 60 GHz "WiGig," this having a range of 30 feet – IEEE 802.11ad and IEEE 802.11ay.) • High adoption; most (but not all) indoor IoT utilize Wi-Fi; good functionality • Free "airtime" • Subject to interference: malicious or non-malicious interference (e.g. too many hot spots) could impair the sensor from send data either on a fine-grain or coarse-grain basis • For outdoors: requires inter-spot connectivity backbone (wired or wireless) (e.g. 802.11ah: distance range up to about 1/2 mile)
Bluetooth	• Low bandwidth (2 Mbps) • Used in medical devices and industrial sensors; low power, good for wearables • Usable for real-time location systems with medium accuracy
ZigBee	• Low data rate • Industrial and some home applications (e.g. home energy monitoring, wireless light switches) • Low transmit power/low battery consumption

Note: A few other legacy IoT wireless technologies exist (e.g. Cat 0, Cat 1, EC-GSM, Weightless) but are not included in this table.

- Globally/worldwide geographically dispersed applications (as supported by satellite-based services, especially for aeronautical and maritime applications, as documented at length in [24]).

Localized applications tend to make use of PAN/LAN technologies (e.g. Wi-Fi, BLE, ZigBee); geographically dispersed applications tend to use LoRa, Sigfox, or cellular solutions discussed below; globally/worldwide geographically dispersed applications use geosynchronous or, better-yet, Low Earth Orbit (LEO) satellites.

4.3.1   Pre-5G Wireless Technologies for the IoT

Local Level   As noted, several connectivity options are available at the PAN/LAN level: IoT networks may use WLAN based on the IEEE 802.11 family of protocols, including Bluetooth Low Energy (BLE) and ZigBee; additionally, Bluetooth Mesh is a newer technology also starting to be utilized for IoT, especially in larger, commercial, or industrial environments. While Wi-Fi is ubiquitous, Wi-Fi transceivers are relatively expensive (in the $25–75 range at press time) and are rather power-intensive, see Figure 4.7. BLE and ZigBee are discussed in Section 4.4 in the context of the discussion on full protocol stacks.

Some less-common (and/or newer) local solutions available at this juncture include the following:

- Hybrid hotspot networks. These networks typically support connectivity to various cloud-based servers when used in an IoT context. A landline-based network usually (but not always) exists to provide backhaul connectivity. These networks also include the newly

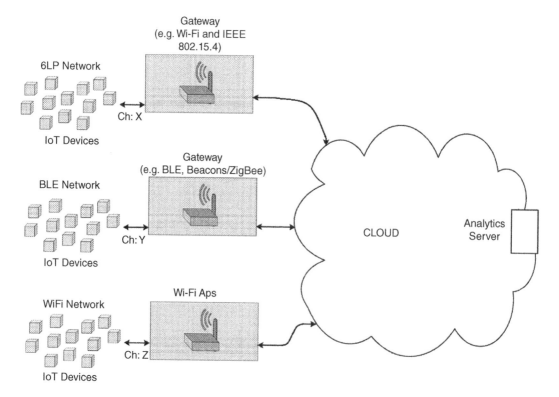

FIGURE 4.7   Example of communication systems in IoT with local aggregation [64].

standardized low-power Wi-Fi known as Wi-Fi HaLow™ – these solutions afford a cost-effective, energy-efficient buildout; they are ideal for e-health and mobility. They generally (but not always) support only relatively low bandwidth. In particular, Wi-Fi HaLow is based on IEEE 802.11ah technology; it operates in the 900 MHz band and affords longer range, lower power connectivity to Wi-Fi-enabled devices. This technology supports power-efficient applications in the smart some, smart venue, connected car, digital healthcare, and other applications; clearly, it is positioned for the low power connectivity needed for wearables and similar use cases. Wi-Fi HaLow's range is approximately twice that of traditional Wi-Fi while also providing more robust connectivity, for example, the ability to penetrate structures, walls, or other barriers more easily. The goal is to support multi-vendor interoperability, strong security, and easy setup (including intrinsic support of IP and IoT). Devices that support Wi-Fi HaLow will also operate in the traditional 2.4 and 5 GHz bands [15].

- Low-Rate Wireless Personal Area Networks (LR-WPANs) such as *IEEE 802.15.4-2020 – Standard for Low-Rate Wireless Networks*. The IEEE 802.15.4-2020 defines the Physical Layer (PHY) and Medium Access Control (MAC) sublayer specifications for low-data rate wireless connectivity with fixed, portable, and moving devices with no battery or very limited battery consumption requirements. In addition, the standard provides modes that allow for precision ranging (some of the PHYs provide precision ranging capability that is accurate to 1 m). PHYs are defined for devices operating in a variety of geographic regions. The standard is a Revision to IEEE Standard 802.15.4-2015; furthermore, six completed amendments were rolled up to the standard – these are 802.15.4n, 802.15.4q, 802.15.4s, 802.15.4t, 802.15.4u, and 802.15.4v. The basic formulation establishes a system for a 10-m communications range with a transfer rate of 250 kbps, but some of the defined PHYs provide different distance-bandwidth trade-offs. IEEE 802.15.4-conformant devices may use one of three possible frequency bands for operation (868, 915, or 2450 MHz). IEEE 802.15.4 is the basis for other PAN systems, including ZigBee, ISA100.11a, 6LoWPAN, WirelessHART, MiWi, Thread and SNAP specifications; these specifications extend the standard by defining the upper layers protocols (typically TCP/IP based). (Additional information on the IEEE 802.15.4 standard is provided in Chapter 7.)

- 6LoWPAN defines a binding for the IPv6 over WPANs and is used by upper layers in Thread.

- Other systems have also emerged or are emerging – e.g. (i) ISO/IEEE 11073 Personal Health Data (PHD) Standards and (ii) ETSI TR 101557.

- Near-Field Communication (NFC) supports contactless communication between devices (e.g. a card or a smartphone) where a user waves the device over an NFC-based reader to transfer information without the devices making contact.

WAN Level   At the WAN level, an LPWAN network can be utilized. Most technologies in this arena aim at supporting large numbers of very low-cost, low-throughput devices with very low power consumption so that even battery-powered devices can be deployed for years. These devices also tend to be constrained in their use of bandwidth, for example, with limited frequencies being allowed to be used within limited duty cycles (usually expressed as a percentage of time per hour that the device is allowed to transmit). Coverage of large (regional) areas is also a common goal. While all constrained networks must balance power consumption /battery life, cost, and bandwidth, LPWANs prioritize power and cost benefits by accepting severe bandwidth and duty cycle constraints when making the required trade-offs. This prioritization is made in order to get the multiple-kilometer radio links [83]. Common LPWANs include: (i) Sigfox, (ii) LoRa™ (Long Range), or (iii) IPv6 over Low Power Wide-Area Networks network based on the Internet Engineering Task Force (IETF)'s relatively recent RFC 8376. See Figures 4.8 and 4.9 for pictorial views of WAN/LPWAN environments; in particular, Figure 4.9

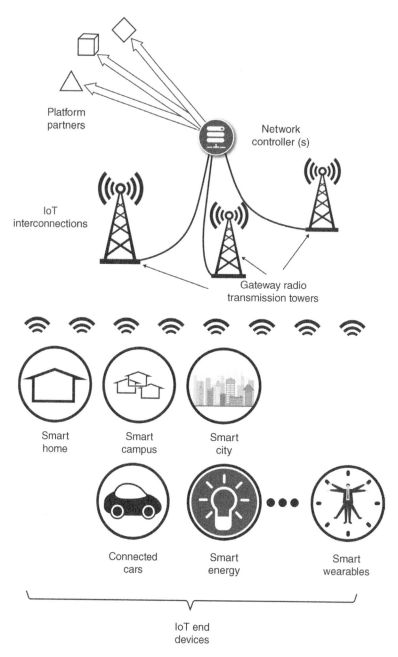

FIGURE 4.8 WAN/LPWAN IoT environment [55].

depicts pictorially an environment where both unlicensed systems (e.g. LoRa) and licensed systems are used. Sigfox supports suburban applications within 50 km from the tower; it provides up to 100 bps. LoRa supports suburban applications within 15–45 km from the tower and 3–8 km in urban environments; it provides up to 50 kbps. A network based on LoRa technology makes use of protocols promulgated by the LoRa Alliance and developed by Semtech.

LoRa is considered the most-adopted technology for IoT applications in smart city environments [84]. LoRa's commercial success is probably due to the use of chirp spread spectrum modulation technique in lieu of conventional modulation techniques; this technique has better

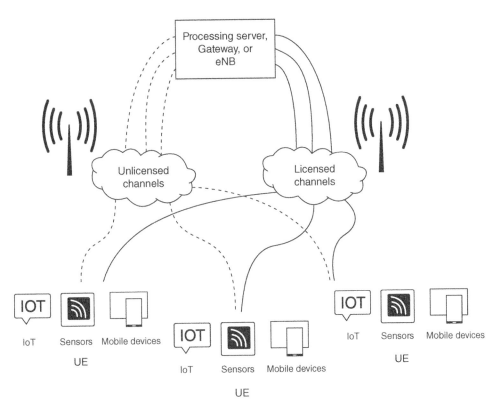

FIGURE 4.9   Dual-mode systems.

performance compared to the former modulation techniques in the presence of multipath fading and interference. The use of orthogonal spreading factors by the dispersed IoT devices provides resilience to packet interference at the receiver end. Performance metrics of interest include message delivery reliability, message delivery delay, power/energy consumption, power efficiency, latency, throughput, path loss, packet delivery rate, delivery ratio, outage probability, spectral efficiency, and coverage. A LoRa network is composed of base stations or gateways generally placed at elevated topographic points to cover a large geographical area. LoRa includes a PHY and MAC protocol structure.

As noted, the PHY protocol utilizes a chirp spread spectrum modulation technique, where a signal is modulated by chirp pulses that continuously vary in frequency, making it robust against interference and resistant to multipath fading. LoRa modulation depends upon five adjustable transmission parameters: bandwidth, coding rate, spreading factor, carrier frequency, and transmission power. LoRa uses multiple chirps to represent each bit of information; the rate of spreading information is called the symbol rate (R_s) [85]. The number of chirps per symbol is calculated as 2^{SF}, i.e. each symbol is expanded by a spreading code of length 2^{SF} chirps with symbol rate of $R_S = BW/2^{SF}$. The spreading factor is defined as the ratio of chirp rate (R_c) to symbol rate, that is, $SF = log_2 (R_c/R_S)$. The system employs multiple orthogonal spreading factors ranging from 7 to 12. Higher SF values result in an increase in receiver sensitivity and transmission range but a decrease in the transmission rate. LoRa operates on different unlicensed Industrial, Scientific and Medical (ISM) bands such as 433, 868, and 915 MHz, depending upon the region of deployment around the world; it operates in three scalable bandwidths of 125, 250, and 500 kHz; higher bandwidth supports higher data rate but reduces the receiver sensitivity [86, 87]. Some recent performance studies show that about three-quarters of total transmitted messages were successfully received by the server, and 99% of messages that were received were received within 10s of transmission [88]. Cooperative communication methods have been employed to further enhance the LoRa network performance; cooperative transmission based on relaying nodes can enhance the network performance by improving reliability, throughput, channel capacity, spectral and power efficiency and reducing outage probability and latency of the wireless network [87].

The MAC layer is known as the LoRaWAN protocol. LoRaWAN utilizes pure ALOHA random access methods to access the channel. The data link has star topology network architecture that consists of three components: IoT end-devices, gateways, and network server(s). In turn, LoRaWAN defines three categories of IoT end-devices based on their method of channel access: Class A, class B, and Class C. A Class A device supports bidirectional communication where, after every uplink transmission, it releases two receive windows at a delay of one second and two seconds in order to be able to receive a downlink transmission from the gateway; thereupon, the device halts in the idle state until the next transmission, optimizing power consumption. Class B devices open extra receive windows, at a scheduled time-synchronized by the gateway. Class C end-devices remain in the reception mode closing only at the time of transmission, thus consuming higher energy [89]. The IoT end-devices communicate the message to the gateway in a single hop communication; the gateway relays the messages received to the network server via IP/TCP/HTTP connections (dual-hop LoRa networks can also be envisioned, where intermediary relays amplify and forward messages [87]).

The LoRa gateways are able to detect messages sent in their zone by equipment or endpoints (end-devices) and to transfer them to at least one LoRa Network Server (LNS). An endpoint wishing to transmit a message (i.e. data) to the LNS transmits this message in a frame, referred to as an uplink LoRa frame, in accordance with the LoRaWAN protocol. The uplink LoRa frame is transmitted in broadcast mode. This uplink LoRa frame is received by at least one LoRa gateway. Each LoRa gateway that has received the uplink LoRa frame decodes it and retransmits the message to the server in an HTTP (HyperText Transfer Protocol) request. If a plurality of LoRa gateways received the uplink LoRa frame, the server receives a plurality of HTTP requests containing the message. The server must then designate, among the LoRa gateways that have received the uplink LoRa frame, the LoRa gateway to be used for relaying a response to the message contained in the uplink LoRa frame. The response is transmitted from the server to the designated LoRa gateway in an HTTP request, and then, in unicast mode from the designated LoRa gateway to the endpoint in a downlink LoRa frame, in accordance with the LoRaWAN protocol [90]. LoRa requires that a dedicated network be deployed; such a network requires sufficient geographic density to provide WAN coverage to the area of interest.

It would be desirable to be able to use (or share) an existing wireless WAN infrastructure. Successive releases of LTE have optimized MTC with improved support for low-power wide-area connectivity. In LTE Release 13, enhanced MTC (eMTC) and NarrowBand Internet of Things (NB-IoT) have been introduced. These protocols have enabled a reduction in device cost and complexity, extended battery life and enhanced coverage.

4.3.2   NB-IoT

NB-IoT is a licensed LPWAN technology designed to coexist with existing LTE specifications and provide cellular-level QoS connectivity for IoT devices. NB-IoT was standardized by 3GPP in LTE Release 13, *but it does not operate in the LTE context* per se [91–93]. NB-IoT has attracted support from Qualcomm, Ericsson and Huawei, among many other vendors and service providers. NB-IoT (also known as LTE Cat-NB1) is based on a Direct Sequence Spread Spectrum (DSSS) modulation in a 200 kHz channel. There are several underutilized 200-kHz GSM spectrum channels, as well as other possible bands such as guard bands. NB-IoT is intended as an alternative to LoRa and Sigfox. This technology can optimize sunken financial investments by service providers and can shorten the service deployment rollout timetable for IoT services since NB-IoT uses existing cellular infrastructure. NB-IoT service goals include: (i) low-complexity end-nodes, (ii) device cost less than $5, (iii) a device battery life expected to last for 10 years if it transmits 200 bytes of data per day, and (iv) uplink latency less than 10 seconds (thus not a true real-time service). NB-IoT operates on 900–1800 MHz frequency bands with coverage of up to about 20 miles; it supports data rates of up to 250 Kbps for uplink and 230 Kbps for downlink communications [94–96]. NB-IoT can be implemented in different ways: (i) in standalone non-cellular licensed bands; (ii) in unused 200 kHz bands in the context of GSM or CDMA; and (iii) in LTE environments where base stations can allocate a resource block to NB-IoT transmissions. Since NB-IoT offers low cost for the device and for the service, it is an optimal choice for large-scale distributed deployment in smart cities and smart grid applications. NB-IoT provides end-to-end security, which entails trusted security and authentication features [97].

As illustrative commercial examples, in 2018 T-Mobile announced a North American NB-IoT plan that costs just $6 a year – one-tenth of Verizon's Cat-M plans – for up to 12 MB per connected device, and several NB-IoT modules based on Qualcomm® MDM9206 LTE IoT modem, certified for use on T-Mobile's network. T-Mobile, in conjunction with Qualcomm and Ericsson, conducted the first trial NB-IoT in the United States in 2017 across multiple sites; T-Mobile and the City of Las Vegas also announced a partnership to deploy IoT technology throughout the city. For applications that require more bandwidth and voice, T-Mobile offers Cat-1 IoT Access Packs [98, 99]. NB-IoT consumes minimal power: while most IoT end-nodes save power when they are quiescent when the node and the modem are running and handling all the signal processing, the systems with simpler waveform (such NB-IoT) consume less overall power. Additionally, chipsets that support a single protocol (such NB-IoT) are cheaper compared to a chipset that supports multiple protocols. Furthermore, prima facie, NB-IoT may provide deeper building penetration than LTE-M (Long-Term Evolution Machine Type Communication).

4.3.3 LTE-M

LTE-M supports low nodal complexity, low nodal power consumption, high deployment density, low latency, and extended geographic coverage while allowing service operators the reuse of the LTE installed base. LTE-M is a power-efficient system; two innovations support battery efficiency: LTE eDRX (Extended Discontinuous Reception) and LTE PSM (Power Saving Mode). LTE-M not only allows the upload of 10 bytes of data a day (LTE-M messages are short compared to NB-IoT messages) but also allows access to Mbps rates. Therefore, LTE-M can support several use cases. In the United States, major carriers such as Verizon and AT&T offer LTE-M services (as noted, Verizon has announced support for NB-IoT – T-Mobile [Sprint] appears to lean in the NB-IoT direction) [100]. Worldwide geographies with GSM deployments will likely offer NB-IoT in the short term. Figure 4.10 depicts some of the IoT compatibility mechanisms to be incorporated into 5G in terms of bandwidth scope; however, the transmission frequencies will be wildly different.

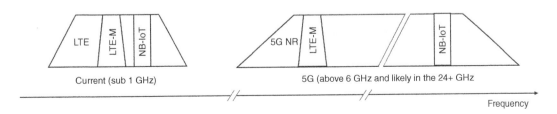

FIGURE 4.10 Support of LTE-M and NB-IoT under 5G.

In summary, LTE-M supports low nodal complexity, high nodal density, low nodal power consumption, low latency, and extended geographic coverage while allowing service operators the reuse of the LTE installed base. NB-IoT aims at improved indoor coverage, high nodal density for low throughput devices, low delay sensitivity, low node cost, low nodal power consumption, and simplified network architecture. NB-IoT and LTE-M are currently providing mobile IoT solutions for smart cities, smart logistics, and smart metering, but only in small deployments to date (as of early 2018, there were 43 commercial NB-IoT and LTE-M networks worldwide [101]). As noted, the commercial success of NB-IoT and LTE-M can serve as a proxy for the eventual success of 5G IoT in a smart city context (compared with non-cellular LPWAN solutions).

Some wireless technologies are better positioned for edge/fog networking than others. For example, LoRa and Sigfox tend to be more city-wide and county/region-wide technologies, possibly with a few centralized antennas serving a wide area; signals are received at these dispersed somewhat remote antennas and then channeled directly to the Internet [102]. Other wireless technologies allow a more hierarchical network tiering where data are aggregated (and possibly summarized) at various points along the way; LTE-M and NB-IoT can operate in both modes, depending on the implementation. Low-power radio systems tend to be more matched to fog/edge-based network designs.

Some less-common solutions include:

- Proprietary short-range (up to a few miles) technologies, with relatively low bandwidth, utilizing unlicensed radio spectrum. These are typically vendor-specific wireless systems; they are subject to channel interference, typically, there is no intrinsic security included, and confidentiality must be implemented by a device-level encryption or tunneling mechanism. Unlicensed non-3GPP LPWAN IoT wireless technologies include the following [103]:
- Platanus (<1 km; 500 kbps)
- OnRamp (4 km; up to 8 kbps)
- Weightless-N (up to 5 km; up to 100 kbps)
- Telensa (up to 8 km; low)
- NWave (10 km; up to 100 bps)
- Amber Wireless (up to 20 km; up to 500 kbps).

4.3.4 5G Technologies for the IoT

As noted in earlier chapters, to meet increasing demands for WAN wireless data traffic after commercialization of the 4G cellular communication system, efforts have been made to develop an improved 5G (or pre-5G) communication system.[5] 5G is the term for the next-generation cellular/wireless service provider network that aims at delivering higher data rates – 100 times faster data speeds than the current 4G LTE technology – lower latency and highly reliable connectivity. In practical terms, it is an evolution of the previous generations of cellular technology. 5G IoT is licensed cellular-based IoT. A 5G system entails devices connected to a 5G access network, which is then connected to a 5G core network. 5G systems subsume important 4G system concepts such as the energy-saving capabilities of NB-IoT radios; secure, low-latency small data transmission for low-power devices – *low latency* is a requirement for making autonomous vehicles *safe*; and devices using energy-preserving dormant states when possible. Bandwidth demand for smart city and smart campus applications is the main driver for mobile broadband-based 5G services in general and new-generation 5G IoT applications.

---

[5]5G or pre-5G communication system is also called a beyond 4G network or a post LTE system.

According to GSMA, 5G is on track to account for 15% (1.4 billion) of global mobile connections by 2025. By early 2019, 11 worldwide operators announced initial 5G service introductions, and several other operators have since activated new base stations with 5G commercial services to follow thereafter [104].

Some 5G systems will use the higher frequency millimeter wave (mmWave) bands to achieve high throughput. To decrease propagation loss of the radio waves and increase the transmission distance, various techniques such as beamforming, massive Multiple-Input Multiple-Output (MIMO), Full Dimensional MIMO (FD-MIMO), array antenna, analog beamforming, and large-scale antenna techniques are being discussed in the 5G communication system. Additionally, approaches and techniques such as but not limited to advanced small cells, Cloud Radio Access Network (cRAN), ultra-dense networks, Device-to-Device (D2D) communication, wireless backhaul, Coordinated Multi-Point (CoMP),[6] reception-end interference cancellation are being investigated and/or applied to enhance overall performance. Hybrid Frequency Shift Keying and Quadrature Amplitude Modulation (FQAM) and sliding window superposition coding (SWSC) are being developed as advanced coding modulation schemes, and Filter Bank Multi-Carrier (FBMC), Non-Orthogonal Multiple Access (NOMA), and Sparse Code Multiple Access (SCMA) are also being developed as advanced access technologies. Efforts are underway to apply 5G communication systems to IoT networks. For example, technologies such as WSN, MTC, and M2M communication may be implemented by beamforming, MIMO, and array antennas [43]. Some see the application of a cRAN in conjunction with Big Data processing technology as an example of 5G IoT.

Standardization efforts for 5G systems have been undertaken by several international bodies in recent years with the goal of establishing one unified global standard. Major standardization bodies included International Telecommunication Union-Radio Communication Sector (ITU-R) and the 3rd Generation Partnership Project (3GPP). The ITU-R has assessed usage scenarios in three classes: Ultra-Reliable and Low-Latency Communications (URLLC), massive MTC (mMTC), and enhanced Mobile Broadband (eMBB). Key performance indicators are identified for each of these classes, such as spectrum efficiency, area traffic capacity, connection density, user-experienced data rate, peak data rate, and latency, among others. The ability to efficiently handle device mobility is also critical. Some examples of eMBB use cases include, but are not limited to, smartphones, home/enterprise/venues applications, and UHD (4K and 8K) broadcast. eMBB will likely not be immediately used for generic IoT applications since traditional IoT applications tend to be at the lower end of the bandwidth scale (although multimedia applications are, in fact, evolving). mMTC use cases include smart buildings, logistics, tracking and fleet management, smart meters. URLLC cases include traffic safety and control, remote surgery, and industrial control. In broad terms, 5G systems are expected to support the following [106] (some aspects of which were already noted in Chapter 3):

- 1000x higher mobile data volume per area than current systems (downlink peak data rate: 20 Gbps; uplink peak data rate: 10 Gbps).
- 10-to-100x higher number of devices than current systems (connection density is 1 000 000 devices per km^2).

---

[6]In a CoMP system, the Base Station (BS) in each serving cell (or sector) is allowed to use not only its own antennas, but also the antennas of neighboring BSs to transmit to mobile terminals in the serving cell to form a floating CoMP cell. The serving BS in each floating CoMP cell computes tentative linear precoding weights for transmissions from the coordinating BSs in the floating CoMP cell to users in the serving cell of the floating CoMP cell. The serving BS determines the power availability for transmit antennas in the floating CoMP cell that are shared with other floating CoMP cells, and scales the tentative precoding weights based on the power availability of the shared transmit antennas to determine final precoding weights so that the power constraints of the shared transmit antennas will not be violated [105].

- 10-to-100x higher user data rate than current systems (downlink "user experienced data rate": 100 Mbps; uplink "user experienced data rate": 50 Mbps).
- 10x longer battery life for low-power IoT devices than current systems (up to a 10-year battery life for mMTC).
- 5x reduced end-to-end latency than current systems (latency for eMBB is 4 ms, latency for URLLC is 1 ms).

The 5G system expands the 4G environment by adding New Radio (NR) capabilities, but in a manner that LTE and NR can evolve in complementary ways. The 5G access network may include 3GPP radio base stations and/or a non-3GPP access network. Either the existing traditional LTE Evolved Packet Core (EPC) can be used (as a transition mechanism) to support the 5G NR, or a new 5G Core (5GC) can be deployed to support 5G services.

4.3.5 WAN-Oriented IoT Connectivity Migration Strategies

Worldwide IoT revenue is expected to increase at an annual rate of 23% to 2025 to reach $1.1 trillion (up from 267 B in 2018). At the end of 2018, there were 83 commercial deployments of LTE-M and NB-IoT worldwide. However, pure connectivity is expected to become increasingly commoditized, making it difficult for operators to compete on the data transmission alone, declining from 9% of total IoT revenue in 2018 to 5% in 2025. Service providers must develop new strategies and business models beyond connectivity services. Applications, platforms, and services (e.g. cloud data analytics and IoT security) are the major growth area of IoT; this segment will be approximately 70% of the market in 2025. Professional services (e.g. consulting, systems integration, also including managed services) will increase in share and will be approximately 25% of the market in 2025 [107].

In the near term, "5G IoT" really equates to NB-IoT and LTE-M technologies. In early 2019, there were nearly one hundred commercial deployments of LTE-M and NB-IoT worldwide. At this juncture, NB-IoT and LTE-M are providing mobile IoT solutions for smart metering, smart logistics, and smart cities applications, but only in small deployments to date (as of 2018, there were in the range of 50 commercial NB-IoT and LTE-M networks worldwide [100]).

Further along, 5G IoT will need to compete with other technologies, both the cellular type (e.g. NB-IoT and LTE-M) as well as the non-cellular type (although NB-IoT and LTE-M are now considered "part of the 5G world"). The economics and availability of these networks in various parts of the world may be such that a level of inertia, frustrating a full migration to truly-novel 5G IoT services, will take hold. Clearly, in principle, 5G is better positioned for city/region-wide applications as contrasted with building or campus applications.

From an end-user perspective, design and implementation questions center around the following issues, which 5G IoT technology must be able to address successfully:

- Availability of equipment.
- Availability of service (geographic coverage in the area of interest).
- Support of required technical details (latency, bandwidth, packet loss, and so on).
- Support of mobility (where needed, e.g. wearables, crowdsensing, Vehicle-to-Vehicle and Vehicle-to-Infrastructure applications, to name a few).
- Adequate reliability (where needed, e.g. physical security, process control, Vehicle-to-Vehicle and Vehicle-to-Infrastructure applications, to name a few).
- Scalability support (functional and geographic/numerical expansion of the application).
- Initial and recurring cost of the equipment.
- Initial and recurring cost of the service.

Recent acceptability and economics of NB-IoT and LTE-M can serve as a proxy for the near-term commercial success of 5G IoT, in particular, truly-novel 5G IoT services in general. Some developers have looked at cellular services for city-wide or region-wide IoT coverage; in some instances, for example, for national truck transportation, a combination of LEO satellite services and cellular services have been and are being used. A current drawback is the cost of the requisite (miniaturized) modems and the cost of the cellular service. New services such as NB-IoT and LTE Cat-M1 (an LTE-based 3GPP-sponsored alternative to NB-IoT, also known as LTE-M) are short-term attempts to address the cost and resource issues. In particular, NB-IoT is seen as providing a pathway to 5G IoT. 5G and truly novel 5G IoT are the target solutions.

NB-IoT, LTE-M, and LTE are 4G standards, but advocates claim that they remain integral parts of early releases of 5G. Proponents make the case that *"enterprises deploying either NB-IoT or LTE-M are futureproofing their IoT projects because when 5G rollouts become common-place these two Mobile IoT standards will continue into foreseeable 5G releases (from 3GPP Release 15 on)"* [98]. In the context of 3GPP Rel 15, it appears, in fact, that NB-IoT and LTE-M will be included as 5G mobile standards. In 2018, the GSMA asserted that *"NB-IoT and LTE-M, as deployed today, are part of the 5G family; with the dawn of the 5G era [...] both NB-IoT and LTE-M technologies are an integral part of 5G, and that 5G from the LPWA perspective, is already here today"* [101]. Including these technologies as initial 5G IoT standards will motivate service providers and vendors to support these implementations for IoT deployments as an evolutionary strategy to 5G. 3GPP Release 16 is recognized as the "second 5G standard" and thereafter transmitted to the ITU for consideration as a global standard. Among other functionality and capabilities, Release 16 aimed at adding standards for connected cars and smart factories (notably, automobile companies have formed the 5G Automotive Association assist 3GPP to set autonomous vehicle standards, such as 5G cellular V2X [C-V2X]).

4.4 EXAMPLES OF SEVEN-LAYER IoT PROTOCOL STACKS

The discussion so far has focused on the PHY and the Data Link layer of the protocol stack. Some well-known protocols of a full IoT stack are shown in Figure 4.11 (other vendor- or organization-specific protocols also exist, e.g. but are not limited to DASH7, Z-Wave, INSTEON, EnOcean.)

Some aspects of these protocol stacks are discussed in the subsections that follow immediately.

4.4.1 UPnP

UPnP[7] was introduced in the early 2000s by Microsoft and is now supported by a large number of vendors, such as AT&T, CableLabs, Canon, Cisco, Epson, HP, Intel, Nokia, Philips, Siemens, Sony, including residential access point/home router vendors [2]. UPnP is a networking process allowing for "zero-configuration" for network resource and service discovery, allowing automatic detection of devices and services in the local home area network. Initially, UPnP was managed by the UPnP Forum; its current form is maintained and promulgated by the Open Connectivity Foundation (OCF). UPnP-based and other home area network-based OCF specifications can be utilized to standardize the control of IoT-enabled appliances and smart home devices such as outlets, smart wall switches, dimmers, thermostats and other devices – for example, the UPnP Forum/OCF has published UPnP device control protocols (DCPs) for lights, thermostats, automatic blinds, and security cameras. The OCF 1.0 specification has been ratified

---

[7]Some portions of the material in subsections 4.4.1–4.4.3 are based on reference [2].

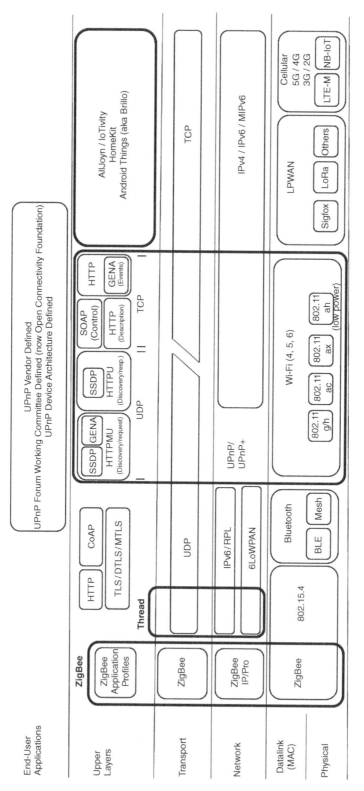

FIGURE 4.11 Key IoT protocols in a full stack.

TABLE 4.6   Features of Some Key Protocol Stacks

Stack	Description
Universal Plug and Play (UPnP)	A network infrastructure allowing for "zero-configuration" for network startup, automatic detection of devices and services in the same local network, and seamless network connection on P2P basis. Key features are on the adoption of Service-oriented Architecture (SOA), Web technologies and Internet standards, i.e. TCP/IP and HTTP, to serve at the lower layer of UPnP protocol stack. UPnP assumes the network runs on IP. It uses HTTP on top of IP, to provide device/service description, actions, data transfer and eventing. Device search requests and advertisements are supported by running HTTP on top of UDP (port 1900) using multicast (known as HTTPMU). Responses to search requests are also sent over UDP but are sent using unicast (known as HTTPU). Web-based applications can be easily implemented on any UPnP devices for performing various tasks, such as interacting with the other networked UPnP devices, supporting UPnP services, or extending these services for improved functionality. UPnP was recently extended (UPnP+) to support cloud connectivity/applications.
Thread	An IPv6-based, low-power mesh networking protocol stack for IoT products that provide security mechanisms, such as AES encryption. Thread uses 6LoWPAN, which utilizes the IEEE 802.15.4 wireless protocol with mesh capabilities. Membership in Thread Group is required to be able to implement, practice, and ship Thread technology and Thread Group specifications. The Thread Group Alliance was launched in 2014, with support from Alphabet/Google, Samsung, Qualcomm, among others – in 2018 Apple also joined the group.
AllJoyn/IoTivity/ Open Connectivity Foundation (OCF)	An open-source software framework (spearheaded by Qualcomm) that facilitates proximal network connectivity (including optional cloud connectivity). The system provides a software framework and basic set of system services that foster interoperability among connected products and software applications. OCF was previously known as the Open Interconnect Consortium.
HomeKit	A software framework by Apple released with iOS 8; it enables users to set up Apple's iOS Device to configure, communicate with, and control smart home appliances. By defining rooms, elements, and actions in the HomeKit service, users can enable automatic actions through voice commands to Siri.
Android Things (aka Brillo)	An Android-based embedded IoT operating system designed to work with small RAM (as low as 32–64 MB) developed by Google in 2015. It is positioned for low-power and memory-constrained IoT devices. It supports Bluetooth Low Energy (BLE) and Wi-Fi.

as an International Standard ISO/IEC 30118. Web-based applications can be easily implemented on any UPnP devices for performing a variety of tasks, such as interacting with the other networked UPnP devices and supporting UPnP services. It is one of the three major environments to support home automation. UPnP was recently extended (UPnP + Cloud) to support cloud connectivity/applications [108–112] (the pre-2015 version did not).

At the lower layers, UPnP typically utilizes Wi-Fi connectivity. However, Wi-Fi is relatively resource-intensive and, in general, IoT sensors have physical power and computing power limitations. Fortunately, sensors associated with stationary home appliances such as refrigerators, washing machines, and HVACs can typically support sensors and affiliated processors of adequate intrinsic capabilities (physical power and computing power). Vendors of high-end

TABLE 4.7 Comparison of Key Features

	UPnP over Wi-Fi	ZigBee	Bluetooth
Frequency Band	2.4 GHz 5 GHz Evolving 6 GHz proposals	2.4 GHz	2.4 GHz

PHY/MAC	Direct Sequence
	  Spread Spectrum
	  (DSSS)
	
	Multiple In Multiple
	  Out (MIMO)
	
	IEEE 802.11n
	IEEE 802.11ac
	IEEE 802.11ax
		Direct Sequence
		  Spread Spectrum
		  (DSSS)
		
		IEEE 802.15.4
			Frequency Hopping Spread
			  Spectrum (FHSS)
			
			IEEE 802.15.1
			• Classical Bluetooth
			• Bluetooth Low Energy (BLE)
			• Bluetooth Basic
			  Rate/Enhanced Data Rate
			  (BR/EDR)
			• Bluetooth Mesh
			• Bluetooth 5.1

Upper layers	TCP/IP stack, SSDP,
	  SOAP
		ZigBee IP/PRO and
		  various application
		  profiles
			TCP/IP stack and/or 6LoWPAN
			CoAP

Range	100 m	30 m	10 m
Data rate	High, depends on standard	250 kbps	Depends on version, 1–2 Mbps
Number of nodes in network	32/AP	High (64000)	Low (7)
Transceiver costs	Medium (but depends on Wi-Fi generation)	Very Low	Low

multi-hundred-dollar home appliances (e.g. refrigerators, smart ovens, washing machines, HVAC systems, outdoor grills) – or appliances requiring high throughput video – can economically use the Wi-Fi technology; however, vendors of lower-end appliances (e.g. clocks, outlets, lighting) may find Wi-Fi to be too expensive, opting instead to utilize the lowest practical cost for the sensor apparatus, thus looking at ZigBee or Bluetooth solutions. As noted in earlier chapters, three generations of Wi-Fi are currently available:

- **Wi-Fi 4** is 802.11n, late 2000s
- **Wi-Fi 5** is 802.11ac, mid 2010s
- **Wi-Fi 6** is 802.11ax, late 2010s.

UPnP stipulates that devices, such as home appliances, support IP-based addresses (particularly IPv4), Extensible Markup Language (XML), and basic Internet protocols such as HyperText Transfer Protocol (HTTP), User Datagram Protocol (UDP), and Transmission Control Protocol (TCP). See Figure 4.11 for the protocol stack. At the top layer, messages contain exclusively vendor-specific UPnP information about their devices. Below that, in the stack, vendor content is complemented by information defined by the Forum's working committees. Below that still is the

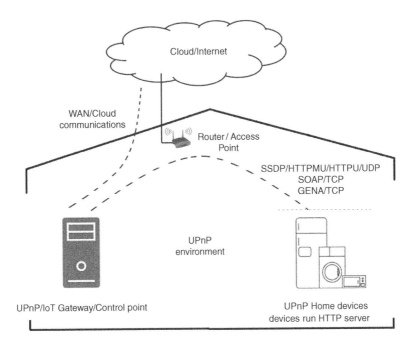

0: DHCP Client or AutoIP
1: Discovery: Device multicasts: SSDP / HTTPMU / UDP
    service advertisements
1a: Discovery response: unicast from control point SSDP / HTTPU / UDP
2: Description: XML / SOAP messaging to Control point
    Architecture schema; Working committees definitions of
    device types and services; vendor information (e.g. name, model)
3. Control from control point: SOAP / HTTP / TCP
4. Eventing: GENA Framework/ XML / TCP against state table

1a: Discovery: sends response to device advertisements
    with SSDP / HTTPU / UDP unicast
2: Solicits service descriptions with SOAP / TCP using XML
3: Issues controls to devices using SOAP / HTTP / TCP
4: If subscribed, gets events changes compared with state tables using GENA / XML

FIGURE 4.12   The UPnP process for device control.

communications stack, as depicted in the figure. UPnP makes use of these protocols to advertise the devices' presence and for data transfers. Some entities act as end-point devices while others act as Control Points – a Control Point is typically a multifunction/data-processing IoT gateway, which resides on the Local Area/Home Area Network and interacts with the rest of the devices in the local environment via a designated router, on a given IP subnet.

The basic mechanisms of the UPnP arrangement are [2]: device addressing, service discovery, device description, device control, and event notification ("eventing"). The UPnP process works as described next (also see Figure 4.12).

In Step 0, when an appliance or device first joins the network, it acts as a Dynamic Host Configuration Protocol (DHCP) client to assign itself an IP address; then it searches for a DHCP server. If no DHCP server is located, the device assigns itself a unique IP address from a set of reserved private addresses, specifically 169.254.0.0/16.

In Step 1, after connectivity is established via Step 0, a process is initiated for (service) discovery. When the appliance or device is added to the network, UPnP allows the appliance to advertise its functional services to other devices on the network using the Simple Service Discovery Protocol (SSDP). Discovery is achieved by broadcasting *SSDP alive* messages utilizing HTTP multicast over UDP (HTTPMU). SSDP also allows an appliance to passively listen to *SSDP alive* messages from other appliances or devices on the network. When two appliances discover each other, typically an appliance and a Control Point entity – almost invariably the IoT gateway – a discovery message is exchanged in a unicast manner, with SSDP over HTTP

Unicast (HTTPU), also over UDP. This message contains basic appliance or device information for that specific device, such as the device type and its services and capabilities.

Step 2 supports the more granular determination of the device/appliance description. After the devices discover each other, they exchange support information that could include, for example, manufacturer identification, device model name, device model number, website URL of the manufacturer where additional device service information can be found, parameters or data to be exchanged between the device and the Control Point for a given service. The information is exchanged in XML format utilizing the Simple Object Access Protocol (SOAP). SOAP is a standardized method to support Remote Procedure Calls (RPCs) in client/server environments; the "object" in question is some appropriate software module. SOAP runs over HTTP and uses XML to describe remote procedure calls to a server and to return results from those procedure calls.

Step 3 supports device "control": a device or program (such as the Control Point) can instruct another device (e.g. a home appliance) to perform a specified action. The control function is achieved by transferring intrinsic commands utilizing SOAP over TCP. The control can be direct, such as the Control Point instructing the device to do something; or the control can be more complex where, for example, after obtaining information about the device and its services, the Control Point endeavors to utilize the service described by the URL provided by the manufacturer and acquired during the device description step (also using SOAP) to acquire additional capabilities to effectuate the control of the device.

Step 4 deals with eventing or status updates. The General Event Notification Architecture (GENA) is the mechanism used for event notification in UPnP; TCP is the underlying transport mechanism, and messages are coded in the XML format. UPnP devices maintain "state variables" utilized for keeping state information in the devices. A Control Point or other entities (programs) can subscribe to state changes for a device, a process known as eventing: when a state variable is changed, the new state is sent to all devices and entities (programs) that have subscribed to the event – note, however, that not all home area network entities are necessarily or automatically subscribed.

When needed, UPnP uses an interdomain or intradomain conversion functionality called *UPnP Bridging*, with the bridging function discussed earlier. OCF has published a bridging specification that specifies a framework for translation between devices in OCF and non-OCF ecosystems [113]. This specification defines requirements for resource discovery, message translation, and security; additionally, it provides requirements for translation between OCF and AllJoyn ecosystems, including mapping of core resources, translation of custom resource types, propagation of errors, and other capabilities. UPnP bridging allows different local networks to interact, for example, for existing device protocols such as Bluetooth or ZigBee. Beyond low-layer bridging, the UPnP device data modeling approach allows devices residing in different ecosystems to set on a common messaging format. Translations to this format ensure common operation between any ecosystems [114, 115].

UPnP-based digital devices utilize a basic architecture for device and services description called UPnP Device Architecture (UDA). The key function of UDA is to control digital devices, to enable a user-friendly and effective home appliance controlling system. UDA describes the protocols for communication between controllers, or Control Points, and devices discussed above, including discovery, description, control, eventing, and presentation. However, there are limitations in UDA 1.0 because the controlling functions are only supported in the local home area network; additionally, because of its architecture formulation, UDA 1.0 also cannot be used for cloud services such as PaaS and SaaS. In order to overcome the limitations, OCF/UPnP Forum recently released the UCA (UPnP Cloud Architecture) in UDA 2.0, and as the name implies is not restricted to the local network [110]. In fact, given the evolving requirement for users to be able to access devices from remote locations using a smartphone, UPnP has recently been extended to directly support, discover, and manage IoT devices with an eye for integration of cloud content and services (however, the OCF recommends its native RESTful architecture in truly greenfield environments). The new framework, UPnP+ Cloud, eliminates previous topological boundaries. It utilizes new and existing UPnP DCPs and UPnP architecture extensions to provide UPnP protocols geared specifically for IoT applications, also considering the predicament of constrained resources.

4.4.2 ZigBee

ZigBee-based systems were early entrants in home automation, and ZigBee is now positioned to capture a major share of the smart home market (there using the ZigBee Home Automation profile) and lighting (there using the ZigBee Light Link profile). ZigBee has been used with several ZigBee home energy systems in the recent past. ZigBee is a low-power, low-data rate, and close-proximity wireless ad hoc networking technology (transmission distances are practically limited to 30–60 ft line-of-sight). ZigBee Home Automation can be used to control thermostats, home energy monitors, window shades, and security appliances, among other appliances. More specifically, applications include but are not limited to building automation (security, HVAC, Automatic Meter Reading [AMR], lighting control, access control); personal healthcare (patient monitoring, fitness monitoring); consumer electronics (TVs, VCRs, DVDs); and residential/commercial lighting (security, HVAC, lighting control, access control, lawn irrigation); and industrial control (process control).

ZigBee is an IEEE 802.15.4-based specification for a set of capabilities that can be used to create wireless PANs (WPANs) with a small footprint, low-power digital radios, making it ideal for low-power low-bandwidth (e.g. 250 kbps) applications, such as domotics and in-home medical device data collection (but typically not video). See Figure 4.11 for a simplified protocol stack. Zigbee operates at 2.4 GHz and supports 128-bit symmetric encryption keys. It supports a star, mesh, or clustered tree topology – ZigBee can make use of mesh routing to identify a reliable route between the source device and the destination, allowing data packets to traverse multiple nodes in a network to transfer data from a source to a destination; however, many basic applications utilize the star topology. ZigBee offers a number of application profiles, typically running over ZigBee PRO at the network layer and, as noted, on IEEE 802.15.4 Media Access Control/Physical layer (MAC/PHY) (manufacturer-specific extensions can be added). Zigbee was standardized in 2003 and revised in 2006 (the latest version being ZigBee 3.0); application profiles have been added in recent years – Home Automation was the first ZigBee application profile, published in the late 2000s. ZigBee's specification and commercial promotion are managed by the ZigBee Alliance.

A ZigBee network is comprised of several types of devices. There are three logical ZigBee device types: coordinator (only one per network), router (one or more per network), and end-device (very often many per network). The network coordinator is a device that activates the network, keeps track of all the nodes within its network, and manages the data about each node along with the information that is being sent over the network; every ZigBee network must contain a network coordinator. There will be 802.15.4-capable Full Function Devices that can serve as network coordinators, network routers, or as appliances; there will be 802.15.4-capable Reduced Function Devices, these just being simple devices/appliances. The differences in functionality among the devices are visible at the network layer of the protocol stack; typically, the coordinator and the routers will be commercial-power-based while the end-devices can be commercial-power-based or, more likely, battery-powered (the latter type of devices can become sleeping-end-devices when not actively monitoring their intrinsic phenomenon or phenomena). The coordinator or router(s) are designated as a Primary Discovery Cache Device whose job is to provide server services to manage discovery data and respond to discovery requests on behalf of the sleeping devices [116]. The network coordinator typically implemented in the ZigBee hub can be seen as a low-end IoT gateway.

As a shortlist, ZigBee is supported by Samsung SmartThings, Philips Hue/Signify lighting systems, Honeywell thermostats, Hive Active Heating accessories, Bosch Security Systems, and Amazon Echo Plus, among others. To interconnect a set of ZigBee devices, a hub is needed to bring them together; this hub, however, is not as complex as a full IoT gateway and typically costs much less (e.g. $50–100 instead of $250–500 for a typical gateway). For example, Amazon Echo Plus can operate as a ZigBee hub: it will scan the home environment for Zigbee devices, without requiring the administrator (homeowner) to set up each device individually; SmartThings and Wink are more distinct hubs that can add and control ZigBee devices from a single app.

4.4.3 Bluetooth

There are several Bluetooth®-based technologies: BLE, Bluetooth Basic Rate/Enhanced Data Rate (BR/EDR), and Bluetooth Mesh. BLE was designed for power-efficient point-to-point communication as well as broadcast data in a one-to-many environments. Bluetooth BR/EDR was developed for the transmission of a predictable or isochronous stream of data (possibly involving low-to-medium resolution video) over point-to-point connections between two devices. Bluetooth Mesh expands the capabilities of Bluetooth, complementing other Bluetooth systems such as BLE and BR/EDR. Bluetooth Mesh facilitates the deployment of scalable smart home and smart building applications with an eye on interoperability, reliability, performance, and security. The new specifications describing Bluetooth Mesh networking were published in 2017. Bluetooth Mesh is a connectivity mechanism that enables the establishment of many-to-many networks of relatively large groups of Bluetooth devices. Bluetooth Mesh makes use of BLE (version 4.0 or better) for wireless radio communications; in the protocol stack, BLE (modulation scheme/air interface) is positioned underneath the Bluetooth Mesh functionality (BLE provides the radio interface fields, such as a preamble, an access address, and a CRC). See Figure 4.11 for the protocol stack. Bluetooth uses the Service Discovery Protocol (SDP) in an analogous manner with SSDP.

In Bluetooth Mesh, messages generated by one node can transit from node to node over the network until these messages reach their intended destination, thus allowing communication to take place significantly beyond the basic radio range of each individual device. Multi-path and multi-hop delivery are intrinsic in the way Bluetooth Mesh works: copies of messages can transit via multiple paths in the network. Bluetooth Mesh networking facilitates the coverage of large spaces making it also applicable to commercial and industrial environments; it is optimized for low energy consumption and makes efficient use of radio resources.

A key consideration is the total system capacity, given that any group of devices within radio range are using the same available (spectrum) resources (but not those further away in view of the concept of space division multiplexing). As long understood from the extensive work on random access techniques, wireless environment packet collisions degrade the overall transmission performance. To address this issue, Bluetooth Mesh Protocol Data Units (PDUs) have been defined to be small: a shorter PDU leads to fewer collisions. Given some bandwidth limitations of Bluetooth, this solution may not be suitable for applications that generate high-bandwidth video (e.g. some video surveillance systems) but is usable for lower quality video (e.g. in some home surveillance security systems).

Mesh-specific messages are utilized to make optimal use of the underlying Bluetooth LE resources, as defined in the bearer layer of the mesh protocol stack. Two bearer mechanisms are defined; of these, the *advertising bearer* is utilized to enable communication between nodes in the mesh network. This bearer defines how PDUs are to be broadcast within BLE advertising packets on the three BLE advertising channels and how they can be received by nodes scanning for PDUs on these channels. Generally, identical copies of each PDU are sent on (up to) three advertising channels to increase transmission reliability.

It is worth noting that Bluetooth Mesh incorporates both network-layer security and application-layer security. Messages may be secured with two independent encryption keys.

This mechanism allows for relay nodes to authenticate a message on a network layer while precluding tampering with the application payload (for example, a light bulb that relays a message to a door lock cannot change the payload from *open to close*, but it can validate if the PDU belongs to its own network).

4.5 GATEWAY-BASED IoT OPERATION

Due to the general low complexity of endpoint elements, IoT gateways are typically used in many IoT-enabled environments. In smart home appliances, these gateways are typically deployed in the home, functioning as a processing hub, as depicted in Figure 4.12. The same broad principles apply to smart buildings, smart campuses, and smart venues. Gateways can also be located at the edge of the fog-core boundary, then serving multiple customers. Gateways operate at two broad protocol levels and for two groups of functions:

- In terms of protocol levels, gateways can be low-level devices that convert protocols up to Open Systems Interconnection Reference Model layer 2 or 3, or possibly up to layer 7; or they can be middleware gateway [117] (e.g. a UPnP Bridge [114, 115, 118]).
- In terms of groups of functions, gateways enable the inclusion of functionally peer entities; or the inclusion of less intelligent and/or legacy appliances such that IP-based functionality is achieved for all devices (in this sense, they are comparable to early "terminal servers" that enabled non-Ethernet-ready devices to communicate with devices on the LAN).

In general, user-level applications, also including home automation functions, operate at three overall levels or layers: (i) the business logic proper; (ii) the middleware, to enable operating across different software/platform environments; and the (iii) the communications apparatus (typically the seven layers of the OSIRM). In a traditional gateway, various layers of the OSIRM are converted: generally, in "mainstream environments" there is sufficient commonality in the two environments that a simple translation suffices (e.g. convert from IPv4 to IPv6). However, some environments, including home automation and/or smart building environments, have "deeper differences" that require middleware-level conversion. Middleware-level gateways can operate as follows [117]:

(i) Transport-level bridging: entails the translation of protocols and also *data types* intrinsic to the environments to be "bridged." Each environment utilizes its own fundamental protocol for element/device communication (e.g. SOAP in UPnP described in the next section). While utilizing these protocols, devices exchange data formatted in their environment-specific types; therefore, the bridging gateway must be able to translate between these distinct protocols and *data representations* to enable devices in different environments to communicate with one another.

(ii) Service-level bridging: entails the translation of *service mechanisms*. For example, when a new device joins the network in one environment, the bridging gateway must be able to dynamically recognize its presence and make the device available for use with other elements/devices. Different environments utilize different discovery approaches; for example, UPnP utilizes one protocol, while Bluetooth utilizes a different one. The bridging gateway must "bridge" the various discovery mechanisms used by the distinct environments.

(iii) Device-level bridging: entails translating the *device semantics*, which are typically represented differently in different environments – such as roles of devices and their compatibility – spanning the different environments. For example, while UPnP provides device types that have states and promulgate state-articulated events, Bluetooth uses device profiles that establish their own protocols over the underlying Bluetooth base protocol. The bridging gateway must translate different representations of device semantics to enable devices in different environments to interact with each other.

As implied above, the term "bridging" has entered the parlance when talking about middleware gatewaying; however, as clearly appreciated, these gateways are much more complex than the very basic MAC-layer bridge defined in the IEEE LAN standards.

Thus, an IoT gateway is not typically an IP router per se or a Wi-Fi Access Point per se, although some vendors may choose to package all these functions in one physical device for convenience. In practical commercial terms, IoT gateways on the market at press time provide logical aggregation of sensor data flows, translation between sensor protocols when or if needed, data computing functions by processing the sensor data before sending it to the analytics system – whether local or in the cloud – and possibly support some encryption and authorization/authentication functions. Many network infrastructure vendors such as Cisco, Dell, HPE, and Huawei offer gateway, along with other lesser-known providers. The Enterprise Management Associates recently found by way of surveys that the basic function performed on IoT gateways is, in fact, support of edge computing and analytics, while other functions documented by them include connectivity to WAN router or the WAN itself (for example in case of LoRa or Sigfox WAN services), encryption/decryption of IoT data given that most systems embedded in "things" do not have sufficient computing power to undertake this function on their own data on their own, IoT device control and management, and, finally, IoT protocol translation [119]. Figure 4.13 depicts a typical environment of an IoT gateway supporting internal protocol conversion (other processing functions such as data aggregation are not shown). Many existing (home) energy systems do not allow consumers to access or control them directly via a smartphone or a tablet, even if they are proximal to the energy system; thus, gateways functionality is needed to support this local or remote access from a standard use device; the same broad principles apply to IoT-based systems in smart buildings, smart campuses, and smart venues.

4.6 EDGE COMPUTING IN THE IoT ECOSYSTEM

Edge computing is also known as fog computing, although some see some nuances in the two terms. Effectively, fog computing entails the "insertion" of a new layer between cloud software computing resources and IoT devices [3]. Fog computing is a paradigm that utilizes a group of fog nodes that are located at the edge, in proximity of the end-nodes, to offload some of the functionality originally provided in data centers resident in the cloud. In some cases, these edge/gateway nodes have relatively low-computing power available locally; in other cases, the edge devices are able to support a substantial amount of computation, storage, and local communication. One can refer to the fog network as the set of device-facing communication links that allow remote, dispersed IoT nodes to communicate with the fog nodes. The use of gateways discussed above supports the edge-computing concept, although other mechanisms can also be employed.

Some IoT applications require fully distributed IoT devices that are broadly dispersed over the open environment; these systems often entail an explicit fog layer where concentration, protocol conversion, processing, transient storage, and other functions may take place prior to having the data transiting a WAN to reach the service-rich cloud. Some IoT applications entail an industry-specific set of IoT devices, for example, in the case of a smart grid or a transportation rail system; these systems may be aggregated at the edge but typically utilize an industry-specific traditional network (e.g. an "extranet"). Other applications are more geographically or

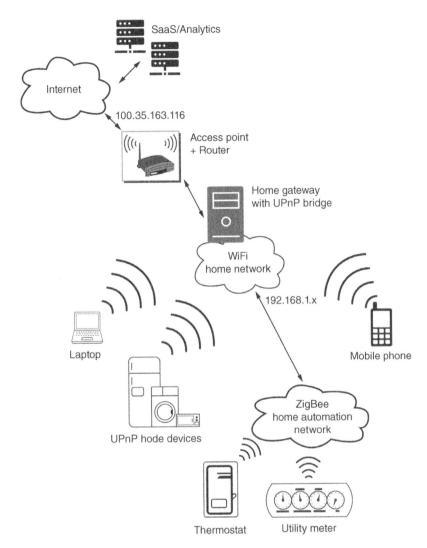

FIGURE 4.13 Example of gateway performing a "bridging" function between non-IP based devices (on ZigBee) and IP-based devices on UPnP (inspired by [120]).

administratively localized, such as an industrial IoT environment or a corporate IoT environment – for example, a Building Management System (BMS); these systems may employ a localized wired or wireless LAN for aggregation and access to the WAN and cloud, and may or may not use a fog layer mechanism per se.

Figures 4.14 and 4.15 provide a pictorial view of the typical core networks that support home (or site) automation at a macroscopic level. Figure 4.14 depicts what might be called a conventional networking arrangement for smart home services, where devices use existing ISP connections to gain access to specific service nodes (e.g. supporting home security value-added services). Figure 4.15 shows a networking arrangement for smart home services, where an IoT-specific edge gateway is used. The edge gateway can perform various supplemental functions, typically in the communications realm, but also possibly support edge-computing/processing functions (e.g. data summation) on behalf of users in some defined local geography. IoT implementers have established that it is often advantageous to analyze some or most of the sensor data at the edge of the networks, thus reducing the amount of bandwidth required in the core network to forward

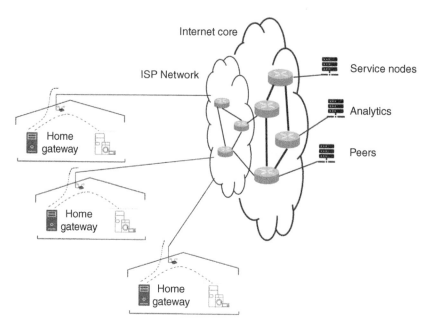

FIGURE 4.14 Typical networking arrangement for smart home services where an IoT home gateway is used.

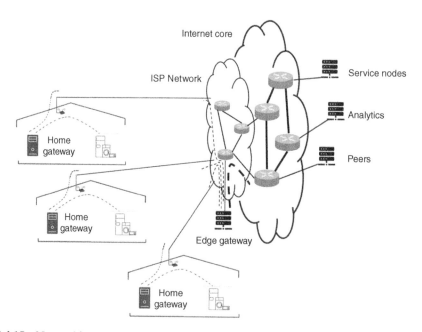

FIGURE 4.15 Networking arrangement for smart home services where an IoT edge gateway is used.

what is often voluminous data generated by the sensors; additionally, this can improve the speed of decisioning and overall reliability by making important, data-based decisions at the edge, without requiring centralized intervention. IoT gateways facilitate the execution of analytics-based decisioning closer to the source of the data and the IoT devices themselves.

In conclusion, as the number of connected devices increases under the impetus of automation, having a multitude of low-end "things" connect on an individual basis to the management or analytics systems becomes impractical. Some end-system devices generate so much transient data in the aggregate that the communication channel in the last mile or even in the core can become saturated; this is detrimental to the bottom line since the raw data per se are not all that valuable to a decision-making or monitoring process. The edge IoT gateways perform many critical functions such as device connectivity, data filtering and processing, authentication and security, protocol translation, and endpoint device software management (updating).

4.7 SESSION ESTABLISHMENT EXAMPLE

When Wi-Fi is used for wireless communication, an IoT device must connect to an Access Point (AP) to conduct wireless communication. In this case, the operation of establishing a connection between the IoT device and the AP is called Wi-Fi setup. Figure 4.16 illustrates a method for connecting an IoT device to an AP [121]. Referring to the figure, an IoT device and an AP may initiate a connection through various connection operations. For example, the IoT device and the AP may be connected via quick response (QR) code recognition or by an input of pressing a button on an application in a mobile terminal, for example, a smartphone or tablet personal computer (PC); the IoT device and the AP may also be connected through an NFC contact of the terminal, for example, an NFC-capable smartphone or tablet PC, or a personal identification number (PIN) input of the terminal, for example, a remote controller. As such, there may be various methods for connecting the IoT device to the AP, depending on the manufacturer, the type of the IoT device, wireless techniques supported by the IoT device, such as QR code, NFC, or BLE. An IoT device may receive information about an AP from a user terminal (e.g. user equipment [UE]) and connect to the AP based on information received about the AP.

4.8 IoT SECURITY

The[8] ubiquity of smarts-enabled embedded systems in practically all types of every-day-life devices, and the mission-critical applications in some cases (e.g. e-health, grid control), makes the concerns for security all-the-more pressing. The deployment of enhanced automation under the IoT leads to an enlarged attack surface: not only can the IoT-enabled devices (e.g. appliances, cameras, HVACs) be compromised by potential intruders, but such compromise may afford the bad actors an unmonitored path into the entity's full Information and Communications Technology (ICT) domain [122–125]. There are widespread concerns of the need for more comprehensive support of security in the IoT[9]; unfortunately, up-to-the-present security mechanisms are generally not put into place as often as it would be advisable.

4.8.1 Challenges

The security apparatus must support an ecosystem-wide state of secure nodes, secure computing components, secure communications, and secure asset access control. Traditionally, cybersecurity has been described as encompassing the following requirements, which are also requirements in IoT environments: Confidentiality (C) – making sure the IoT data flow in not intercepted; Integrity (I) – making sure that the IoT information received has not been altered; and, that the resident data have not been changed in an unauthorized manner; Availability (A) – making sure that IoT devices are not incapacitated and/or not properly performing their function, or that the devices are not hijacked to become rough devices, or that the IoT analytics

---

[8]Portions of these suctions are based on material from references [9, 15, 19, 20].
[9]The field of IoT security is also known as *IoTSec*.

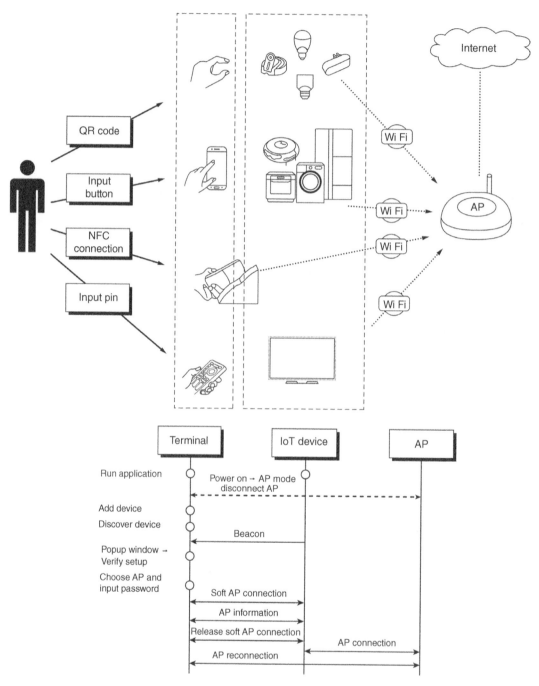

FIGURE 4.16   IoT Session establishment example [121].

systems do not become flooded with spurious traffic, or that the communication channels or Network Element are not intentionally jammed (see Table 4.8). These requirements are known as a group as CIA. These three categories of security concerns are applicable to the IoT as a whole, and they, individually, require explicit attention; trust and privacy are also ubiquitous security concerns in this arena. The security requirements of various applications may differ; therefore, not all IoT end-nodes must necessarily comply with the tightest security strictures;

TABLE 4.8   CIA Security Goals

Requirement	Description
Confidentiality (C)	Ascertaining that the IoT data flow is not intercepted and read; and, that the endsystem is not corrupted such that an intruder can steal data, credentials, or configuration parameters.
Integrity (I)	Ascertaining that the IoT information received (or stored) has not been compromised and/or altered in an unauthorized manner. Infection caused by viruses and worms can be utilized by an inimical agent to alter the original data, thus impacting integrity (among other possible damage). Integrity can also be perceived in the context of authentication: making sure that entities/devices are who they claim to be, and that the identity of the systems or users is not compromised (misappropriated).
Availability (A)	Ascertaining that:
	IoT devices are not precluded from functioning and/or performing their function in an improper or compromised manner (because they might have been infected with viruses, worms, and other debilitating intrusions and/or have been exploited in terms of various Operating System, software utilities, packaged microcode, or applications).
	IoT devices are not hijacked to become rogue devices with camouflaged identity marshaled to compromise other IoT devices and systems either as distributor of viruses or worms, or as sources of flooding traffic (this impacting availability).
	The IoT analytics systems do not become flooded with useless traffic which would cause practical shutdowns and denial of service (DoS).
	The data channels or Network Element (NE) used for IoT are not intentionally swamped (especially when using unlicensed spectrum), or the (normal) traffic overwhelms the system to cause buffer, packet, or message loss.

security policies should be consistent with the application, the risks, and the practical limits of the device in question. In some applications (e.g. healthcare), there typically is a stringent privacy requirement dictated by regulatory regiments; thus, strong security mechanisms are needed. In some instances, the IoT is used to control physical infrastructure (e.g. grids, process control); here, again, strict control is (almost invariably) a requirement [20].

The inclusion of various security functions to protect IoT devices and the networks in the IoT ecosystem can intrinsically have a negative impact on efficiency; also, the power consumption associated with executing security protocols presents a significant challenge. The challenges to the consistent incorporation of security mechanisms are fueled by the following considerations [20]:

- IoT technology and systems are relatively new, less broadly understood than traditional ICT systems.
- IoT systems are widely scattered geographically, often in open, uncontrolled environments.
- IoT systems are widely scattered administratively, where multiple, often heterogeneous, environments, processes, technologies, and security mechanisms exist.
- IoT systems are widely splintered across vertical applications, many of which tend to be silos of their own.
- IoT systems being deployed in the short term (including M2M systems) tend to be vendor-specific; wide-ranging, comprehensive standards have not been developed, matured, or implemented.

- IoT systems being deployed at this time tend not to follow an accepted layered architecture, which would enable concept/function simplicity/standardization and the ability to integrate systems (including security) from various vendors, all resulting in a fragmented environment.
- IoT endpoints may use different addressing models and addressing formats, each optimized for a specific application.
- IoT systems tend to employ relatively low complexity (low cost) endpoint platforms with limited memory and computational power.
- IoT endpoint systems tend to consume limited electrical power (battery-driven).

Unfortunately, these conditions and operating limitations exist and must be dealt with in order to deploy a security mechanism for the IoT. In particular, the lack of agreed-upon architectures and standards have hindered not only the IoT deployment per se up to the present, but have also hampered the symbiotic integration of security mechanisms. Interoperability of multiple levels is a key concern and a factor impacting deployment penetration, deployment velocity, and system cost.

The security requirements of various IoT systems and applications may differ in several ways; therefore, not all IoT end nodes necessarily comply with the tightest security measures; security policies should be consistent with the application, the risks, and the practical limits of the devices in question. In some applications (e.g. healthcare, assisted living, physical patient monitoring), there is almost invariably a stringent privacy requirement dictated by regulatory regimens; thus, strong security mechanisms are needed [126]. In some instances, the IoT is used to control physical infrastructure (e.g. grids, process control) – mobility may also be a factor in some of these applications; here, again, strict control is (almost invariably) a requirement. Furthermore, in some applications, especially in the case of wireless "fog" connectivity, there is a need for the IoT to enjoy "network autonomy," that is, having the ability for distributed ad hoc self-organization to support routing, "opportunistic routing," forwarding, and relaying of the content and topology information. In these cases, the security considerations are even more critical. In autonomous operation, individual endpoints or even clusters of endpoints are not dependent on a remote central entity; perhaps back in the cloud, to enable them to fulfill their sensing and data collection roles. Thus, a security compromise may not be detected by a centrally located control and audit system. The expectation is that as more M2M applications emerge, the trend toward autonomous operation of the endpoint devices and toward self-organizing network paradigms will accelerate. Appropriate security mechanisms are thus required. In an autonomous distributed environment, decisions may be made locally; one example might be the streetlight intensity during the night, based on, say, the sensed number of people or vehicles on that particular street; (not all) the data need travel to a central point for decision-making (although some "accounting" information may). On the other hand, a centralized IoT architecture has a different risk profile than an autonomous distributed environment. The point of integration, where all data for decision-making, analytics, or storage is aggregated, is a target-rich environment for hackers. Fortunately, more powerful network elements will be typically present (e.g. full-fledged firewalls); thus, it is both critical, as well as more doable to implement strong security measures at those locations.

A layered architecture approach comes into play both as a requirement for a better understanding of what security measures are needed and defining what security functions must be implemented at various physical or logical points in the overall IoT ecosystem. Security benefits from the availability of underlying system architecture, upon which predictable and reliable functionality can be defined [65]. As noted earlier, several IoT architectural models have been advanced in recent years, each focusing on some specific formulation or abstraction IoT ecosystem. Some IoT architectures are advanced by prominent vendors (e.g. see [75, 76]), while other architectures are proposed by industry consortia or standardization bodies. Aspects such as effective connectivity (both unicast and multicast), end-node management and administration (configuration and identity management), data representation, data collection and analysis, and, certainly, reliable security are key features of an effective and useful (IoT) architecture. Unfortunately, some of the existing reference models (just) make passing reference to security.

4.8.2 Applicable Security Mechanisms

Many existing security measures in the cloud or on-premise can be reused in the IoT environment, but they must be applied consistently, granularly (at each layer of the model), and in consideration of the IoT nodal constraints listed earlier (e.g. low computing complexity, limited storage, limited power, etc.). In practice, different techniques or mechanisms (possibly of increasing complexity and sophistication) may be utilized at each layer. In fact, this is the way some of the communications links already operate today: thus, a wireless LAN may include data link layer encryption; however, a mobile user on a wireless laptop working from home with a VPN tunnel to a corporate office will also get layer 3 encryption. Furthermore, the worker could download a Word file that had been encrypted on an intranet folder, for example. In broad terms, system designs based on decentralized architectures are preferable from an IoT security perspective; for example, DoS attacks are better weathered by a decentralized model, and overall operational functioning is more impervious to breaches and incapacitation; however, many applications require a centralized paradigm. Some of the proposed IoT architectures do include security considerations, e.g. [79]. For example, the model in [79] includes Authorization (supporting the "I" requirement); Key Exchange and Management (supporting the "C" requirement); Trust and Reputation (I); Identity Management (I); and Authentication (I). This model, like many others, includes security as a homogenous vertical stack.

The application of the OSiRM discussed above may at least provide some pragmatic simplification and low-level standardization for the deployment of security safeguards. Security mechanisms covering confidentiality, integrity, and availability are needed at each of these layers. In practical terms, layer-specific mechanisms are needed. As a minimum, encrypted tunnels, encryption of data at rest, and key management are key IoT/IoTSec imperatives. Capabilities such as security and management are often included in architectures as vertical stacks that cut across multiple layers; in other environments, the capabilities are included at each layer. There are advantages and disadvantages to either model. In the former, the security is tacked-on at the side, as a separable product or functionality, say by a specialized vendor; also, there is minimal redundancy of functions replicated at each layer. This is generally the approach that has been used in the ICT industry to date; unfortunately, this model has weaknesses, as the chronic, daily breaches attest (with billions of business records compromised each year). The latter (baking-in the capabilities at each layer) is more complex, requires that each layer-vendor implements the features, and may require more processing power, but intuitively one can see that this is a tighter and more reliable model; it also provides "security in depth" by implementing redundant safety checks at several points in the system/architecture/model. Some architectures do endeavor to start to place some security functionality in various places in the model; for example, Intel's IoT architecture places a security element in their "things" block, in their "network" block (in the context of the gateway), and in their "cloud" block [76].

OSiRM cited earlier includes three security-related mechanisms that exist independently at each layer and may be designed differently as needed:

- Authorization and Authentication – this supports part of the Integrity requirement (who is the "user" and what kind of data can this user read/write/modify).
- Encryption and Key Management – this supports the Confidentiality requirement mentioned earlier.
- Trust and Identity Management – this supports part of the Integrity requirement (e.g. can the data/user be trusted).

(Other realms can be added to the model if deemed appropriate.) In the OSiRM, there will be optimized differences for a given security function at different layers, as well as specializations that may occur with the type of thing and/or type of application. For example, in the Fog Networking layer one might use a 64-bit encryption algorithm, while in the "Data Aggregation" or "Data Centralization" layer one might use a 256-bit encryption algorithm. As another example, for authentication at a "Fog Networking" layer, one might use a MAC-address list of VLANs managed by the access point, while at "Data Aggregation" or "Data Centralization" layer one might use a RADIUS or a DIAMETER mechanism, and so on [127]. An important aspect of the security mechanism in IoT must be implemented at the "Data Aggregation" layer, where the IoT gateway is positioned. The gateway supports the connection between the sensing realm and the data aggregation, forwarding, processing, and analysis realm. Its basic functions include protocol conversion, address mapping, and security, especially authentication and firewalling.

This approach is consistent with the IoT predicament discussed earlier, where IoT systems tend to employ relatively low complexity endpoint platforms with limited memory and limited computational power; and also tend to have limited endpoint electrical (battery) power, limiting "number crunching." As embodied by the OSiRM, some IoT nodes might utilize a basic software firewall with few packet inspection states, while other more complex or strategic endpoints might employ a more sophisticated full-packet inspection firewall. While security is always a trade-off with cost and convenience, the widespread deployment of connectivity-enabled devices with high intersection with people's daily life requires more diligence in building-in security defenses upfront and in more places and instances to avoid always trying to catch up.

Obviously, the IoT is not a completely new technology since sensor networks have been around for several years [26], and power grid control, for example, (also including meter reading) has been practiced in one form or another for a couple of decades. However, broad deployment of sensors is now being contemplated, also using cheaper, possibly Commercial off the Shelf (COTS) hardware. Unfortunately, it is clear that existing security techniques leave a lot to be desired, considering the billions of customer records being breached and/or compromised yearly and the daily press on breaches and infractions at the business and even governmental level; but, basically, to address the IoT security issue, these techniques can, in principle, be adopted, extended, and more affirmatively integrated and disseminated in the IoT ecosystem. As noted, some of the limitations of the IoT sensors, also including a possibly large endpoint population that speaks to the scalability of existing security mechanisms, need to be taken into consideration and addressed. Using existing approaches, however, can also be problematic. As an example, [128] notes that many embedded systems have the practice of reusing keys: devices share their built-in keys for HTTPS/SSH sessions, exposing them to exploitation, where intruders gain access to the keys and not only can eavesdrop on the encrypted stream but also can interfere with the node's operation (e.g. by changing the parameters). According to trade press, 4.5 million appliances, embedded systems, and IoT devices were at, some point, actually "net-accessible vulnerable devices" [128]. Accordingly, some of the "well known existing security measures" need to be re-tweaked or re-designed for the IoT environment.

Within the framework of IoT, the cryptographic algorithms for securing the data must be adapted to the needs of embedded devices with limited resources and, therefore, must meet limited area and power demands. Different security mechanisms may be used in PAN/LAN-based systems, LPWAN-based systems, or cellular-based systems. The implementation of these algorithms can be done at both the hardware and software levels. While software implementations enable the reuse of resources and support to different cellular standards, these are typically slower and less secure than their hardware counterparts. For example, a processor subsystem in an NB-IoT node would become too overloaded with security processing if implemented as software functions.

In the context of cellular-based systems, there is a recognized need for an improved crypto-processor that can process the security algorithms such as, for example, SNOW 3G,[10] Advanced Encryption Standard (AES) and ZUC,[11] or other security algorithms [97]. The user plane *confidentiality* mechanisms for LTE systems have been defined in Technical Specification System Architecture Evolution (SAE) of 3GPP under the Security Architecture specification. Three Evolved Packet System (EPS) Encryption Algorithms (EEAs) and EPS Integrity Algorithms (EIAs) are defined for the protection of data against malicious attacks: (i) the 128-EEA1 based on SNOW 3G, (ii) the 128-EEA2 based on Advanced Encryption Standard (AES), and (iii) 128-EEA3 based on ZUC ciphering algorithm; see Table 4.9 [97]. Table 4.10 depicts algorithms that protect the *integrity* of the information, that is, to authenticate the data with a Message Authentication Code (MAC) as being the original source of information without any tampered data (the keystream generators described above are used to produce a 32-bit MAC).

In the context of local area security, some of the local area protocols listed above (notably ZigBee, BLE, and Wi-Fi HaLow) offer MAC-layer encryption in support of first-hop confidentiality, while others do not, and, thus, the developer or technology provider must provide encryption tools. Even when providing first-hop confidentiality, end-to-end confidentiality must be assured. As implied in the OSiRM discussed above, strong security measures for authorization and authentication, encryption and key management, and trust and identity management, must be implemented at each layer of the model and end-to-end. Table 4.11 (partially inspired by [131]) depicts an OSiRM-assisted transition from "*As Is*" IoT security environment to a target "*To Be*" environment.

4.8.3   Hardware Considerations

Some security capabilities are best handled at the hardware level. Such capabilities have emerged in recent years. The Trusted Execution Environment (TEE) is of particular importance to (IoT) applications that deal with sensitive user data, including e-health-related data such as user real estate locations, user real estate contents (e.g. jewelry), medical claims, and so on. In addition, Intel Trusted Execution Technology (Intel TXT) is a system hardware technology that deals with verifying and/or maintaining a trusted Operating System (OS). Both technologies are discussed in Section 4.8.3.1 and 4.8.3.2 since they have relevance to the processing environment that may be utilized by e-health companies.

Trusted Execution Environment   The TEE is a secure ecosystem of the IoT device processor or of the application system ("Trusted Applications [TAs]"), offering the capability of isolated execution of authorized security software. In doing so, TEE provides end-to-end security by enforcing protected execution of authenticated code; it also ensures confidentiality, authenticity, privacy, system integrity, and data access rights [15, 132]. In general, IoT devices are single-purpose devices and may have a specifically chosen OS or application script. However, a user may also utilize his or her smartphone as a device and/or to access IoT-related data repositories.

---

[10](Apparently an obscure acronym.) The SNOW family are word-based synchronous stream ciphers developed by T. Johansson and P. Ekdahl at Lund University, Sweden. SNOW 3G is a word-oriented stream cipher that generates a sequence of 32-bit words under the control of a 128-bit key and a 128-bit initialization variable; these words can be used to mask the plaintext: first a key initialization is performed, that is the cipher is clocked without producing output; then with every clock tick it produces a 32-bit word of output.

[11](Apparently an obscure acronym.) ZUC is a relatively new stream cipher included in the LTE standards specified in Release 11 [129]. ZUC is a word-oriented stream cipher and is the basic building block for both 128-EEA3 and 128-EIA3. ZUC is a stream cipher designed by the Data Assurance and Communication Security Research Center (DACAS) of the Chinese Academy of Sciences. The cipher forms the core of the 3GPP mobile standards 128-EEA3 (for encryption) and 128-EIA3 (for message integrity). It was advanced for inclusion in LTE. ZUC is Linear-Feedback Shift Register (LFSR)-based and uses a 128-bit key and a 128-bit Initialization Vector (IV).

TABLE 4.9   User Plane Confidentiality Protection Mechanisms for LTE

User Plane	Description
Confidentiality Protection Mechanisms for LTE Systems 128-EEA1 SNOW 3G Algorithm	Note: LTE introduced a new set of cryptographic algorithms: there are three sets of algorithms for both confidentiality and integrity: they are the EPS Encryption Algorithms (EEA) and EPS Integrity Algorithms (EIA). EEA1/EIA1 are based on SNOW 3G, EEA2/EIA2 are based on the Advanced Encryption Standard (AES) with EEA2 defined by AES in Counter mode (CTR mode) and EIA2 defined by AES ciphered Message Authentication Code (MAC). EEA3/EIA3 are both based on a Chinese cipher ZUC. The keystream generator for 128-EEA1 is based on SNOW 3G, which is composed of three primary modules: (i) a bit reorganization mechanism; (ii) a Finite State Machine (FSM); and a (iii) Feedback Mechanism. The keystream produced by the keystream generator is used to mask the plaintext data using an exclusive OR operation. The Linear Feedback Shift Register (LFSR) initialization is specified in ETSI/SAGE, "Document 1: UEA2 and UIA2 Specification."
128-EEA2 Advanced Encryption Standard (AES) Algorithm	AES is a symmetric block cipher algorithm specified by the National Institute of Standards and Technology (NIST) that processes the input data arranged as a fixed block size of 128 bits. The principle of operation of AES cipher is to pass a block of data to encrypt, defined as the state, through a network of substitutions and permutations. AES encryption flow 200 is shown in Fig. 2. The 128-EEA2 algorithm uses AES in the Counter (CTR) mode of operation, with a key size of 128 bits. The cipher generates a series of output blocks that are XOR-ed with the input plain text data. AES encryption is performed over a set of input blocks, called counters, which are obtained as specified in Annex B of 3GPP, "Technical Specification TS 33.401: Security Architecture." To recover a plaintext block of data, the same XOR operation is performed between the ciphertext data and the output of the AES cipher, which is obtained using the same counter generated during encryption as an input block.
128-EEA3 ZUC Algorithm	ZUC is a stream cipher that forms the heart of the 3GPP confidentiality algorithm 128-EEA3 and the 3GPP integrity algorithm 128-EIA3. ZUC is a word-oriented stream cipher. It takes a 128-bit initial key and a 128-bit initial vector (IV) as input, and outputs a keystream of 32-bit words (where each 32-bit word is hence called a keyword). This keystream can be used for encryption/decryption. The execution of ZUC has two stages: initialization stage and working stage. The keystream generator for 128-EEA3 algorithm has the same general structure as SNOW 3G algorithm, but with different feedback and Finite State Machine (FSM) logic. The Linear Feedback Shift Register (LFSR) consists of 16 registers of 31 bits each, and the feedback path is constructed by a primitive polynomial in $GF(2^{31}-1)$. This is the primary difference from the SNOW 3G algorithm, since ZUC produces sequences over the prime field $GF(2^{31}-1)$ instead of $GF(2^m)$. Described in "Specification of the 3GPP Confidentiality and Integrity Algorithms 128-EEA3 & 128-EIA3. Document 2: ZUC Specification - June 28, 2011."

In this instance, the situation may occur where several other, often poorly secured, applications are downloaded, which can cause cross-domain contamination. Mechanisms are thus needed to allow trusted access to devices, systems, or data resources. Service providers and device manufacturers (for example, OEMs) have the challenge of protecting applications at various concentric domains, such as attacks against (i) the device's OS; (ii) device-resident application; (iii) device (and/or user's) credentials (e.g. authenticating the correct user to the correct service); (iv) transmission integrity and privacy; and (v) the data-at-rest content. The TEE isolates secure

TABLE 4.10   User Plane Integrity Protection Mechanisms for LTE

User Plane Integrity	Description
Protection Mechanisms for LTE Systems 128-EIA1 SNOW 3G Based Algorithm	See Note in Previous Table 3GPP's 128-EIA1 algorithm is implemented the same way as ETSI/SAGE UIA2 Integrity algorithm, specified in 3GPP, "Technical Specification TS 35.215: Document 1: UEA2 and UIA2 specifications."
128-EIA2 AES Based Algorithm	The 128-EIA2 algorithm also uses AES cipher to perform integrity protection. This mode of operation for integrity protection is called Cipher-based MAC (CMAC) mode. The message data are divided into blocks of 128 bits, which is the input size of data for the AES keystream generator. The resulting data from the cipher are XOR-ed with the next block of the message and encrypted in another round of AES cipher. This is repeated until the entire message has been processed.
128-EIA3 ZUC Based Algorithm	This algorithm will process the keystreams produced by ZUC stream cipher as specified in to produce an output word MAC. See, 3GPP, "Technical Specification TS 35.221: Document 1: EEA3 and EIA3 specifications."

Note: Recently P. Ekdahl, T. Johansson et al proposed a new member in the SNOW family of stream ciphers, called SNOW-V. The motivation is to meet an industry demand for very high-speed encryption in a virtualized environment, such as in 5G/5G-IoT environments. The SNOW 3G architecture was reassessed to be competitive in a pure software environment, making use of both existing acceleration instructions for the AES encryption round function as well as the ability of modern CPUs to handle large vectors of integers (e.g. Single Instruction/Multiple Data [SIMD] instructions). SNOW-V kept the general design from SNOW 3G in terms of linear feedback shift register (LFSR) and Finite State Machine (FSM), but both entities are updated to better align with vectorized implementations; the LFSR part is new and operates eight times the speed of the FSM. SNOW-V (i) increases the total state size by using 128-bit registers in the FSM, (ii) uses the full AES encryption round function in the FSM update, and (iii) the initialization phase includes a masking with key bits at its end. The result is an algorithm generally much faster than AES-256 and with expected security not worse than AES-256 [130].

applications and screens them from malware and viruses that might be injected (or downloaded) inadvertently. TEE-based approaches may be leveraged to address these concerns.

Naturally, a key desideratum for TEE is to minimize the device or application processing overhead while at the same time providing controlled access to a large amount of processor memory. In particular, TEE ensures that sensitive data, such as insurance-related transactions, are processed, stored, and protected in a trusted, isolated environment. Developers have defined the main operating system on an IoT device or an application server as the "Rich OS" (e.g. Windows, Android, etc.). The TEE is a lean OS-like environment that resides aside the Rich OS (this is also known as the Rich Execution Environment [REE]). It is designed to extend the level of protection against attacks that may have been generated in the Rich OS (e.g. from malware), for example, by taking control of access rights. TEE assumes that nothing coming from the Rich OS is trustworthy. TEE domiciles sensitive applications that are best isolated from the Rich OS, and it maintains all cybersecurity credentials and data manipulation in the lean TEE rather than in a larger Rich OS. In this paradigm, sensitive functions are meticulously defined and assigned to the TEE in the form of TAs (while also integrating physical tamper-resistant mechanisms).

The TEE concept was at first defined in the 2007 timeframe by the Open Mobile Terminal Platform (OMTP) Forum. TEE standardization is critical to avoid application, hardware, and industry fragmentation; standardization enables simplified implementation, improves interoperability, and reduced costs. GlobalPlatform is an organization of over 130 member firms developing standards. In support of an effort to bring some industry standardization to the TEE environment, the GlobalPlatform organization has defined two sets of APIs: TEE Internal APIs (1.0) and TEE Client APIs (1.0). The TEE internal API is utilized by a TA; the TEE Client APIs

TABLE 4.11   OSiRM-assisted Transition from "As Is" to a "To Be" Environment

Layers	Device	Typical Status quo ("As is")	Target ("to Be")/OSiRM-assisted
Lower Layers	Sensors (sensor to base station communication)	• No encryption • Weak Encryption • Weak protocols • No passwords • Weak passwords • Weak Operating Systems • Weak applications	• Strong encryption • Robust protocols • Use of TLS • Device health checks • Stronger Operating System (OS) • Stronger applications • Strong ID • Strong User Interface • Memory Isolation • Firmware over the Air (FOTA) • Hardware RoT (Root of Trust) • Trusted Execution Environment (TEE)
	Base Station/ Gateway	• All the communication issues listed above • Hacked device keys • Side-channel attacks • Weak Network Element OS/memory leakage	• Hardware Root of Trust • Secure boot • Trusted Execution Environment • Trusted Firmware • Secure clocks and counters • Anti-rollback mechanisms • Secure key storage • Strong encryption/cryptography
Upper Layers	Data Servers/cloud (communication with base station)	• No encryption • Weak Encryption • Weak protocols • Weak OS • Weak application	• Strong encryption • Strong Protocols • Strong OS
	Key Server (PKI/ Public Key infrastructure)	• Weak Encryption • Weak protocols • Weak OS • Disclosure ok keys	• Secure key provisioning • Key rotation

support the communication interfaces that Rich OS software can use to interact with its TAs. Ericson and Nokia are two of several vendors that have developed the TEE. The GlobalPlatform also promulgated a compliance-testing process and issued a Protection Profile, certifying that a TEE meets the target security level. The advantage of working within a community, rather than "going it alone" is that multiple vendors can contribute to the security assurance efforts, and products are interoperable and interchangeable. Open-source TEE supporting standard interfaces have become available recently, sometimes known as OP(en)-TEE [15, 133].

Chipmakers and devices using these chipsets utilize TEEs to deliver platforms that have trust (that is, the assurance that the device is only running legitimate, uncorrupted firmware/software) built-in from the get-go; in turn, service and content providers rely on integral trust to build critical applications (e.g. financial, insurance, e-health). Typical TEE-oriented applications to date have included Digital Rights Management (e.g. digital content, films, music); mCommerce and mPayments credentials and transactions; and enterprise data (which can include insurance data). While these concepts are described here as supporting the end-user device (e.g. the medical monitoring gear), they can also apply to the processing end of the path (e.g. the analytics engines, portals, or SaaS cloud-based applications). Looking specifically at security, the TEE environment can increase the level of assurance of the medical devices as related to the following: User Authentication, Trusted Processing and Isolation, Transaction Validation,

uintanilla, F.G., Cardin, O., L'Anton, A. et al (2016). Implementation Framework For Cloud-Based olonic Control Of Cyber-Physical Production Systems. In: *IEEE Proceedings of the 14th International onference on Industrial Informatics*. Poitiers, France. (19–21 July 2016).

ing, C., Zhu, Y., Shi, W. et al. (September 2018). A dependable time series analytic framework for ber-physical systems of IoT-based smart grid. ACM Transactions on Cyber-Physical Systems 3 (1): 7, CM New York, NY, USA, doi:https://doi.org/10.1145/3145623.

ramboulidis, K. and Christoulakis, F. (October 2016). UML4IoT – A UML-based approach to ploit IoT in cyber-physical manufacturing systems. Computers in Industry, Elsevier 82: 259–272. ps://doi.org/10.1016/j.compind.2016.05.010.

rg, A., Chattopadhyay, A., and Lam, K.Y. (2018). Wireless communication and security issues for ber–physical systems and the internet-of-things. Proceedings of the IEEE 106 (1): 38–60. https://doi. /10.1109/JPROC.2017.2780172.

tonino, P.O., Morgenstern, A., Kallweit, B., et al. (2018). Straightforward Specification of Adaptation-chitecture-Significant Requirements of IoT-enabled Cyber-Physical Systems. *2018 IEEE ernational Conference on Software Architecture Companion (ICSA-C)*, 30 April-4 May 2018, Seattle, A, USA, doi:10.1109/ICSA-C.2018.00012.

, H., Maple, C., Watson, T., et al. (2016). The Security Challenges in the IoT Enabled Cyber-Physical tems And Opportunities For Evolutionary Computing & Other Computational Intelligence. 6 *IEEE Congress on Evolutionary Computation (CEC)*, 24–29 July 2016, Vancouver, BC, Canada, :10.1109/CEC.2016.7743900.

sch, E., Kadar, I., Grewe, L.L., et al. (2017). Panel Summary Of Cyber-Physical Systems (CPS) and ernet of Things (IoT) Opportunities With Information Fusion. *Proceedings Volume 10200, Signal cessing, Sensor/Information Fusion, and Target Recognition XXVI; Event: SPIE Defense + Security*, 7, Anaheim, CA (2017), doi.org/10.1117/12.2264683.

ffor, E. (2017). Reference Architecture for Cyber-Physical Systems. NIST (National Institute of dards and Technology), 12 December 2017, 100 Bureau Drive, Gaithersburg, MD 20899.

Baheti, H. Gill, "Cyber-physical systems", Chapter in The Impact of Control Technology, T. Samad A.M. Annaswamy (eds.), IEEE Control Systems Society, 2011, www.ieeecss.org. 161–166

g, H., Rawat, D.B., Jeschke, S. et al. (2017). Cyber-Physical Systems: Foundations, Principles and plications. London: Academic Press/Elsevier ISBN: 978-0-12-803801-7.

anovsky, A. and Ishikawa, F. (eds.) (2017). Trustworthy Cyber-Physical Systems Engineering. a Roton, FL: CRC.

chevsky, A.A., Letychevsky, O.O. et al. (November 2017). Cyber-physical systems. Cybernetics and ems Analysis 53 (6) https://doi.org/10.1007/s10559-017-9984-9.

kovic, J.A. (February 2017). Research directions for cyber physical Systems in Wireless and Mobile lthcare. ACM Transactions on Cyber-Physical Systems – Inaugural Issue 1 (1) https://doi. 10.1145/2899006.

er, H.A., Litoiu, M., and Mylopoulos, J. (2016). Engineering Cybersecurity into Cyber Physical ems. *CASCON 2016 Markham*, Ontario Canada, 2016 ACM. ISBN 123-4567-24-567/08/06, 0.475/123 4.

ing, T. Lei, Zhang, L., and Hsu, C.-H. "Offloading Mobile data traffic for QoS-aware service provi-in vehicular cyber-physical systems", Future Generation Computer Systems, 61 August 2016, s 118–127. doi: https://doi.org/10.1016/j.future.2015.10.004.

soukos, X., Karsai, G., Laszka, A. et al. SURE: a Modeling and simulation integration platform for ation of secure and resilient cyber-physical systems. Proceedings of the IEEE PP (99): 1–20. ://doi.org/10.1109/JPROC.2017.2731741.

Y.-S., Kwon, Y.-H., Yoon, S.-Y., and Hwang, S.-S. (2020). Apparatus and method for transmitting eceiving signals in [a] wireless communication system. US Patent 10, 693, 696; 23 June 2020; filed ovember 2017. Uncopyrighted.

.A. (February 2015). The past, present and future of cyber-physical systems: a focus on models. rs 15 (3): 4837–4869. https://doi.org/10.3390/s150304837.

o, S. (April 2017). Cyber-physical systems. Computer 50 (4): 14–16. https://doi.org/10.1109/ 017.105.

umar, M., Sadagopan, C., and Baskaran, M. (August 2016). Wireless sensor network to cyber al systems: addressing mobility challenges for energy efficient data aggregation using dynamic Sensor Letters 14 (8): 852–857. (6), DOI: https://doi.org/10.1166/sl.2016.3624.

Usage of Secure Resources, and Certification. For example, a TEE-based environment ensures that a rogue device that presents into the environment is not accepted as a legitimate user – the rogue device may have the nefarious intent to inject some malware at some point in time. Also, the TEE makes sure that a device (or user) does not escalate its privileges beyond what is the intent of the system administrator (by providing isolation that limits functionality or access to data or some function – for example, a function controlling automatic medication delivery from a body-worn pump).

As a side note, TEEs can increase the capabilities of Secure Elements (SEs). SEs are secure components that consist of autonomous, tamper-resistant hardware within which secure applications and related confidential cryptographic data (e.g. key management) are stored and executed. While enjoying a high level of security, these devices have limited functionality; an example includes an NFC device. SEs can work in conjunction with the TEE to enhance their capabilities.

Intel TXT Intel® TXT is a scalable architecture that specifies hardware-based security protection [134]. These methodologies are built into Intel's chipsets to address threats across physical and virtual infrastructures; the technology is designed to harden platforms to better deal with threats of hypervisor attacks, BIOS, or other firmware attacks, malicious rootkit installations, or other software-based attacks. TXT aims at increased protection by allowing greater control of the launch stack through a Measured Launch Environment (MLE) and enabling isolation in the boot process. It extends the Virtual Machine Extensions (VMX) environment of Intel Virtualization Technology to provide a verifiably secure installation, launch, and use of a hypervisor or OS. MLE enables an accurate comparison of all critical elements of the launch environment against a known good source. TXT creates a cryptographically unique identifier for each approved launch-enabled component and then provides hardware-based enforcement mechanisms to block the launch of code that does not match approved code. This hardware-based solution provides the foundation on which trusted platform solutions can be built to protect against software-based attacks; the technology is broadly applicable to servers and to IoT devices. More specifically, TXT provides [15, 134]:

- *Verified Launch:* A hardware-based chain of trust that enables the launch of the MLE into a "known good" state. Changes to the MLE can be detected through cryptographic (hash-based or signed) measurements.
- *Launch Control Policy (LCP):* A policy engine for the creation and implementation of enforceable lists of "known good" or approved executable code.
- *Secret Protection:* Hardware-assisted methods that remove residual data at an improper MLE shutdown, protecting data from memory-snooping software and reset attacks.
- *Attestation:* The ability to provide platform measurement credentials to local or remote users or systems to complete the trust verification process and support compliance and audit activities.

After the system goes thought a boot-up stage, various state variables are compared with pre-stored data; only if a match in state parameters is established, the device is allowed to move to the next stage that would entail loading the application code, at which point other state checks are made. If the checks fail, reporting of the condition occurs (to some administrator); there would be an indication that "trust" of the platform cannot be established. TXT can be used in IoT environments, particularly at the server end, to ascertain that the medical analytics engines and the storage systems (e.g. storing patient health data) are secure. These processes can also be embedded in the medical IoT device to ascertain, among other validations, that the device is legitimate – in the sense that it runs only official code, performing only specific functions, and not exceeding its level of functional authority. The TPM is an ISO (International

Organization for Standardization) standard for a secure cryptoprocessor; a secure cryptoprocessor is a dedicated microcontroller designed to secure hardware by integrating cryptographic keys into devices (ISO/IEC 11889 published in 2009 – the technical specification was TPM was developed by Trusted Computing Group [TCG]).

4.8.4 Other Approaches: Blockchains

Some have sought to implement security mechanisms at a much higher level, both in an IoT context and in the broader IT content. One such mechanism is the "blockchain." A blockchain is a form of distributed ledger (a distributed database) that retains an expanding list of records while precluding revision or tampering (a tamper-proof ledger). The blockchain encompasses a data structure of "child" (aka successor) blocks; each block includes sets of transactions, timestamps, and links to a "parent" (aka predecessor) block; the linked blocks constitute a chain. It intrinsically provides universal accessibility, incorruptibility, openness, and the ability to store and transfer data in a secure manner. The original application was as a ledger for bitcoins. Users can add transactions, verify transactions, and add new blocks. Proponents see opportunities for the use of blockchains for e-health companies (and banks and many other industries), as it allows a replacement of the common centralized data paradigm, thus fostering additional process disintermediation [2, 3, 8, 9, 15, 135]. Possible applications of interest to the e-health industry include claims filing/processing; claim fraud detection, for example, spot multiple claims from a claimant (medical office) for the same procedure; data decentralization; and cybersecurity management (e.g. data integrity). Given the expected distributed nature of an IoT-based ecosystem (also in the e-health industry context), blockchains may play an important role in the future.

REFERENCES

1. D. Minoli, B. Occhiogrosso, "IoT-driven advances in commercial and industrial building lighting and in street lighting", in the book Industrial IoT: Challenges, Design Principles, Applications, and Security, Ismail Butun Editor. Springer, 2020, ISBN 978-3-030-42500-5 Pages 97–159.
2. Minoli, D. (November 2019). Positioning of Blockchain mechanisms in IoT-powered smart home systems: a gateway-based approach. Elsevier IoT Journal, Special Issue on IoT Blockchains https://www.sciencedirect.com/science/article/pii/S2542660519302525.
3. D. Minoli, B. Occhiogrosso, "Blockchain-enabled fog and edge computing: concepts, architectures and smart city applications". Chapter in book: Blockchain-Enabled Fog and Edge Computing: Concepts, Architectures and Applications, Editors: M. H. Rehmani, M. M. Rehan, CRC Press, Taylor & Francis Group, 2020. Boca Raton, Fla. Pages 20–32
4. D. Minoli, B. Occhiogrosso, "Practical aspects for the integration of 5G networks and IoT applications in smart cities environments", Special Issue titled "Integration of 5G Networks and Internet of Things for Future Smart City," Wireless Communications and Mobile Computing". Vol. 2019, Article ID 5710834, 30 pages, August 2019. Hindawi & John Wiley & Sons (https://doi.org/10.1155/2019/5710834).
5. D. Minoli, B. Occhiogrosso, W. Wang, "MIPv6 in Crowdsensing applications for SIoT environments", Chapter in book Towards Social Internet of Things: Enabling Technologies, Architectures and Applications, Hassanien A., Bhatnagar R., Khalifa N., Taha M. (eds), Chapter 3, Springer, 2020. pp. 31–49, Part of the Studies in Computational Intelligence book series (SCI, volume 846). https://link.springer.com/chapter/10.1007/978-3-030-24513-9_3.
6. D. Minoli, B. Occhiogrosso, "Constrained average design method for QoS-based traffic engineering at the edge/gateway boundary in VANETs and cyber-physical environments", Chapter in book Managing Resources for Futuristic Wireless Networks, IGI Global, 2020. Mamata Rath, Editor.
7. Minoli, D. and Occhiogrosso, B. (2018). Ultrawideband (UWB) Technology For Smart Cities IoT Applications. *2018 IEEE International Smart Cities Conference (ISC2) – IEEE ISC2 2018- Buildings, Infrastructure, Environment Track*, Kansas City, 16–19 September 2018.
8. Minoli, D. and Occhiogrosso, B. (November 2019). Editor-in-chief, special issue cations in IoT environments, overview. Elsevier IoT Journal https://www.science article/pii/S2542660519302537.
9. Minoli, D. and Occhiogrosso, B. (Summer 2018). Blockchain mechanisms for IoT s Journal 1 (1): 1–13.
10. D. Minoli, B. Occhiogrosso, "IoT Applications to Smart Campuses and a Case Stu Alliance for Innovation) Endorsed Transactions on Smart Cities, European Un ISSN 2518–3893 Volume 2. Published 19 December 2017. http://eudl.eu/dc 2017.153483
11. D. Minoli, B. Occhiogrosso, "Implementing the IoT for renewable energy", Cha Things a to Z: Technologies and Applications, Editor: Q. Hassan, Chapter 15, IEE 2018, ISBN-13: 978-1119456742. (With L. Finco). 425–442, 881–895
12. Minoli, D., Occhiogrosso, B. et al. (2017). A review of wireless and satellite-based in Support of smart grids. Mobile Networks and Applications Journal, Springer.
13. Minoli, D. and Occhiogrosso, B. (2017). Mobile IPv6 Protocols and High Efficienc Smart City IoT Applications. *CEWIT2017 Conference*, Nov. 8-9-2017, Stony Brook
14. Minoli, D., Occhiogrosso, B., Sohraby, K., et al. (2017). A Review of Wireless and S Services in Support of Smart Grids. *1st EAI International Conference on Smart G of Things (SGIoT 2017)*, 11–13 July 2017, Sault Ste. Marie, Ontario, Canada.
15. Minoli, D., Occhiogrosso, B., Sohraby, K., et al. (2017). IoT Security (IoTSec) Mecha and Ambient Assisted Living Applications. The *Second IEEE/ACM International Energy-Aware, & Reliable Connected Health (SEARCH 2017) (collocated with CHA on Connected Health: Applications, Systems, and Engineering Technologies)*, Philadel
16. Minoli, D., Occhiogrosso, B., Sohraby, K., et al. (2017). Multimedia IoT System *Global IoT Summit, GIoTS-2017, Organized by Mandat International, IEEE IoT T Service Computing), the IoT Forum and IPv6 Forum (collocated with the IoT We* Geneva, Switzerland.
17. D. Minoli, B. Occhiogrosso, "The emerging "energy internet of things", Chapter in I to Z: Technologies and Applications, Editor: Q. Hassan, Chapter 14, IEEE Pres ISBN-13: 978-1119456742 385–417.
18. D. Minoli, B. Occhiogrosso, K. Sohraby, "IoT Considerations, Requirements, and Smart Buildings – Energy Optimization and Next Generation Building Managem Internet of Things Journal, Year: 2017, Volume: 4, Issue: 1, Pages: 269–283. DOI: http JIOT.2017.2647881.
19. D. Minoli, B. Occhiogrosso, Sohraby, K., et al., "Security considerations for IoT s applications", Chapter in Internet of Things: Challenges, Advances and Applic Hassan, A. R. Khan, S. A. Madani, Chapter 16, Chapman & Hall/CRC Compute Science Series, CRC Press, Taylor & Francis, 2018, ISBN 9781498778510. 321–347
20. D. Minoli, B. Occhiogrosso, Sohraby, K., et al. (2017). IoT Security (IoTSec Requirements, and Architectures. *2017 14th IEEE Annual Consumer Communicati Conference (CCNC)*, 8–11 January 2017 in Las Vegas, NV, USA. IEEE catalog nur CDR, ISBN:978–1–5090-6195-2, ISSN: 2331–9860.
21. D. Minoli, B. Occhiogrosso, and Sohraby, K. (2017). Internet Of Things (IoT)-Base Method For Rail Crossing Alerting Of Static Or Dynamic Rail Track Intrusior Proceedings of JRC (Joint Rail Conference), April 4–7, 2017, Philadelphia, PA.
22. D. Minoli, B. Occhiogrosso, "Internet of things applications for smart cities". Chap Things a to Z: Technologies and Applications, Editor: Q. Hassan, Chapter 12, IEEE 2018, ISBN-13: 978-1119456742. 319–350
23. D. Minoli, B. Occhiogrosso, Sohraby, K. et al., "IoT considerations, requirements, ani insurance applications". Chapter in Internet of Things: Challenges, Advances and App Q. Hassan, A. R. Khan, S. A. Madani, Chapter 17, Chapman & Hall/CRC Computer Science Series, CRC Press, Taylor & Francis, 2018, ISBN 9781498778510. 347–362
24. Minoli, D. (2015). Innovations in Satellite Communications and Satellite Technol Implications of DVB-S2X, High Throughput Satellites, Ultra HD, M2M, and IP. Wile
25. Minoli, D. (2013). Building the Internet of Things with IPv6 and MIPv6. Wiley.
26. Minoli, D. and Sohraby, K. (2007). Wireless Sensor Networks. Wiley.

47. Hua, F., Lua, Y., Vasilakos, A.V. et al. (March 2016). Robust cyber–physical systems: concept, models, and implementation. Future Generation Computer Systems 56: 449–475. https://doi.org/10.1016/j.future.2015.06.006.

48. Monostori, L. (2014). Cyber-physical production systems: roots, expectations and R&D challenges. Procedia CIRP 17: 9–13. https://doi.org/10.1016/j.procir.2014.03.115.

49. Yu, X. and Xue, Y. (May 2016). Smart grids: a cyber–physical systems perspective. Proceedings of the IEEE 104 (5): 1058–1070. https://doi.org/10.1109/JPROC.2015.2503119.

50. Swierk, T.E., Cox, T.R., and Hammons, M.R. (2020). Internet-of-things (IoT) gateway tampering detection and management. US Patent 10, 671, 765; 2 June 2020; filed 03 March 2017. Uncopyrighted.

51. Patent Trial and Appeal Board (PTAB). (2020). Petitioner's Notice of Deposition of Mr. Daniel Minoli. GMG Products LLC v. Traeger Pellet Grills LLC – http://Law360www.law360.com, PGR2019–00034 &00035 21 Febuary 2020.

52. Massey, D. (2017). Applying Cybersecurity Challenges to Medical and Vehicular Cyber Physical Systems. Proceeding of SafeConfig '17, 2017 Workshop on Automated Decision Making for Active Cyber Defense, Dallas, Texas, USA – 03 November 2017, 39–39, ISBN: 978-1-4503-5203-1 doi:10.1145/3140368.3140379.

53. Buntz, B. (2020). COVID-19 Driving Data-Integration Projects in IoT. IoT World Today, 10 August 2020. Available online on 17 October 2020 at https://urgentcomm.com/2020/08/10/covid-19-driving-data-integration-projects-in-iot

54. Matthews, K. (2020). 5 IoT Projects Aiming to Help Fight Against Coronavirus. The IoT Magazine, 15 April 2020. Available online on 17 October 2020 at https://theiotmagazine.com/5-iot-projects-aiming-to-help-fight-against-coronavirus-3e912043f88a.

55. J. J. Corbett, D. L. Kjendal, Woodhead, J.R. (2020). System and method for low power wide area virtual network for IoT. US Patent 10, 778, 752; 15 September 2020; Filed 06 May 2019. Uncopyrighted.

56. Guinard, D., Fischer, M., and Trifa, V. (2010). Sharing Using Social Networks In A Composable Web Of Things. *Proc. 8th IEEE Int. Conf. Pervasive Comput. Commun. Workshops (PERCOM)*, 702–707.

57. Atzori, L., Iera, A., and Morabito, G. (2012). The social internet of things (SIoT) – when social networks meet the internet of things: concept, architecture and network characterization. Computer Networks 56 (16): 3594–3608.

58. Atzori, L., Iera, A., and Morabito, G. (November 2011). SIoT: giving a social structure to the internet of things. IEEE Communications Letters 15 (11) https://doi.org/10.1109/LCOMM.2011.090911.111340.

59. Nitti, M., Girau, R., Atzori, L., et al. (2012). A Subjective Model For Trustworthiness Evaluation In The Social Internet of Things. *Proc. IEEE 23rd Int. Symp. Pers. Indoor Mobile Radio Communications (PIMRC)*, 18–23.

60. Lee, G.M., Rhee, W.S., and Crespi, N. (2013). Proposal of a New Work Item on Social and Device Networking. *ITU Telecommunications Standard, Sector, SG13 Rapporteur Group Meeting*, Geneva, Switzerland.

61. Ciortea, A., Boissier, O., Zimmermann, A., et al. (2013). Reconsidering The Social Web Of Things: Position Paper. *Proc. ACM Conf. Pervasive Ubiquitous Computing, UbiComp'13 Adjunct*, 1535–1544.

62. Xu, L.D., He, W., and Li, S. (2014). Internet of things in industries: a survey. IEEE Transactions on Industrial Informatics 10 (4) https://doi.org/10.1109/TII.2014.2300753.

63. Ortiz, A.M., Hussein, D., Park, S. et al. (June 2014). The cluster between internet of things and social networks: review and research challenges. IEEE Internet of Things Journal 1 (3) https://doi.org/10.1109/JIOT.2014.2318835.

64. Sarwar, U. and Chun, A.L. (2020). Distributed Adaptive Heterogeneous Wireless Communications Management. US Patent 10, 791, 560; 29 September 2020; filed 28 September 2017. Uncopyrighted.

65. Minoli, D. (2008). Enterprise Architecture a to Z: Frameworks, Business Process Modeling, SOA, and Infrastructure Technology. Auerbach.

66. Weyrich, M. and Ebert, C. (2016). Reference Architectures For The Internet of Things. IEEE Software, IEEE Computer Society, Jan/Feb 2016, p.112 ff.

67. Hu, L., Xie, N., Kuang Z, and Zhao K. (2012). Review of cyber-physical system architecture. *2012 IEEE 15th International Symposium on Object/Component/Service-Oriented Real-Time Distributed Computing Workshops*, 11–11 April 2012, Shenzhen, Guangdong, China. doi:10.1109/ISORCW.2012.15.

68. Guo, S. and Zeng, D. (Eds.) (2019). *Cyber-Physical Systems: Architecture, Security and Application*, *EAI/Springer Innovations in Communication and Computing*. 2019. Part of the EAI/Springer Innovations in Communication and Computing book series (EAISICC). ISBN 978-3-319-92564-6.

69. Jiang, J.-R. (2018). An improved cyber-physical systems architecture for industry 4.0 smart factories. Advances in Mechanical Engineering https://doi.org/10.1177/1687814018784192.

70. International Society of Automation (ISA), *ISA-95, Enterprise-control System Integration*, 2018, https://www.isa.org/isa95 (accessed November 2018).

71. Lee, J. and Bagheri, B. (2015). A cyber-physical systems architecture for industry 4.0-based manufacturing systems. Manufacturing Letters 3: 18–23.

72. Ahmadi, A., Cherifi, C., Cheutet, V., and Ouzrout, Y. (2017). A Review of CPS 5 Components Architecture for Manufacturing Based on Standards. *11th IEEE International Conference on Software, Knowledge, Information Management and Applications (SKIMA 2017)*, December 2017, Colombo, Sri Lanka. 6 p. ffhal-01679977f.

73. Tiburski, R., Amaral, L.A., de Matos, E., and Hessel, F.P. (2015). The importance of a standard security architecture for SOA-based Iot middleware. IEEE Communications Magazine 53 (12): 20–26.

74. ISO/IEC. (2014). Study Report on IoT Reference Architectures/Frameworks, 2014.

75. Cisco Systems. (2014). Cisco Whitepaper: The internet of things reference model. 2014, Cisco Systems, Inc., 170 West Tasman Dr., San Jose, CA 95134, USA.

76. Intel Staff. (2015). Whitepaper: The Intel IoT platform, architecture specification. Santa Clara, California, 2015.

77. Walewski, J.W. (ed.) (2013). *Internet-of-Things Architecture (IoT-A), Project Deliverable D1.2 – Initial Architectural Reference Model for IoT*. https://cocoa.ethz.ch/downloads/2014/01/1360_D1%202_Initial_architectural_reference_model_for_IoT.pdf

78. ETSI. *ETSI TS 102 690: Machine-to-Machine communications (M2M); Functional architecture*. October 2011.

79. Bauer, M., Boussard, M., Bui, N. et al. (2013). IoT reference architecture. In: Enabling Things to Talk (eds. A. Bassi, M. Bauer, M. Fiedler, et al.). Berlin, Heidelberg: Springer https://doi.org/10.1007/978-3-642-40403-0_8.

80. Singh, N.F. (2020). Active and Passive Asset Monitoring System. US Patent 10, 783, 419; 22 September 2020; filed 2 July 2019. Uncopyrighted.

81. Sedjelmaci, H., Senouci, S.M., and Feham, M. (2012). Intrusion Detection Framework Of Cluster-Based Wireless Sensor Network. *2012 IEEE Symposium on Computers and Communications (ISCC)*, Cappadocia, Turkey, 1–4.

82. Komal, K. and Mohasin, T. (2015). Secure and economical information transmission for clustered wireless sensor network. International Journal of Science and Research (IJSR). ISSN (Online) 4 (12): 2319–7064.

83. IETF. RFC 8376: Low-Power Wide Area Network (LPWAN) Overview, 2018–05.

84. Andrade, R.O. and Yoo, S.G. (Nov. 2019). A comprehensive study of the use of LoRa in the development of smart cities. Applied Sciences 9 (22): 4753. https://doi.org/10.3390/app9224753.

85. Elshabrawy, T. and Robert, J. (Apr. 2019). Interleaved chirp spreading LoRa-based modulation. IEEE Internet of Things Journal 6 (2): 3855–3863. https://doi.org/10.1109/JIOT.2019.2892294.

86. Sallum, E., Pereira, N., Alves, M., and Santos, M.M. (2020). Improving quality-of-service in LoRa low-power wide-area networks through optimized radio resource management. Journal of Sensor and Actuator Networks 9 (1): 10. https://doi.org/10.20944/preprints201909.0243.v1.

87. Kaura, G., Guptab, S.H., and Kaurc, H. Enhancing LoRa Network Performance for Smart City IoT Applications. Sustainable Cities and Society. Elsevier.

88. Basford, P.J., Bulot, F.M.J., Apetroaie-Cristea, M. et al. (2020). LoRaWAN for smart city IoT deployments: a long term evaluation. Sensors 20 (3): 648. https://doi.org/10.3390/s20030648.

89. Wu, W., Li, Y., Zhang, Y. et al. (2020). Distributed queueing-based random access protocol for LoRa networks. IEEE Internet of Things Journal 7 (1): 763–772. https://doi.org/10.1109/JIOT.2019.2945327.

90. Teboulle, H., Le Gourrierec, M. Franck Harnay, (2020). Device for Transporting LoRa Frames on a PL Network. US Patent 10, 742, 266; 11 August 2020; filed 15 July 2019. Uncopyrighted.

91. Saleem, Y., Crespi, N., Rehmani, M.H., and Copeland, R. (2019). Internet of things-aided smart grid: technologies, architectures, applications, prototypes, and future research directions. IEEE Access: 62962–63003.

92. Li, Y., Cheng, X., Cao, Y. et al. (2018). Smart choice for the smart grid: Narrowband internet of things (NB-IoT). IEEE Internet of Things Journal 5 (3): 1505–1515.

93. Reininger, P. (2016). 3GPP Standards for the Internet-of-Things. Published online: ftp://http://www.3gpp.org/Information/presentations/presentations 2016/2016 11 3gpp Standards for IoT.pdf

94. 3GPP. Cellular System Support For Ultra-Low Complexity And Low Throughput Internet of Things (CIoT). Published online https://portal.3gpp.org/desktopmodules/Specifications/SpecificationDetails. aspx?specificationId=2719

95. Ratasuk, R., Vejlgaard, B., Mangalvedhe, N., and Ghosh, A. (2016). NB-IoT System for M2M Communication. *IEEE Wireless Communications and Networking Conference (WCNC)*, pp. 1–5.

96. Wang, Y.-P.E., Lin, X., Adhikary, A. et al. (2017). A primer on 3GPP Narrowband internet of things. IEEE Communications Magazine 55 (3): 117–123.

97. Nunes, L.C. and Fuhrmann, S.F.R. (2020). Low Area Optimization for NB-IoT Applications. US Patent 10, 797, 859; 6 October 2020; filed 22 March 2018. Uncopyrighted.

98. T-Mobile NB-IoT plan, https://iot.t-mobile.com/pricing

99. Verizon wireless Cat-M Plan, https://www.verizonwireless.com/biz/plans/m2m-business-plans

100. GSMA. (2018). Mobile IoT in the 5G Future: NB-IoT and LTE-M in the Context of 5G. https://www.gsma.com/iot/wp-content/uploads/2018/05/GSMAIoT_MobileIoT_5G_Future_May2018.pdf

101. Contento, M. (2019). 5G and IoT – Emerging Tech with Endless Use Cases.19 February 2019. Published online at https://www.telit.com/blog/state-of-5g-and-iot-current-future-applications

102. Lin, J., Shen, Z., Zhang, A., Chai, Y. (2017). Using Blockchain Technology To Build Trust In Sharing LoRaWAN IoT. *Proceeding, ICCSE'17 Proceedings of the 2nd International Conference on Crowd Science and Engineering*, Beijing, China, 06–09 July 2017, doi: 10.1145/3126973.3126980.

103. Moyer, B. (Sept. 2015). Low power, wide area: a survey of longer-range IoT wireless protocols. Electronic Engineering Journal.

104. GOS World Staff. Skyworks Unveils Sky5 Ultra Platform for 5G Architecture. 27 February 2019. Published online at https://www.gpsworld.com/skyworks-unveils-sky5-ultra-platform-for-5g-architecture

105. Hui, D. and Zangi, K. (2012). Distributed Computation Of Precoding Weights For Coordinated Multipoint Transmission On The Downlink. US Patent 8, 107, 965; 31 January 2012; filed August 26, 2009. Uncopyrighted.

106. ITU-R SG05 Contribution 40. *Minimum Requirements Related To Technical Performance For IMT-2020 Radio Interface(s)*, February 2017.

107. Global System for Mobile Communications Association (GSMA). *The Mobile Economy 2019, Report*, 2019. Published online at https://www.gsmaintelligence.com/research/?file=b9a6e6202ee1d5f 787cfebb95d3639c5&download

108. Vui, Y. and Lee, H. (2016). A Novel Digital Device Monitoring System Using UPnP Cloud Architecture. International Conference on Engineering Technologies and Big Data Analytics (ETBDA'2016) 21–22 January 2016 Bangkok (Thailand).

109. Rajalakshmi, Narayanan, M., Venkatesh, R., and Scholar, U.G. (2015). An exclusive study on unstructured data mining with big data. International Journal of Applied Engineering Research 10 (4): 3875–3886. ISSN 0973-4562.

110. Cui, Y., Kim, M., and Lee, H. (2013). Cloud-based home media system model: providing a novel media streaming service using UPnP Technology in a Home Environment. International Journal of Software Engineering and Its Applications 7 (4): 127–136.

111. Open Connectivity Forum, UPnP-resources, https://openconnectivity.org/developer/specifications/upnp-resources/upnp

112. Sony Corporation. (2004). Method of and Apparatus For Bridging a UPnP Network And A Rendezvous Network. US20050160172A1. https://patents.google.com/patent/US20050160172A1/en?oq=US20050160172

113. OCF. *OCF 2.0.5 Bridging Specification*, https://openconnectivity.org/specs/OCF_Bridging_Specification_v2.0.5.pdf

114. UPnP Forum/OCF. (2015). UPnP: The Discovery & Service Layer For The Internet of Things. April 2015. http://upnp.org/resources/whitepapers/UPnP_Internet_of_Things_Whitepaper_2015.pdf

115. UPnP Forum/OCF. *UDA V2.0 with Cloud Annex*, http://upnp.org/specs/arch/UPnP-arch-Device Architecture-v2.0.pdf

116. Tomar, A. (2011). Introduction to ZigBee Technology, Global Technology Centre, Volume 1 July 2011, https://www.cs.odu.edu/~cs752/papers/zigbee-001.pdf

117. Nakazawa, J., Tokuda, H., Keith Edwards, W., Umakishore Ramachandran (2006). A Bridging Framework For Universal Interoperability In Pervasive Systems. *26th IEEE International Conference on Distributed Computing Systems (ICDCS'06)*, Lisboa, Portugal, July 2006.

118. Baek, Y.M., Ahn, S.C., and Kwon, Y.-M. (2010). UPnP network bridge for supporting interoperability through non-IP channels. IEEE Transactions on Consumer Electronics 56 (4): 2226–2232. http://dx.doi.org/10.1109/TCE.2010.5681094.

119. Treadway, J. (2018). Using an IoT Gateway To Connect The 'Things' To The Cloud. TechTarget – IoT Agenda, 1 October 2018, https://internetofthingsagenda.techtarget.com/feature/Using-an-IoT-gateway-to-connect-the-Things-to-the-cloud

120. Lu, J. and Scott Kelly, T. (2011). Network integration system and method. US Patent 8, 761, 050; 04 October 2011. Uncopyrighted.

121. Lee, M.-G., Lee, B.-M., Patil, M.-M. et al. (2020). Method and apparatus for setup of wireless communication. US Patent 10, 609, 581; 31 March 2020; filed 05 June 2017. Uncopyrighted.

122. Lai, C., Lu, R., Zheng, D. et al. (2015). Toward Secure Large-Scale Machine-to-Machine Communications in 3GPP Networks: Challenges and Solutions. *IEEE Communications Magazine – Communications Standards Supplement*, December 2015.

123. ISO, Introduction to ISO JTC1/WG10, June 2015. ISO materials.

124. K. Komal, T. Mohasin, "Secure and economical information transmission for clustered wireless sensor network", IJSR, Vo. 4, Issue 12 Dec. 2015, pp. 1033ff.

125. Tiburski, R., Amaral, A., L., de Matos, E., and Hessel, F.P. (2015). The importance of a standard security architecture for SOA-based IoT middleware. IEEE Communications Magazine 53 (12): 20–26.

126. Cornejo-García, R. Strengthening Elder's Social Networks through Ambient Information Systems and SNS. Computer Science Department, CICESE. (Online).

127. Chen, G., Gong, Y., Xiao, P. et al. (2015). Physical layer network security in the full duplex relay system. IEEE Transactions on Information Forensics and Security 10 (3): 574–583.

128. Nichols, S. (2016). Internet of Sins: Million More Devices Sharing Known Private Keys for HTTPS, SSH admin. The Register (http://theregister.co.uk), 07 September 2016.

129. Madani, M., Benkhaddra, I., Tanougast, C., and Chitroub, S. Digital implementation of an improved LTE stream cipher Snow-3G based on Hyperchaotic PRNG. Security and Communication Networks 2017: 5746976. https://doi.org/10.1155/2017/5746976.

130. Ekdahl, P., Johansson, T., Maximov, A., and Yang, J. (2019). A new SNOW stream cipher called SNOW-V. IACR Transactions on Symmetric Cryptology 2019 (3): 1–42. https://doi.org/10.13154/tosc.v2019.i3.1-42.

131. Wallace, J. (2016). Securing the Embedded IoT World. Guest Blog, IoT Security Foundation Website.

132. GlobalPlatform. *Trusted Execution Environment (TEE) Guide*, White Paper, GlobalPlatform, Inc., 544 Hillside Road, Redwood City, CA 94062, USA.

133. Bech, J. (2014). OP-TEE, Open-Source Security For The Mass-Market. Core Dump online magazine, http://www.linaro.org/blog/core-dump/op-tee-open-source-security-mass-market.

134. Intel Corporation. (2012). Intel® Trusted Execution Technology: White Paper, 2012. Intel, 2200 Mission College Blvd., Santa Clara, CA 95054-1549, USA.

135. Piesse, D., Kemp, L., Gellan, C. et al. (2016). Insurers, new kids on the Blockchain?. Whitepaper, FC Business Intelligence Ltd, 2016, 7–9 Fashion Street, London, E1 6PX, UK.

5 Evolved Campus Connectivity

The previous chapters covered three basic technologies that generally tend to give rise to three overlay networks: (i) on-site Wireless Local Area Networks (WLANs), usually for intranet or Internet access; (ii) cellular networks, especially in the context of building-level access with Distributed Antenna Systems (DASs); and (iii) networks to support distributed devices in an Internet of Things (IoT) context. Additionally, the chapters that follow describe technologies used for in-building/in-campus localization using Real-Time Location Systems (RTLSs), which may comprise a fourth communication overlay. Traditionally, these systems have employed and deployed different technologies, different facilities, different implementations, different administrative support, and different security mechanisms. Using applications-specific solutions might give the impression that a better technical fit is achievable and perhaps some savings in hardware expenditure; however, the better financial measure is Total Cost of Ownership (TCO). When the TCO is taken into account, using one solution across multiple requirements may be cheaper overall rather than using individual solutions for individual requirements, due, typically, to the higher operations cost (including training, sparing, software upgrade management) associated with a larger set of technology solutions.

This chapter reviews some of the latest discrete technologies applicable to the campus environment, but, in the end, it makes a case for an integrated solution, both at the technology and at the administrative level. It is not the goal of this chapter to provide an all-inclusive review of these technologies or a self-contained tutorial – readers are referred to numerous available references to address any such goal.

Figure 5.1 shows a pedagogically illustrative example. In the example, there are two applications requiring support and two technological choices; for this example, it is assumed that either solution could support the entire application set and that the operations cost scales linearly with the number of end-devices (although it usually grows less sharply, for example, as a function of the square-root of the number of devices). If one were to focus just on the hardware costs, the example would indicate that solution A is better for application 1 and solution B is better for application 2; with the respective operations costs for the selected solutions based on the minimal hardware cost criterion, the TCO would be $1085. In this illustrative example, using solution A for both requirements would lead to a TCO of $1300, while using solution B for both requirements would lead to a TCO of $870, which is really the true minimum.

Wi-Fi can be used for on-site (e.g. airport) WLANs for Internet access (for both traditional ISM and CBRS bands); (ii) VoWi-Fi/Voice over IP (VoIP) cellular access/service with voice Wi-Fi-enabled smartphones; (iii) Wireless Sensor Networks (WSNs) for IoT in the context of Wi-Fi-enabled sensors, and RTLSs with Wi-Fi-enabled tags. One solution in lieu of multiple overlay networks may be cheaper, overall, when the TCO is considered. And, while at a particular point in time, some Wi-Fi-based solutions may still be a bit "pricey" (e.g. for IoT and/or RTLS), the expectation is that these costs will continue to drop in the near future.

High-Density and De-Densified Smart Campus Communications: Technologies, Integration, Implementation, and Applications, First Edition. Daniel Minoli and Jo-Anne Dressendofer.
© 2022 John Wiley & Sons, Inc. Published 2022 by John Wiley & Sons, Inc.

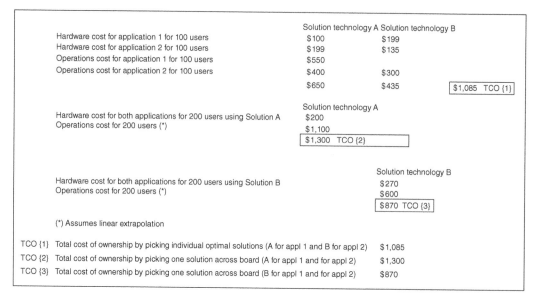

FIGURE 5.1 TCO illustrative example.

5.1 ADVANCED SOLUTIONS

An institution looking to deploy a state-of-the-art local/campus infrastructure at this juncture would likely consider deploying a system based on the IEEE 802.11ax specification [1–3]. Wireless Nodes (WNs) (also commonly known as Stations [STAs], or also known as User Equipment [UE] in some contexts) and Access Points (APs) constitute a Basic Service Set (BSS). We discussed the Medium Access Control (MAC) and Physical (PHY) layer operation of a generic Wireless LAN (WLAN) in Chapter 2. A brief summary follows, to position the need for new mechanisms in the emerging solutions.

Figure 5.2 (top) illustrates schematically a process in which a WN/STA establishes a link with an AP [4]. Figure 5.2 (middle) illustrates a Carrier Sense Multiple Access (CSMA)/Collision Avoidance (CA) method used in WLAN communications. Figure 5.2 (bottom) illustrates a method for performing a Distributed Coordination Function (DCF) using a Request to Send (RTS) frame and a Clear to Send (CTS) frame. Interframe spaces (IFSs) are waiting periods between transmission of frames operating in the MAC sublayer. These waiting periods are used to prevent collisions as defined in IEEE 802.11-based WLAN standards; they represent the time period between completion of the transmission of the last frame and starting transmission of the next frame, apart from the variable backoff period. Table 2.3 in Chapter 2 listed the different types of IFSs spacing, ordered by duration.

Traditionally, at the PHY level, the 802.11 protocol uses a CSMA channel management method: WNs first sense the channel and endeavor to avoid collisions by transmitting a packet only when they sense the channel to be idle; if the WN detects the transmission of another node, it waits for a random amount of time for that other WN to stop transmitting before sensing again to assess if the channel is free. The process is based on the AP or the WN establishing signal detection energy on an given channel, specifically the Received Signal Strength Index (RSSI) of the received PLCP (Physical Layer Convergence Protocol) Protocol Data Unit (PPDU): if the signal detection energy is less than a Clear Channel Assessment (CCA) threshold, the AP or the WN then contends for the channel, and transmits its data [5–8].

Traditionally, at the frame level, Wi-Fi WNs use RTS/CTS frames to avail itself of the shared medium. The AP only issues a CTS packet to one WN at a time. When the WN receives the CTS,

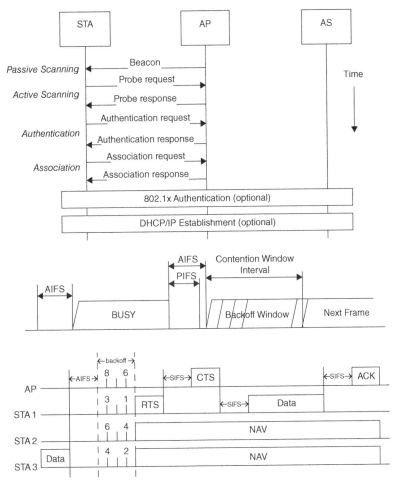

FIGURE 5.2   Basic channel access/management in recent 802.11 specifications [4].

it sends its entire frame to the AP; the WN then waits for an acknowledgement (ACK) frame from the AP indicating that it received the packet correctly; if the WN does not get the ACK in a specified amount of time, it postulates that the packet in fact collided with some other WN transmission – at that point the WN transitions into a period of binary exponential backoff; it will try to access the medium and retransmit its packet after the backoff time expires.

More specifically, when[1] it is determined that the radio channel is idle, each terminal having data to be transmitted performs a backoff procedure after an IFS time interval, depending on the state of the terminal. Each terminal stands by while locally decreasing the slot time(s) counter, such time counter being a random number allocated to the terminal during an idle state of the channel. When a terminal completely exhausts the slot time(s), it will attempt to access the channel under consideration. The interval during which each terminal performs the backoff procedure is referred to as a *Contention Window (CW) interval*. When a specific terminal successfully assesses the channel as being available, that terminal may transmit data over that channel. However, when the terminal that attempts the access collides with another terminal, the terminals that collide with each other are allocated new random slot time(s) counters, respectively, to

---

[1]This discussion partially based on reference [4].

perform the backoff procedure again; the new random number allocated to each terminal is typically within a range that is twice larger than the random number that was previously allocated to that terminal. Meanwhile, each terminal attempts the access to the channel by performing the backoff procedure again in the next CW interval, starting the backoff procedure from slot time(s) that remained in the previous CW interval.

As described, a terminal in a WLAN checks whether a channel is busy or not by performing carrier/channel sensing before transmitting data. Such a process is referred to as *CCA*, and a signal level used to decide whether the corresponding signal is sensed is referred to as a *CCA threshold*. When a radio signal is received by a terminal, it is processed to determine if it has a value exceeding the CCA threshold. When a radio signal having a predetermined or higher-strength value is sensed, it is determined that the channel under consideration is physically busy and the terminal delays its access to that channel. When a radio signal is not sensed in the channel under consideration or a radio signal is sensed having a strength smaller than the CCA, then the terminal determines that the channel is idle.

Even when the radio channel is physically available, the AP and STAs in the BSS must contend for higher-level system resources in order to obtain authority to transmit any data. When data transmission at the previous step is completed, each terminal having data to be transmitted performs a backoff procedure by decreasing a backoff counter based on a random number allocated to each terminal after an Arbitration Inter-frame Space (AIFS) time. A transmitting station in which the backoff counter has expired transmits the RTS frame to notify the AP that it has data to transmit. In Figure 5.2 (bottom), STA1 holds a lead in contention with minimum backoff; hence, it may transmit the RTS frame after the backoff counter has expired. The RTS frame includes information on a receiver address, a transmitter address, and duration. A receiving terminal (i.e. the AP in bottom part of the figure) that receives the RTS frame transmits the CTS frame after waiting for a Short Inter-frame Space (SIFS) time to notify that the data transmission is available to terminal STA1. The CTS frame includes the information on a receiver address and duration. In this case, the receiver address of the CTS frame may be set identically to a transmitter address of the RTS frame corresponding thereto, that is, an address of the transmitting terminal STA1. Upon receiving the CTS frame, STA1 transmits the data after a SIFS time.

When the data transmission is completed, the receiving AP transmits an ACK frame after a SIFS time, to notify that the data transmission is completed. When the transmitting terminal receives the ACK frame within a predetermined time, the transmitting terminal regards that the data transmission is successful. However, when the transmitting terminal does not receive the ACK frame within the predetermined time, the transmitting terminal regards that the data transmission is failed. Meanwhile, adjacent terminals that receive at least one of the RTS frame and the CTS frame in the course of the transmission procedure set a Network Allocation Vector (NAV) and do not perform data transmission until the set NAV is terminated. In this case, the NAV of each terminal may be set based on a duration field of the received RTS frame or CTS frame. In the course of the aforementioned data transmission procedure, when the RTS frame or CTS frame of the terminals is not transferred successfully to a target due to a situation such as interference or a collision, a subsequent process is temporarily suspended; the transmitting terminal STA1 that transmitted the RTS frame regards that the data transmission is unavailable and participates in a follow-up contention by being allocated with a new random number. In this case, the newly allocated random number may be determined within a range twice larger than a previous predetermined random number range.

These PHY layer and data link layer protocols described above work well for a reasonably small population of users in proximity; however, the system efficiency decreases when the number of users is large. In addition to the intrinsic protocol limitations, overlapping signal areas that may exist where there might be a high number of APs to support a large number of users in a confined space can impact throughput, since a WN in an overlapping area can trigger backoff procedures in one or more signal areas. Yet another factor to consider is the shared use of wider channels (note, for example, that for 802.11ac operation in North America there is only one 160 MHz channel available). Designing dense coverage when using a small number of channels is arduous because it forces planners to reuse channels in nearby cells – unless power management is meticulously taken into consideration in that case, users will experience degraded performance due to co-channel interference, especially for cases where high Modulation and Coding Schemes (MCS) index (say 8–11) are used. High MCS environments are more susceptible to low Signal-to-Noise Ratio (SNR). High system throughput can be achieved by using a wider radio frequency bandwidth (e.g. a maximum of 160 MHz – also at the 60 GHz band); a higher number of Multiple Input(s) Multiple Output(s) (MIMO) spatial streams (for example, eight in a major band and four in another major band); multi-user MIMO (MU-MIMO); beamforming technology; and high-grade modulation (e.g. a maximum of 1024 Quadrature Amplitude Modulation [QAM]). Although millimeter waves (mmWave) offer wider radio frequency bandwidth, recall from previous chapters that transmission of mmWave signals requires line-of-sight and that transmission is subject to significant attenuation, so that these systems can be used only among devices in a short-distance space.

5.1.1 802.11ax Basics

802.11ax, also called High-Efficiency Wireless (HEW) or High-Efficiency (HE), aims at improving the average throughput per user fourfold in dense user environments; additionally, the standard implements a number of mechanisms to support a higher number of users with consistent and reliable data throughput in crowded wireless environments. As of late 2020, 802.11ax was an IEEE draft amendment with expected full ratification to follow in 2021. 802.11ax defines modifications to the 802.11 PHY layer and the MAC sublayer for high-efficiency operation in the sub-6 GHz frequency bands between 1 and 6 GHz. While 802.11ac operates at 5 GHz only, 802.11ax radios can operate either at the 2.4 or 5 GHz frequency bands and, in the future, it will also operate in the 6 GHz band.

Higher data rates and wider channels were not the principal goal of 802.11ax standardization development; instead the goal was to achieve better and more efficient 802.11 traffic management; a related goal was to increase the average throughput 4x per user in high-density WLAN environments. Recent amendments to the 802.11 specification (e.g. 802.11ac) defined higher data rates and wider channels but did not address efficiency. Since the majority of the typical 802.11 data frames (75–80%) have been shown to be small, and under 256 bytes, a WLAN experiences excessive overhead at the MAC sublayer and medium contention overhead for each small frame [9]. In many (but not all) high-density deployment situations, overall system efficiency may become more important than higher data rates; for example, in high-density hotspot and cellular offloading scenarios, many devices competing for the wireless medium may have low-to-moderate data rate requirements [10]. 802.11ax does indeed provide the efficiency improvement. Specifically, besides being backwards compatible with 802.11a/b/g/n/ac, key 802.11ax features include [9, 11]:

- Increase 4x the average throughput per user in high-density scenarios, such as train stations, airports, and stadiums (see Figure 5.3 for an example of a high-density venue).
- Data rates and channel widths similar to 802.11ac, with the exception of new MCSs (MCS 10 and 11) with 1024-QAM.
- Specified for DL and UL multi-user operation by means of MU-MIMO and Orthogonal Frequency Division Multiple Access (OFDMA) technology. With this technology, an AP can coordinate upstream client transmissions using a Trigger Frame (TF) to allocate transmission resources and set transmit timing for each client.

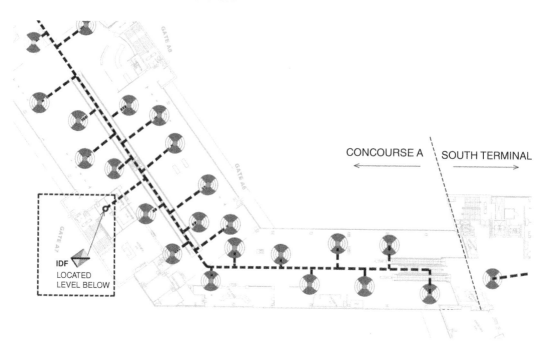

CONCOURSE A   SOUTH TERMINAL

IDF
LOCATED
LEVEL BELOW

FIGURE 5.3 Example scenario of an airport with high user density targeted for 802.11ax deployment.

- Larger OFDM FFT (Fast Fourier Transform) sizes (4x larger), narrower subcarrier spacing (4x tighter), and longer symbol time (4x) for improved robustness and performance in multipath fading environments and outdoors.
- Improved traffic flow and channel access.
- Better power management for longer battery life.
- Supporting Cellular data offloading as well as supporting environments with many APs and a high concentration of users with heterogeneous devices.
- Outdoors/indoors mixed environments (the outdoor mode is particularly useful for campus environments such as stadiums, smart cities, and so on).

The 802.11ax specification incorporated important changes to the PHY; however, it maintains backward compatibility with 802.11a/b/g/n and /ac devices[2]. However, 802.11ax APs will not improve the performance of legacy Wi-Fi clients (802.11a/b/g/n/ac): 802.11ax-based clients are required to take full advantage of 802.11ax HE capabilities. Nonetheless, as a larger number of 802.11ax clients are mixed into the client population in a given environment, the efficiency improvements gained by 802.11ax client devices will free valuable channel resources for those older clients, therefore improving the overall efficiency of the system [9]. Table 5.1 highlights some key features of 802.11ax and Table 5.2 depicts some the key parametric changes in 802.11ax (loosely inspired by [9, 12]). As noted in previous chapters, recently the Wi-Fi Alliance adopted a new naming convention for Wi-Fi technologies, the goal being that the new naming convention should be easier to understand for the average consumer. Because 802.11ax technology has many new technical features, it has been given the generational name of Wi-Fi 6. Major chipset

---

[2]For example, an 802.11ax WN can send and receive data to legacy WNs; legacy clients are able to demodulate and decode 802.11ax packet headers – but not the entire 802.11ax packet – and initiate a backoff when an 802.11ax WN is transmitting.

Usage of Secure Resources, and Certification. For example, a TEE-based environment ensures that a rogue device that presents into the environment is not accepted as a legitimate user – the rogue device may have the nefarious intent to inject some malware at some point in time. Also, the TEE makes sure that a device (or user) does not escalate its privileges beyond what is the intent of the system administrator (by providing isolation that limits functionality or access to data or some function – for example, a function controlling automatic medication delivery from a body-worn pump).

As a side note, TEEs can increase the capabilities of Secure Elements (SEs). SEs are secure components that consist of autonomous, tamper-resistant hardware within which secure applications and related confidential cryptographic data (e.g. key management) are stored and executed. While enjoying a high level of security, these devices have limited functionality; an example includes an NFC device. SEs can work in conjunction with the TEE to enhance their capabilities.

Intel TXT   Intel® TXT is a scalable architecture that specifies hardware-based security protection [134]. These methodologies are built into Intel's chipsets to address threats across physical and virtual infrastructures; the technology is designed to harden platforms to better deal with threats of hypervisor attacks, BIOS, or other firmware attacks, malicious rootkit installations, or other software-based attacks. TXT aims at increased protection by allowing greater control of the launch stack through a Measured Launch Environment (MLE) and enabling isolation in the boot process. It extends the Virtual Machine Extensions (VMX) environment of Intel Virtualization Technology to provide a verifiably secure installation, launch, and use of a hypervisor or OS. MLE enables an accurate comparison of all critical elements of the launch environment against a known good source. TXT creates a cryptographically unique identifier for each approved launch-enabled component and then provides hardware-based enforcement mechanisms to block the launch of code that does not match approved code. This hardware-based solution provides the foundation on which trusted platform solutions can be built to protect against software-based attacks; the technology is broadly applicable to servers and to IoT devices. More specifically, TXT provides [15, 134]:

- *Verified Launch:* A hardware-based chain of trust that enables the launch of the MLE into a "known good" state. Changes to the MLE can be detected through cryptographic (hash-based or signed) measurements.
- *Launch Control Policy (LCP):* A policy engine for the creation and implementation of enforceable lists of "known good" or approved executable code.
- *Secret Protection:* Hardware-assisted methods that remove residual data at an improper MLE shutdown, protecting data from memory-snooping software and reset attacks.
- *Attestation:* The ability to provide platform measurement credentials to local or remote users or systems to complete the trust verification process and support compliance and audit activities.

After the system goes thought a boot-up stage, various state variables are compared with pre-stored data; only if a match in state parameters is established, the device is allowed to move to the next stage that would entail loading the application code, at which point other state checks are made. If the checks fail, reporting of the condition occurs (to some administrator); there would be an indication that "trust" of the platform cannot be established. TXT can be used in IoT environments, particularly at the server end, to ascertain that the medical analytics engines and the storage systems (e.g. storing patient health data) are secure. These processes can also be embedded in the medical IoT device to ascertain, among other validations, that the device is legitimate – in the sense that it runs only official code, performing only specific functions, and not exceeding its level of functional authority. The TPM is an ISO (International

Organization for Standardization) standard for a secure cryptoprocessor; a secure cryptoprocessor is a dedicated microcontroller designed to secure hardware by integrating cryptographic keys into devices (ISO/IEC 11889 published in 2009 – the technical specification was TPM was developed by Trusted Computing Group [TCG]).

4.8.4 Other Approaches: Blockchains

Some have sought to implement security mechanisms at a much higher level, both in an IoT context and in the broader IT content. One such mechanism is the "blockchain." A blockchain is a form of distributed ledger (a distributed database) that retains an expanding list of records while precluding revision or tampering (a tamper-proof ledger). The blockchain encompasses a data structure of "child" (aka successor) blocks; each block includes sets of transactions, timestamps, and links to a "parent" (aka predecessor) block; the linked blocks constitute a chain. It intrinsically provides universal accessibility, incorruptibility, openness, and the ability to store and transfer data in a secure manner. The original application was as a ledger for bitcoins. Users can add transactions, verify transactions, and add new blocks. Proponents see opportunities for the use of blockchains for e-health companies (and banks and many other industries), as it allows a replacement of the common centralized data paradigm, thus fostering additional process disintermediation [2, 3, 8, 9, 15, 135]. Possible applications of interest to the e-health industry include claims filing/processing; claim fraud detection, for example, spot multiple claims from a claimant (medical office) for the same procedure; data decentralization; and cybersecurity management (e.g. data integrity). Given the expected distributed nature of an IoT-based ecosystem (also in the e-health industry context), blockchains may play an important role in the future.

REFERENCES

1. D. Minoli, B. Occhiogrosso, "IoT-driven advances in commercial and industrial building lighting and in street lighting", in the book Industrial IoT: Challenges, Design Principles, Applications, and Security, Ismail Butun Editor. Springer, 2020, ISBN 978-3-030-42500-5 Pages 97–159.
2. Minoli, D. (November 2019). Positioning of Blockchain mechanisms in IoT-powered smart home systems: a gateway-based approach. Elsevier IoT Journal, Special Issue on IoT Blockchains https://www.sciencedirect.com/science/article/pii/S2542660519302525.
3. D. Minoli, B. Occhiogrosso, "Blockchain-enabled fog and edge computing: concepts, architectures and smart city applications". Chapter in book: Blockchain-Enabled Fog and Edge Computing: Concepts, Architectures and Applications, Editors: M. H. Rehmani, M. M. Rehan, CRC Press, Taylor & Francis Group, 2020. Boca Raton, Fla. Pages 20–32
4. D. Minoli, B. Occhiogrosso, "Practical aspects for the integration of 5G networks and IoT applications in smart cities environments", Special Issue titled "Integration of 5G Networks and Internet of Things for Future Smart City," Wireless Communications and Mobile Computing". Vol. 2019, Article ID 5710834, 30 pages, August 2019. Hindawi & John Wiley & Sons (https://doi.org/10.1155/2019/5710834).
5. D. Minoli, B. Occhiogrosso, W. Wang, "MIPv6 in Crowdsensing applications for SIoT environments", Chapter in book Towards Social Internet of Things: Enabling Technologies, Architectures and Applications, Hassanien A., Bhatnagar R., Khalifa N., Taha M. (eds), Chapter 3, Springer, 2020. pp. 31–49, Part of the Studies in Computational Intelligence book series (SCI, volume 846). https://link.springer.com/chapter/10.1007/978-3-030-24513-9_3.
6. D. Minoli, B. Occhiogrosso, "Constrained average design method for QoS-based traffic engineering at the edge/gateway boundary in VANETs and cyber-physical environments", Chapter in book Managing Resources for Futuristic Wireless Networks, IGI Global, 2020. Mamata Rath, Editor.
7. Minoli, D. and Occhiogrosso, B. (2018). Ultrawideband (UWB) Technology For Smart Cities IoT Applications. *2018 IEEE International Smart Cities Conference (ISC2) – IEEE ISC2 2018- Buildings, Infrastructure, Environment Track*, Kansas City, 16–19 September 2018.

8. Minoli, D. and Occhiogrosso, B. (November 2019). Editor-in-chief, special issue on Blockchain applications in IoT environments, overview. Elsevier IoT Journal https://www.sciencedirect.com/science/article/pii/S2542660519302537.

9. Minoli, D. and Occhiogrosso, B. (Summer 2018). Blockchain mechanisms for IoT security. Elsevier IoT Journal 1 (1): 1–13.

10. D. Minoli, B. Occhiogrosso, "IoT Applications to Smart Campuses and a Case Study", EAI (European Alliance for Innovation) Endorsed Transactions on Smart Cities, European Union Digital Library, ISSN 2518–3893 Volume 2. Published 19 December 2017. http://eudl.eu/doi/10.4108/eai.19-12-2017.153483

11. D. Minoli, B. Occhiogrosso, "Implementing the IoT for renewable energy", Chapter in Internet of Things a to Z: Technologies and Applications, Editor: Q. Hassan, Chapter 15, IEEE Press/Wiley, June 2018, ISBN-13: 978-1119456742. (With L. Finco). 425–442, 881–895

12. Minoli, D., Occhiogrosso, B. et al. (2017). A review of wireless and satellite-based M2M/IoT Services in Support of smart grids. Mobile Networks and Applications Journal, Springer.

13. Minoli, D. and Occhiogrosso, B. (2017). Mobile IPv6 Protocols and High Efficiency Video Coding For Smart City IoT Applications. *CEWIT2017 Conference*, Nov. 8-9-2017, Stony Brook University, NY.

14. Minoli, D., Occhiogrosso, B., Sohraby, K., et al. (2017). A Review of Wireless and Satellite-based M2M Services in Support of Smart Grids. *1st EAI International Conference on Smart Grid Assisted Internet of Things (SGIoT 2017)*, 11–13 July 2017, Sault Ste. Marie, Ontario, Canada.

15. Minoli, D., Occhiogrosso, B., Sohraby, K., et al. (2017). IoT Security (IoTSec) Mechanisms For e-Health and Ambient Assisted Living Applications. The *Second IEEE/ACM International Workshop on Safe, Energy-Aware, & Reliable Connected Health (SEARCH 2017) (collocated with CHASE 2017, Conference on Connected Health: Applications, Systems, and Engineering Technologies)*, Philadelphia, July 2017.

16. Minoli, D., Occhiogrosso, B., Sohraby, K., et al. (2017). Multimedia IoT Systems and Applications. *Global IoT Summit, GIoTS-2017, Organized by Mandat International, IEEE IoT TsC (Transactions on Service Computing), the IoT Forum and IPv6 Forum (collocated with the IoT Week)*, 6–9 June 2017, Geneva, Switzerland.

17. D. Minoli, B. Occhiogrosso, "The emerging "energy internet of things", Chapter in Internet of Things a to Z: Technologies and Applications, Editor: Q. Hassan, Chapter 14, IEEE Press/Wiley, June 2018, ISBN-13: 978-1119456742 385–417.

18. D. Minoli, B. Occhiogrosso, K. Sohraby, "IoT Considerations, Requirements, and Architectures for Smart Buildings – Energy Optimization and Next Generation Building Management Systems", IEEE Internet of Things Journal, Year: 2017, Volume: 4, Issue: 1, Pages: 269–283. DOI: https://doi.org/10.1109/JIOT.2017.2647881.

19. D. Minoli, B. Occhiogrosso, Sohraby, K., et al., "Security considerations for IoT support of E-health applications", Chapter in Internet of Things: Challenges, Advances and Applications, Editors: Q. Hassan, A. R. Khan, S. A. Madani, Chapter 16, Chapman & Hall/CRC Computer and Information Science Series, CRC Press, Taylor & Francis, 2018, ISBN 9781498778510. 321–347

20. D. Minoli, B. Occhiogrosso, Sohraby, K., et al. (2017). IoT Security (IoTSec) Considerations, Requirements, and Architectures. *2017 14th IEEE Annual Consumer Communications & Networking Conference (CCNC)*, 8–11 January 2017 in Las Vegas, NV, USA. IEEE catalog number: CFP17CCN-CDR, ISBN:978-1-5090-6195-2, ISSN: 2331–9860.

21. D. Minoli, B. Occhiogrosso, and Sohraby, K. (2017). Internet Of Things (IoT)-Based Apparatus And Method For Rail Crossing Alerting Of Static Or Dynamic Rail Track Intrusions. JRC2017–2304, Proceedings of JRC (Joint Rail Conference), April 4–7, 2017, Philadelphia, PA.

22. D. Minoli, B. Occhiogrosso, "Internet of things applications for smart cities". Chapter in Internet of Things a to Z: Technologies and Applications, Editor: Q. Hassan, Chapter 12, IEEE Press/Wiley, June 2018, ISBN-13: 978-1119456742. 319–350

23. D. Minoli, B. Occhiogrosso, Sohraby, K. et al., "IoT considerations, requirements, and architectures for insurance applications". Chapter in Internet of Things: Challenges, Advances and Applications, Editors: Q. Hassan, A. R. Khan, S. A. Madani, Chapter 17, Chapman & Hall/CRC Computer and Information Science Series, CRC Press, Taylor & Francis, 2018, ISBN 9781498778510. 347–362

24. Minoli, D. (2015). Innovations in Satellite Communications and Satellite Technology, the Industry Implications of DVB-S2X, High Throughput Satellites, Ultra HD, M2M, and IP. Wiley.

25. Minoli, D. (2013). Building the Internet of Things with IPv6 and MIPv6. Wiley.

26. Minoli, D. and Sohraby, K. (2007). Wireless Sensor Networks. Wiley.

27. Quintanilla, F.G., Cardin, O., L'Anton, A. *et al* (2016). Implementation Framework For Cloud-Based Holonic Control Of Cyber-Physical Production Systems. In: *IEEE Proceedings of the 14th International Conference on Industrial Informatics*. Poitiers, France. (19–21 July 2016).

28. Wang, C., Zhu, Y., Shi, W. et al. (September 2018). A dependable time series analytic framework for cyber-physical systems of IoT-based smart grid. ACM Transactions on Cyber-Physical Systems 3 (1): 7, ACM New York, NY, USA, doi:https://doi.org/10.1145/3145623.

29. Thramboulidis, K. and Christoulakis, F. (October 2016). UML4IoT – A UML-based approach to exploit IoT in cyber-physical manufacturing systems. Computers in Industry, Elsevier 82: 259–272. https://doi.org/10.1016/j.compind.2016.05.010.

30. Burg, A., Chattopadhyay, A., and Lam, K.Y. (2018). Wireless communication and security issues for cyber–physical systems and the internet-of-things. Proceedings of the IEEE 106 (1): 38–60. https://doi.org/10.1109/JPROC.2017.2780172.

31. Antonino, P.O., Morgenstern, A., Kallweit, B., et al. (2018). Straightforward Specification of Adaptation-Architecture-Significant Requirements of IoT-enabled Cyber-Physical Systems. *2018 IEEE International Conference on Software Architecture Companion (ICSA-C)*, 30 April-4 May 2018, Seattle, WA, USA, doi:10.1109/ICSA-C.2018.00012.

32. He, H., Maple, C., Watson, T., et al. (2016). The Security Challenges in the IoT Enabled Cyber-Physical Systems And Opportunities For Evolutionary Computing & Other Computational Intelligence. *2016 IEEE Congress on Evolutionary Computation (CEC)*, 24–29 July 2016, Vancouver, BC, Canada, doi:10.1109/CEC.2016.7743900.

33. Blasch, E., Kadar, I., Grewe, L.L., et al. (2017). Panel Summary Of Cyber-Physical Systems (CPS) and Internet of Things (IoT) Opportunities With Information Fusion. *Proceedings Volume 10200, Signal Processing, Sensor/Information Fusion, and Target Recognition XXVI; Event: SPIE Defense + Security*, 2017, Anaheim, CA (2017), doi.org/10.1117/12.2264683.

34. Griffor, E. (2017). Reference Architecture for Cyber-Physical Systems. NIST (National Institute of Standards and Technology), 12 December 2017, 100 Bureau Drive, Gaithersburg, MD 20899.

35. R. Baheti, H. Gill, "Cyber-physical systems", Chapter in The Impact of Control Technology, T. Samad and A.M. Annaswamy (eds.), IEEE Control Systems Society, 2011, www.ieeecss.org. 161–166

36. Song, H., Rawat, D.B., Jeschke, S. et al. (2017). Cyber-Physical Systems: Foundations, Principles and Applications. London: Academic Press/Elsevier ISBN: 978-0-12-803801-7.

37. Romanovsky, A. and Ishikawa, F. (eds.) (2017). Trustworthy Cyber-Physical Systems Engineering. Boca Roton, FL: CRC.

38. Letichevsky, A.A., Letychevsky, O.O. et al. (November 2017). Cyber-physical systems. Cybernetics and Systems Analysis 53 (6) https://doi.org/10.1007/s10559-017-9984-9.

39. Stankovic, J.A. (February 2017). Research directions for cyber physical Systems in Wireless and Mobile Healthcare. ACM Transactions on Cyber-Physical Systems – Inaugural Issue 1 (1) https://doi.org/10.1145/2899006.

40. Müller, H.A., Litoiu, M., and Mylopoulos, J. (2016). Engineering Cybersecurity into Cyber Physical Systems. *CASCON 2016 Markham*, Ontario Canada, 2016 ACM. ISBN 123-4567-24-567/08/06, doi:10.475/123 4.

41. S. Wang, T. Lei, Zhang, L., and Hsu, C.-H. "Offloading Mobile data traffic for QoS-aware service provision in vehicular cyber-physical systems", Future Generation Computer Systems, 61 August 2016, Pages 118–127. doi: https://doi.org/10.1016/j.future.2015.10.004.

42. Koutsoukos, X., Karsai, G., Laszka, A. et al. SURE: a Modeling and simulation integration platform for evaluation of secure and resilient cyber-physical systems. Proceedings of the IEEE PP (99): 1–20. https://doi.org/10.1109/JPROC.2017.2731741.

43. Kim, Y.-S., Kwon, Y.-H., Yoon, S.-Y., and Hwang, S.-S. (2020). Apparatus and method for transmitting and receiving signals in [a] wireless communication system. US Patent 10, 693, 696; 23 June 2020; filed 21 November 2017. Uncopyrighted.

44. Lee, E.A. (February 2015). The past, present and future of cyber-physical systems: a focus on models. Sensors 15 (3): 4837–4869. https://doi.org/10.3390/s150304837.

45. Zanero, S. (April 2017). Cyber-physical systems. Computer 50 (4): 14–16. https://doi.org/10.1109/MC.2017.105.

46. Sivakumar, M., Sadagopan, C., and Baskaran, M. (August 2016). Wireless sensor network to cyber physical systems: addressing mobility challenges for energy efficient data aggregation using dynamic nodes. Sensor Letters 14 (8): 852–857. (6), DOI: https://doi.org/10.1166/sl.2016.3624.

47. Hua, F., Lua, Y., Vasilakos, A.V. et al. (March 2016). Robust cyber–physical systems: concept, models, and implementation. Future Generation Computer Systems 56: 449–475. https://doi.org/10.1016/j.future.2015.06.006.

48. Monostori, L. (2014). Cyber-physical production systems: roots, expectations and R&D challenges. Procedia CIRP 17: 9–13. https://doi.org/10.1016/j.procir.2014.03.115.

49. Yu, X. and Xue, Y. (May 2016). Smart grids: a cyber–physical systems perspective. Proceedings of the IEEE 104 (5): 1058–1070. https://doi.org/10.1109/JPROC.2015.2503119.

50. Swierk, T.E., Cox, T.R., and Hammons, M.R. (2020). Internet-of-things (IoT) gateway tampering detection and management. US Patent 10,671,765; 2 June 2020; filed 03 March 2017. Uncopyrighted.

51. Patent Trial and Appeal Board (PTAB). (2020). Petitioner's Notice of Deposition of Mr. Daniel Minoli. GMG Products LLC v. Traeger Pellet Grills LLC – http://Law360www.law360.com, PGR2019–00034 &00035 21 Febuary 2020.

52. Massey, D. (2017). Applying Cybersecurity Challenges to Medical and Vehicular Cyber Physical Systems. Proceeding of SafeConfig '17, 2017 Workshop on Automated Decision Making for Active Cyber Defense, Dallas, Texas, USA – 03 November 2017, 39–39, ISBN: 978-1-4503-5203-1 doi:10.1145/3140368.3140379.

53. Buntz, B. (2020). COVID-19 Driving Data-Integration Projects in IoT. IoT World Today, 10 August 2020. Available online on 17 October 2020 at https://urgentcomm.com/2020/08/10/covid-19-driving-data-integration-projects-in-iot

54. Matthews, K. (2020). 5 IoT Projects Aiming to Help Fight Against Coronavirus. The IoT Magazine, 15 April 2020. Available online on 17 October 2020 at https://theiotmagazine.com/5-iot-projects-aiming-to-help-fight-against-coronavirus-3e912043f88a.

55. J. J. Corbett, D. L. Kjendal, Woodhead, J.R. (2020). System and method for low power wide area virtual network for IoT. US Patent 10,778,752; 15 September 2020; Filed 06 May 2019. Uncopyrighted.

56. Guinard, D., Fischer, M., and Trifa, V. (2010). Sharing Using Social Networks In A Composable Web Of Things. *Proc. 8th IEEE Int. Conf. Pervasive Comput. Commun. Workshops (PERCOM)*, 702–707.

57. Atzori, L., Iera, A., and Morabito, G. (2012). The social internet of things (SIoT) – when social networks meet the internet of things: concept, architecture and network characterization. Computer Networks 56 (16): 3594–3608.

58. Atzori, L., Iera, A., and Morabito, G. (November 2011). SIoT: giving a social structure to the internet of things. IEEE Communications Letters 15 (11) https://doi.org/10.1109/LCOMM.2011.090911.111340.

59. Nitti, M., Girau, R., Atzori, L., et al. (2012). A Subjective Model For Trustworthiness Evaluation In The Social Internet of Things. *Proc. IEEE 23rd Int. Symp. Pers. Indoor Mobile Radio Communications (PIMRC)*, 18–23.

60. Lee, G.M., Rhee, W.S., and Crespi, N. (2013). Proposal of a New Work Item on Social and Device Networking. *ITU Telecommunications Standard, Sector, SG13 Rapporteur Group Meeting*, Geneva, Switzerland.

61. Ciortea, A., Boissier, O., Zimmermann, A., et al. (2013). Reconsidering The Social Web Of Things: Position Paper. *Proc. ACM Conf. Pervasive Ubiquitous Computing, UbiComp'13 Adjunct*, 1535–1544.

62. Xu, L.D., He, W., and Li, S. (2014). Internet of things in industries: a survey. IEEE Transactions on Industrial Informatics 10 (4) https://doi.org/10.1109/TII.2014.2300753.

63. Ortiz, A.M., Hussein, D., Park, S. et al. (June 2014). The cluster between internet of things and social networks: review and research challenges. IEEE Internet of Things Journal 1 (3) https://doi.org/10.1109/JIOT.2014.2318835.

64. Sarwar, U. and Chun, A.L. (2020). Distributed Adaptive Heterogeneous Wireless Communications Management. US Patent 10,791,560; 29 September 2020; filed 28 September 2017. Uncopyrighted.

65. Minoli, D. (2008). Enterprise Architecture a to Z: Frameworks, Business Process Modeling, SOA, and Infrastructure Technology. Auerbach.

66. Weyrich, M. and Ebert, C. (2016). Reference Architectures For The Internet of Things. IEEE Software, IEEE Computer Society, Jan/Feb 2016, p.112 ff.

67. Hu, L., Xie, N., Kuang Z, and Zhao K. (2012). Review of cyber-physical system architecture. *2012 IEEE 15th International Symposium on Object/Component/Service-Oriented Real-Time Distributed Computing Workshops*, 11–11 April 2012, Shenzhen, Guangdong, China. doi:10.1109/ISORCW.2012.15.

68. Guo, S. and Zeng, D. (Eds.) (2019). *Cyber-Physical Systems: Architecture, Security and Application, EAI/Springer Innovations in Communication and Computing*. 2019. Part of the EAI/Springer Innovations in Communication and Computing book series (EAISICC). ISBN 978-3-319-92564-6.

69. Jiang, J.-R. (2018). An improved cyber-physical systems architecture for industry 4.0 smart factories. Advances in Mechanical Engineering https://doi.org/10.1177/1687814018784192.

70. International Society of Automation (ISA), *ISA-95, Enterprise-control System Integration*, 2018, https://www.isa.org/isa95 (accessed November 2018).

71. Lee, J. and Bagheri, B. (2015). A cyber-physical systems architecture for industry 4.0-based manufacturing systems. Manufacturing Letters 3: 18–23.

72. Ahmadi, A., Cherifi, C., Cheutet, V., and Ouzrout, Y. (2017). A Review of CPS 5 Components Architecture for Manufacturing Based on Standards. *11th IEEE International Conference on Software, Knowledge, Information Management and Applications (SKIMA 2017)*, December 2017, Colombo, Sri Lanka. 6 p. ffhal-01679977f.

73. Tiburski, R., Amaral, L.A., de Matos, E., and Hessel, F.P. (2015). The importance of a standard security architecture for SOA-based Iot middleware. IEEE Communications Magazine 53 (12): 20–26.

74. ISO/IEC. (2014). Study Report on IoT Reference Architectures/Frameworks, 2014.

75. Cisco Systems. (2014). Cisco Whitepaper: The internet of things reference model. 2014, Cisco Systems, Inc., 170 West Tasman Dr., San Jose, CA 95134, USA.

76. Intel Staff. (2015). Whitepaper: The Intel IoT platform, architecture specification. Santa Clara, California, 2015.

77. Walewski, J.W. (ed.) (2013). *Internet-of-Things Architecture (IoT-A), Project Deliverable D1.2 – Initial Architectural Reference Model for IoT*. https://cocoa.ethz.ch/downloads/2014/01/1360_D1%202_Initial_architectural_reference_model_for_IoT.pdf

78. ETSI. *ETSI TS 102 690: Machine-to-Machine communications (M2M); Functional architecture.* October 2011.

79. Bauer, M., Boussard, M., Bui, N. et al. (2013). IoT reference architecture. In: Enabling Things to Talk (eds. A. Bassi, M. Bauer, M. Fiedler, et al.). Berlin, Heidelberg: Springer https://doi.org/10.1007/978-3-642-40403-0_8.

80. Singh, N.F. (2020). Active and Passive Asset Monitoring System. US Patent 10, 783, 419; 22 September 2020; filed 2 July 2019. Uncopyrighted.

81. Sedjelmaci, H., Senouci, S.M., and Feham, M. (2012). Intrusion Detection Framework Of Cluster-Based Wireless Sensor Network. *2012 IEEE Symposium on Computers and Communications (ISCC)*, Cappadocia, Turkey, 1–4.

82. Komal, K. and Mohasin, T. (2015). Secure and economical information transmission for clustered wireless sensor network. International Journal of Science and Research (IJSR). ISSN (Online) 4 (12): 2319–7064.

83. IETF. RFC 8376: Low-Power Wide Area Network (LPWAN) Overview, 2018–05.

84. Andrade, R.O. and Yoo, S.G. (Nov. 2019). A comprehensive study of the use of LoRa in the development of smart cities. Applied Sciences 9 (22): 4753. https://doi.org/10.3390/app9224753.

85. Elshabrawy, T. and Robert, J. (Apr. 2019). Interleaved chirp spreading LoRa-based modulation. IEEE Internet of Things Journal 6 (2): 3855–3863. https://doi.org/10.1109/JIOT.2019.2892294.

86. Sallum, E., Pereira, N., Alves, M., and Santos, M.M. (2020). Improving quality-of-service in LoRa low-power wide-area networks through optimized radio resource management. Journal of Sensor and Actuator Networks 9 (1): 10. https://doi.org/10.20944/preprints201909.0243.v1.

87. Kaura, G., Guptab, S.H., and Kaurc, H. Enhancing LoRa Network Performance for Smart City IoT Applications. Sustainable Cities and Society. Elsevier.

88. Basford, P.J., Bulot, F.M.J., Apetroaie-Cristea, M. et al. (2020). LoRaWAN for smart city IoT deployments: a long term evaluation. Sensors 20 (3): 648. https://doi.org/10.3390/s20030648.

89. Wu, W., Li, Y., Zhang, Y. et al. (2020). Distributed queueing-based random access protocol for LoRa networks. IEEE Internet of Things Journal 7 (1): 763–772. https://doi.org/10.1109/JIOT.2019.2945327.

90. Teboulle, H., Le Gourrierec, M. Franck Harnay, (2020). Device for Transporting LoRa Frames on a PL Network. US Patent 10, 742, 266; 11 August 2020; filed 15 July 2019. Uncopyrighted.

91. Saleem, Y., Crespi, N., Rehmani, M.H., and Copeland, R. (2019). Internet of things-aided smart grid: technologies, architectures, applications, prototypes, and future research directions. IEEE Access: 62962–63003.

92. Li, Y., Cheng, X., Cao, Y. et al. (2018). Smart choice for the smart grid: Narrowband internet of things (NB-IoT). IEEE Internet of Things Journal 5 (3): 1505–1515.

93. Reininger, P. (2016). 3GPP Standards for the Internet-of-Things. Published online: ftp://http://www.3gpp.org/Information/presentations/presentations 2016/2016 11 3gpp Standards for IoT.pdf

94. 3GPP. Cellular System Support For Ultra-Low Complexity And Low Throughput Internet of Things (CIoT). Published online https://portal.3gpp.org/desktopmodules/Specifications/SpecificationDetails. aspx?specificationId=2719

95. Ratasuk, R., Vejlgaard, B., Mangalvedhe, N., and Ghosh, A. (2016). NB-IoT System for M2M Communication. *IEEE Wireless Communications and Networking Conference (WCNC)*, pp. 1–5.

96. Wang, Y.-P.E., Lin, X., Adhikary, A. et al. (2017). A primer on 3GPP Narrowband internet of things. IEEE Communications Magazine 55 (3): 117–123.

97. Nunes, L.C. and Fuhrmann, S.F.R. (2020). Low Area Optimization for NB-IoT Applications. US Patent 10, 797, 859; 6 October 2020; filed 22 March 2018. Uncopyrighted.

98. T-Mobile NB-IoT plan, https://iot.t-mobile.com/pricing

99. Verizon wireless Cat-M Plan, https://www.verizonwireless.com/biz/plans/m2m-business-plans

100. GSMA. (2018). Mobile IoT in the 5G Future: NB-IoT and LTE-M in the Context of 5G. https://www.gsma.com/iot/wp-content/uploads/2018/05/GSMAIoT_MobileIoT_5G_Future_May2018.pdf

101. Contento, M. (2019). 5G and IoT – Emerging Tech with Endless Use Cases.19 February 2019. Published online at https://www.telit.com/blog/state-of-5g-and-iot-current-future-applications

102. Lin, J., Shen, Z., Zhang, A., Chai, Y. (2017). Using Blockchain Technology To Build Trust In Sharing LoRaWAN IoT. *Proceeding, ICCSE'17 Proceedings of the 2nd International Conference on Crowd Science and Engineering*, Beijing, China, 06–09 July 2017, doi: 10.1145/3126973.3126980.

103. Moyer, B. (Sept. 2015). Low power, wide area: a survey of longer-range IoT wireless protocols. Electronic Engineering Journal.

104. GOS World Staff. Skyworks Unveils Sky5 Ultra Platform for 5G Architecture. 27 February 2019. Published online at https://www.gpsworld.com/skyworks-unveils-sky5-ultra-platform-for-5g-architecture

105. Hui, D. and Zangi, K. (2012). Distributed Computation Of Precoding Weights For Coordinated Multipoint Transmission On The Downlink. US Patent 8, 107, 965; 31 January 2012; filed August 26, 2009. Uncopyrighted.

106. ITU-R SG05 Contribution 40. *Minimum Requirements Related To Technical Performance For IMT-2020 Radio Interface(s)*, February 2017.

107. Global System for Mobile Communications Association (GSMA). *The Mobile Economy 2019, Report*, 2019. Published online at https://www.gsmaintelligence.com/research/?file=b9a6e6202ee1d5f787cfebb95d3639c5&download

108. Vui, Y. and Lee, H. (2016). A Novel Digital Device Monitoring System Using UPnP Cloud Architecture. International Conference on Engineering Technologies and Big Data Analytics (ETBDA'2016) 21–22 January 2016 Bangkok (Thailand).

109. Rajalakshmi, Narayanan, M., Venkatesh, R., and Scholar, U.G. (2015). An exclusive study on unstructured data mining with big data. International Journal of Applied Engineering Research 10 (4): 3875–3886. ISSN 0973-4562.

110. Cui, Y., Kim, M., and Lee, H. (2013). Cloud-based home media system model: providing a novel media streaming service using UPnP Technology in a Home Environment. International Journal of Software Engineering and Its Applications 7 (4): 127–136.

111. Open Connectivity Forum, UPnP-resources, https://openconnectivity.org/developer/specifications/upnp-resources/upnp

112. Sony Corporation. (2004). Method of and Apparatus For Bridging a UPnP Network And A Rendezvous Network. US20050160172A1. https://patents.google.com/patent/US20050160172A1/en?oq=US20050160172

113. OCF. *OCF 2.0.5 Bridging Specification*, https://openconnectivity.org/specs/OCF_Bridging_Specification_v2.0.5.pdf

114. UPnP Forum/OCF. (2015). UPnP: The Discovery & Service Layer For The Internet of Things. April 2015. http://upnp.org/resources/whitepapers/UPnP_Internet_of_Things_Whitepaper_2015.pdf

115. UPnP Forum/OCF. *UDA V2.0 with Cloud Annex*, http://upnp.org/specs/arch/UPnP-arch-Device Architecture-v2.0.pdf

116. Tomar, A. (2011). Introduction to ZigBee Technology, Global Technology Centre, Volume 1 July 2011, https://www.cs.odu.edu/~cs752/papers/zigbee-001.pdf

117. Nakazawa, J., Tokuda, H., Keith Edwards, W., Umakishore Ramachandran (2006). A Bridging Framework For Universal Interoperability In Pervasive Systems. *26th IEEE International Conference on Distributed Computing Systems (ICDCS'06)*, Lisboa, Portugal, July 2006.

118. Baek, Y.M., Ahn, S.C., and Kwon, Y.-M. (2010). UPnP network bridge for supporting interoperability through non-IP channels. IEEE Transactions on Consumer Electronics 56 (4): 2226–2232. http://dx.doi.org/10.1109/TCE.2010.5681094.

119. Treadway, J. (2018). Using an IoT Gateway To Connect The 'Things' To The Cloud. TechTarget – IoT Agenda, 1 October 2018, https://internetofthingsagenda.techtarget.com/feature/Using-an-IoT-gateway-to-connect-the-Things-to-the-cloud

120. Lu, J. and Scott Kelly, T. (2011). Network integration system and method. US Patent 8, 761, 050; 04 October 2011. Uncopyrighted.

121. Lee, M.-G., Lee, B.-M., Patil, M.-M. et al. (2020). Method and apparatus for setup of wireless communication. US Patent 10, 609, 581; 31 March 2020; filed 05 June 2017. Uncopyrighted.

122. Lai, C., Lu, R., Zheng, D. et al. (2015). Toward Secure Large-Scale Machine-to-Machine Communications in 3GPP Networks: Challenges and Solutions. *IEEE Communications Magazine – Communications Standards Supplement*, December 2015.

123. ISO, Introduction to ISO JTC1/WG10, June 2015. ISO materials.

124. K. Komal, T. Mohasin, "Secure and economical information transmission for clustered wireless sensor network", IJSR, Vo. 4, Issue 12 Dec. 2015, pp. 1033ff.

125. Tiburski, R., Amaral, A., L., de Matos, E., and Hessel, F.P. (2015). The importance of a standard security architecture for SOA-based IoT middleware. IEEE Communications Magazine 53 (12): 20–26.

126. Cornejo-García, R. Strengthening Elder's Social Networks through Ambient Information Systems and SNS. Computer Science Department, CICESE. (Online).

127. Chen, G., Gong, Y., Xiao, P. et al. (2015). Physical layer network security in the full duplex relay system. IEEE Transactions on Information Forensics and Security 10 (3): 574–583.

128. Nichols, S. (2016). Internet of Sins: Million More Devices Sharing Known Private Keys for HTTPS, SSH admin. The Register (http://theregister.co.uk), 07 September 2016.

129. Madani, M., Benkhaddra, I., Tanougast, C., and Chitroub, S. Digital implementation of an improved LTE stream cipher Snow-3G based on Hyperchaotic PRNG. Security and Communication Networks 2017: 5746976. https://doi.org/10.1155/2017/5746976.

130. Ekdahl, P., Johansson, T., Maximov, A., and Yang, J. (2019). A new SNOW stream cipher called SNOW-V. IACR Transactions on Symmetric Cryptology 2019 (3): 1–42. https://doi.org/10.13154/tosc.v2019.i3.1-42.

131. Wallace, J. (2016). Securing the Embedded IoT World. Guest Blog, IoT Security Foundation Website.

132. GlobalPlatform. *Trusted Execution Environment (TEE) Guide*, White Paper, GlobalPlatform, Inc., 544 Hillside Road, Redwood City, CA 94062, USA.

133. Bech, J. (2014). OP-TEE, Open-Source Security For The Mass-Market. Core Dump online magazine, http://www.linaro.org/blog/core-dump/op-tee-open-source-security-mass-market.

134. Intel Corporation. (2012). Intel® Trusted Execution Technology: White Paper, 2012. Intel, 2200 Mission College Blvd., Santa Clara, CA 95054-1549, USA.

135. Piesse, D., Kemp, L., Gellan, C. et al. (2016). Insurers, new kids on the Blockchain?. Whitepaper, FC Business Intelligence Ltd, 2016, 7–9 Fashion Street, London, E1 6PX, UK.

5 Evolved Campus Connectivity

The previous chapters covered three basic technologies that generally tend to give rise to three overlay networks: (i) on-site Wireless Local Area Networks (WLANs), usually for intranet or Internet access; (ii) cellular networks, especially in the context of building-level access with Distributed Antenna Systems (DASs); and (iii) networks to support distributed devices in an Internet of Things (IoT) context. Additionally, the chapters that follow describe technologies used for in-building/in-campus localization using Real-Time Location Systems (RTLSs), which may comprise a fourth communication overlay. Traditionally, these systems have employed and deployed different technologies, different facilities, different implementations, different administrative support, and different security mechanisms. Using applications-specific solutions might give the impression that a better technical fit is achievable and perhaps some savings in hardware expenditure; however, the better financial measure is Total Cost of Ownership (TCO). When the TCO is taken into account, using one solution across multiple requirements may be cheaper overall rather than using individual solutions for individual requirements, due, typically, to the higher operations cost (including training, sparing, software upgrade management) associated with a larger set of technology solutions.

This chapter reviews some of the latest discrete technologies applicable to the campus environment, but, in the end, it makes a case for an integrated solution, both at the technology and at the administrative level. It is not the goal of this chapter to provide an all-inclusive review of these technologies or a self-contained tutorial – readers are referred to numerous available references to address any such goal.

Figure 5.1 shows a pedagogically illustrative example. In the example, there are two applications requiring support and two technological choices; for this example, it is assumed that either solution could support the entire application set and that the operations cost scales linearly with the number of end-devices (although it usually grows less sharply, for example, as a function of the square-root of the number of devices). If one were to focus just on the hardware costs, the example would indicate that solution A is better for application 1 and solution B is better for application 2; with the respective operations costs for the selected solutions based on the minimal hardware cost criterion, the TCO would be $1085. In this illustrative example, using solution A for both requirements would lead to a TCO of $1300, while using solution B for both requirements would lead to a TCO of $870, which is really the true minimum.

Wi-Fi can be used for on-site (e.g. airport) WLANs for Internet access (for both traditional ISM and CBRS bands); (ii) VoWi-Fi/Voice over IP (VoIP) cellular access/service with voice Wi-Fi-enabled smartphones; (iii) Wireless Sensor Networks (WSNs) for IoT in the context of Wi-Fi-enabled sensors, and RTLSs with Wi-Fi-enabled tags. One solution in lieu of multiple overlay networks may be cheaper, overall, when the TCO is considered. And, while at a particular point in time, some Wi-Fi-based solutions may still be a bit "pricey" (e.g. for IoT and/or RTLS), the expectation is that these costs will continue to drop in the near future.

High-Density and De-Densified Smart Campus Communications: Technologies, Integration, Implementation, and Applications, First Edition. Daniel Minoli and Jo-Anne Dressendofer.
© 2022 John Wiley & Sons, Inc. Published 2022 by John Wiley & Sons, Inc.

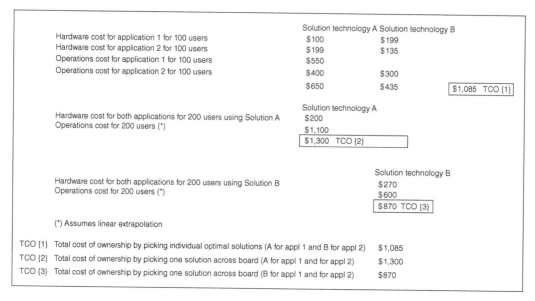

FIGURE 5.1   TCO illustrative example.

5.1   ADVANCED SOLUTIONS

An institution looking to deploy a state-of-the-art local/campus infrastructure at this juncture would likely consider deploying a system based on the IEEE 802.11ax specification [1–3]. Wireless Nodes (WNs) (also commonly known as Stations [STAs], or also known as User Equipment [UE] in some contexts) and Access Points (APs) constitute a Basic Service Set (BSS). We discussed the Medium Access Control (MAC) and Physical (PHY) layer operation of a generic Wireless LAN (WLAN) in Chapter 2. A brief summary follows, to position the need for new mechanisms in the emerging solutions.

Figure 5.2 (top) illustrates schematically a process in which a WN/STA establishes a link with an AP [4]. Figure 5.2 (middle) illustrates a Carrier Sense Multiple Access (CSMA)/Collision Avoidance (CA) method used in WLAN communications. Figure 5.2 (bottom) illustrates a method for performing a Distributed Coordination Function (DCF) using a Request to Send (RTS) frame and a Clear to Send (CTS) frame. Interframe spaces (IFSs) are waiting periods between transmission of frames operating in the MAC sublayer. These waiting periods are used to prevent collisions as defined in IEEE 802.11-based WLAN standards; they represent the time period between completion of the transmission of the last frame and starting transmission of the next frame, apart from the variable backoff period. Table 2.3 in Chapter 2 listed the different types of IFSs spacing, ordered by duration.

Traditionally, at the PHY level, the 802.11 protocol uses a CSMA channel management method: WNs first sense the channel and endeavor to avoid collisions by transmitting a packet only when they sense the channel to be idle; if the WN detects the transmission of another node, it waits for a random amount of time for that other WN to stop transmitting before sensing again to assess if the channel is free. The process is based on the AP or the WN establishing signal detection energy on an given channel, specifically the Received Signal Strength Index (RSSI) of the received PLCP (Physical Layer Convergence Protocol) Protocol Data Unit (PPDU): if the signal detection energy is less than a Clear Channel Assessment (CCA) threshold, the AP or the WN then contends for the channel, and transmits its data [5–8].

Traditionally, at the frame level, Wi-Fi WNs use RTS/CTS frames to avail itself of the shared medium. The AP only issues a CTS packet to one WN at a time. When the WN receives the CTS,

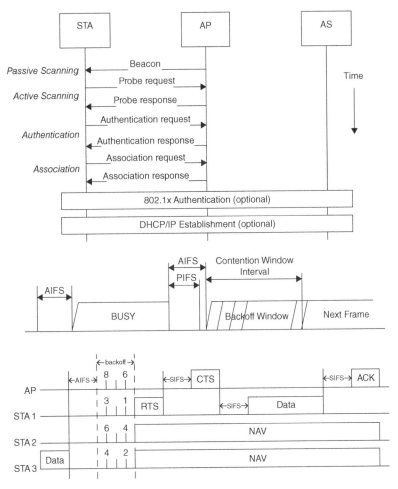

FIGURE 5.2   Basic channel access/management in recent 802.11 specifications [4].

it sends its entire frame to the AP; the WN then waits for an acknowledgement (ACK) frame from the AP indicating that it received the packet correctly; if the WN does not get the ACK in a specified amount of time, it postulates that the packet in fact collided with some other WN transmission – at that point the WN transitions into a period of binary exponential backoff; it will try to access the medium and retransmit its packet after the backoff time expires.

More specifically, when[1] it is determined that the radio channel is idle, each terminal having data to be transmitted performs a backoff procedure after an IFS time interval, depending on the state of the terminal. Each terminal stands by while locally decreasing the slot time(s) counter, such time counter being a random number allocated to the terminal during an idle state of the channel. When a terminal completely exhausts the slot time(s), it will attempt to access the channel under consideration. The interval during which each terminal performs the backoff procedure is referred to as a *Contention Window (CW) interval*. When a specific terminal successfully assesses the channel as being available, that terminal may transmit data over that channel. However, when the terminal that attempts the access collides with another terminal, the terminals that collide with each other are allocated new random slot time(s) counters, respectively, to

---

[1]This discussion partially based on reference [4].

perform the backoff procedure again; the new random number allocated to each terminal is typically within a range that is twice larger than the random number that was previously allocated to that terminal. Meanwhile, each terminal attempts the access to the channel by performing the backoff procedure again in the next CW interval, starting the backoff procedure from slot time(s) that remained in the previous CW interval.

As described, a terminal in a WLAN checks whether a channel is busy or not by performing carrier/channel sensing before transmitting data. Such a process is referred to as *CCA*, and a signal level used to decide whether the corresponding signal is sensed is referred to as a *CCA threshold*. When a radio signal is received by a terminal, it is processed to determine if it has a value exceeding the CCA threshold. When a radio signal having a predetermined or higher-strength value is sensed, it is determined that the channel under consideration is physically busy and the terminal delays its access to that channel. When a radio signal is not sensed in the channel under consideration or a radio signal is sensed having a strength smaller than the CCA, then the terminal determines that the channel is idle.

Even when the radio channel is physically available, the AP and STAs in the BSS must contend for higher-level system resources in order to obtain authority to transmit any data. When data transmission at the previous step is completed, each terminal having data to be transmitted performs a backoff procedure by decreasing a backoff counter based on a random number allocated to each terminal after an Arbitration Inter-frame Space (AIFS) time. A transmitting station in which the backoff counter has expired transmits the RTS frame to notify the AP that it has data to transmit. In Figure 5.2 (bottom), STA1 holds a lead in contention with minimum backoff; hence, it may transmit the RTS frame after the backoff counter has expired. The RTS frame includes information on a receiver address, a transmitter address, and duration. A receiving terminal (i.e. the AP in bottom part of the figure) that receives the RTS frame transmits the CTS frame after waiting for a Short Inter-frame Space (SIFS) time to notify that the data transmission is available to terminal STA1. The CTS frame includes the information on a receiver address and duration. In this case, the receiver address of the CTS frame may be set identically to a transmitter address of the RTS frame corresponding thereto, that is, an address of the transmitting terminal STA1. Upon receiving the CTS frame, STA1 transmits the data after a SIFS time.

When the data transmission is completed, the receiving AP transmits an ACK frame after a SIFS time, to notify that the data transmission is completed. When the transmitting terminal receives the ACK frame within a predetermined time, the transmitting terminal regards that the data transmission is successful. However, when the transmitting terminal does not receive the ACK frame within the predetermined time, the transmitting terminal regards that the data transmission is failed. Meanwhile, adjacent terminals that receive at least one of the RTS frame and the CTS frame in the course of the transmission procedure set a Network Allocation Vector (NAV) and do not perform data transmission until the set NAV is terminated. In this case, the NAV of each terminal may be set based on a duration field of the received RTS frame or CTS frame. In the course of the aforementioned data transmission procedure, when the RTS frame or CTS frame of the terminals is not transferred successfully to a target due to a situation such as interference or a collision, a subsequent process is temporarily suspended; the transmitting terminal STA1 that transmitted the RTS frame regards that the data transmission is unavailable and participates in a follow-up contention by being allocated with a new random number. In this case, the newly allocated random number may be determined within a range twice larger than a previous predetermined random number range.

These PHY layer and data link layer protocols described above work well for a reasonably small population of users in proximity; however, the system efficiency decreases when the number of users is large. In addition to the intrinsic protocol limitations, overlapping signal areas that may exist where there might be a high number of APs to support a large number of users in a confined space can impact throughput, since a WN in an overlapping area can trigger backoff procedures in one or more signal areas. Yet another factor to consider is the shared use of wider channels (note, for example, that for 802.11ac operation in North America there is only one 160 MHz channel available). Designing dense coverage when using a small number of channels is arduous because it forces planners to reuse channels in nearby cells – unless power management is meticulously taken into consideration in that case, users will experience degraded performance due to co-channel interference, especially for cases where high Modulation and Coding Schemes (MCS) index (say 8–11) are used. High MCS environments are more susceptible to low Signal-to-Noise Ratio (SNR). High system throughput can be achieved by using a wider radio frequency bandwidth (e.g. a maximum of 160 MHz – also at the 60 GHz band); a higher number of Multiple Input(s) Multiple Output(s) (MIMO) spatial streams (for example, eight in a major band and four in another major band); multi-user MIMO (MU-MIMO); beamforming technology; and high-grade modulation (e.g. a maximum of 1024 Quadrature Amplitude Modulation [QAM]). Although millimeter waves (mmWave) offer wider radio frequency bandwidth, recall from previous chapters that transmission of mmWave signals requires line-of-sight and that transmission is subject to significant attenuation, so that these systems can be used only among devices in a short-distance space.

5.1.1   802.11ax Basics

802.11ax, also called High-Efficiency Wireless (HEW) or High-Efficiency (HE), aims at improving the average throughput per user fourfold in dense user environments; additionally, the standard implements a number of mechanisms to support a higher number of users with consistent and reliable data throughput in crowded wireless environments. As of late 2020, 802.11ax was an IEEE draft amendment with expected full ratification to follow in 2021. 802.11ax defines modifications to the 802.11 PHY layer and the MAC sublayer for high-efficiency operation in the sub-6 GHz frequency bands between 1 and 6 GHz. While 802.11ac operates at 5 GHz only, 802.11ax radios can operate either at the 2.4 or 5 GHz frequency bands and, in the future, it will also operate in the 6 GHz band.

Higher data rates and wider channels were not the principal goal of 802.11ax standardization development; instead the goal was to achieve better and more efficient 802.11 traffic management; a related goal was to increase the average throughput 4x per user in high-density WLAN environments. Recent amendments to the 802.11 specification (e.g. 802.11ac) defined higher data rates and wider channels but did not address efficiency. Since the majority of the typical 802.11 data frames (75–80%) have been shown to be small, and under 256 bytes, a WLAN experiences excessive overhead at the MAC sublayer and medium contention overhead for each small frame [9]. In many (but not all) high-density deployment situations, overall system efficiency may become more important than higher data rates; for example, in high-density hotspot and cellular offloading scenarios, many devices competing for the wireless medium may have low-to-moderate data rate requirements [10]. 802.11ax does indeed provide the efficiency improvement. Specifically, besides being backwards compatible with 802.11a/b/g/n/ac, key 802.11ax features include [9, 11]:

- Increase 4x the average throughput per user in high-density scenarios, such as train stations, airports, and stadiums (see Figure 5.3 for an example of a high-density venue).
- Data rates and channel widths similar to 802.11ac, with the exception of new MCSs (MCS 10 and 11) with 1024-QAM.
- Specified for DL and UL multi-user operation by means of MU-MIMO and Orthogonal Frequency Division Multiple Access (OFDMA) technology. With this technology, an AP can coordinate upstream client transmissions using a Trigger Frame (TF) to allocate transmission resources and set transmit timing for each client.

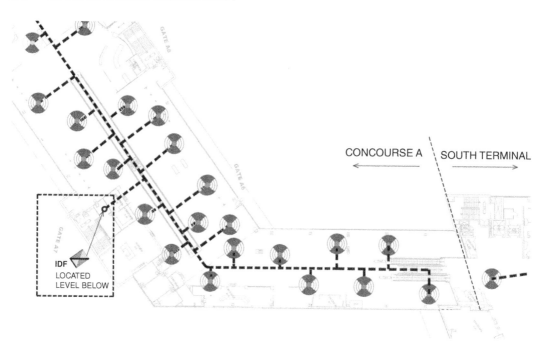

FIGURE 5.3 Example scenario of an airport with high user density targeted for 802.11ax deployment.

- Larger OFDM FFT (Fast Fourier Transform) sizes (4x larger), narrower subcarrier spacing (4x tighter), and longer symbol time (4x) for improved robustness and performance in multipath fading environments and outdoors.
- Improved traffic flow and channel access.
- Better power management for longer battery life.
- Supporting Cellular data offloading as well as supporting environments with many APs and a high concentration of users with heterogeneous devices.
- Outdoors/indoors mixed environments (the outdoor mode is particularly useful for campus environments such as stadiums, smart cities, and so on).

The 802.11ax specification incorporated important changes to the PHY; however, it maintains backward compatibility with 802.11a/b/g/n and /ac devices[2]. However, 802.11ax APs will not improve the performance of legacy Wi-Fi clients (802.11a/b/g/n/ac): 802.11ax-based clients are required to take full advantage of 802.11ax HE capabilities. Nonetheless, as a larger number of 802.11ax clients are mixed into the client population in a given environment, the efficiency improvements gained by 802.11ax client devices will free valuable channel resources for those older clients, therefore improving the overall efficiency of the system [9]. Table 5.1 highlights some key features of 802.11ax and Table 5.2 depicts some the key parametric changes in 802.11ax (loosely inspired by [9, 12]). As noted in previous chapters, recently the Wi-Fi Alliance adopted a new naming convention for Wi-Fi technologies, the goal being that the new naming convention should be easier to understand for the average consumer. Because 802.11ax technology has many new technical features, it has been given the generational name of Wi-Fi 6. Major chipset

---

[2]For example, an 802.11ax WN can send and receive data to legacy WNs; legacy clients are able to demodulate and decode 802.11ax packet headers – but not the entire 802.11ax packet – and initiate a backoff when an 802.11ax WN is transmitting.

TABLE 5.1   Higher Efficiency and Capacity Features Under 802.11ax/Wi-Fi 6

Feature	Description
Operating bands	Wi-Fi 6 extends the new features to the 2.4 GHz band.
MU-MIMO	Employs spatial multiplexing (spatial diversity) to increase the number of simultaneous users of the same frequency resources, increasing capacity (specifically, system throughput), time efficiency, and spectral efficiency. Wi-Fi 6 has 12 spatial streams available for MU-MIMO (Wi-Fi 5 had only four). In Wi-Fi 6 (802.11ax), MU-MIMO can make use of all 12 spatial streams in both the UL and DL directions; Wi-Fi 5 (802.11ac) can use MU-MIMO only in the 5 GHz band and in the DL direction only. In Wi-Fi 6, an AP can use as many as eight spatial streams in the 5 GHz band using eight antennas to simultaneously address eight 1×1 WNs or four 2×2 WNs.
OFDMA	Subdivides the total available spectrum among WNs. It reduces overhead and contention, reduces packet-to-packet latency, and increases efficiency in congested conditions. It can split bandwidth among low data rate clients. Most Wi-Fi 6 product development activity up to the present has concentrated on OFDMA.
Triggering	An AP can utilize triggers to inform WNs how to make use of time and frequency resources, modulation, coding, and spatial streams to minimize contention and improve capacity.
1024-QAM	Higher-order modulation that delivers increased throughput in low noise environments.
Target Wake Time (TWT)	Reduces power consumption, useful for IoT sensors.
"20-MHz-only channels"	Allows IoT sensors to use spectrum more efficiently.

vendors including Broadcom, Qualcomm, and Intel were already manufacturing $2 \times 2 : 2$ Wi-Fi 6 radios for smartphones, tablets, and laptops at press time, with an expectation that 1 billion Wi-Fi 6 chipsets will ship annually by 2022; Samsung and Apple already have Wi-Fi 6-ready smartphones. The PPDU in an 802.11ax format includes a data frame; a control frame such as RTS, CTS, or TF; or a management frame. See Figure 5.4 for an example of a data frame.

The 802.11ax specification defines a four times larger Fast Fourier Transform (FFT): 2048 versus 512, multiplying the number of subcarriers (but in conjunction with this, the subcarrier spacing has been reduced to 1/4 the subcarriers spacing of previous 802.11 specifications, preserving the existing channel bandwidths). The OFDM symbol duration and Cyclic Prefix (CP) also increased fourfold, keeping the raw link data rate the same as 802.11ac, but improving efficiency and robustness in indoor/outdoor and mixed environments; 1024-QAM and smaller CP ratios can also be utilized in indoor environments, which will increase the maximum data rate [11]. The basic concept of OFDM is to use a large number of parallel narrow-band subcarriers instead of a single wide-band carrier. Key advantages include simplicity and efficiency in dealing with multipath and other interference; however, the modulation scheme is sensitive to frequency offset and phase noise. This modulation technique has been used for various standards going back to 802.11a as well as telco Digital Subscriber Line (DSL) services.

802.11ax employs an explicit beamforming procedure (similar to that of 802.11ac), where the beamformer initiates a channel sounding procedure with a Null Data Packet (NDP) [14–18]. The beamformee measures the channel and responds with a beamforming feedback frame, containing a compressed feedback matrix. The beamformer uses this information to compute the channel matrix, H. A narrowband time-invariant wireless channel with n_t transmit and n_r receive antennas is described by a n_r by n_t matrix H; the entries h_{ij} represent the channel gain from transmit antenna j to receive antenna i. The beamformer can then use this channel matrix to

TABLE 5.2   802.11ax Changes Compared with 802.11ac

Feature	802.11ac (Published in December 2013 – Also Known as Wi-Fi 5)	802.11ax (Final Approval in Early 2021 – Also Known as Wi-Fi 6)
Spectrum bands	5 GHz 5170–5330 MHz (eight 20 MHz subchannels, or four 40 MHz subchannels, or two 80 MHz subchannels, or one 160 MHz subchannel) 5490–5730 MHz (twelve 20 MHz subchannels, or six 40 MHz subchannels, or three 80 MHz subchannels, or one 160 MHz subchannel) 5735–5835 MHz (four 20 MHz subchannels, or two 40 MHz subchannels, or one 80 MHz subchannels)	2.4 GHz and 5 GHz; 6 GHz in the future
Channel bandwidth	20, 40, 80, 80 + 80, 160 MHz	Same as 802.11ac
Fast Fourier Transform (FFT) sizes	64, 128, 256, 512	256, 512, 1024, 2048
Subcarrier spacing	312.5 kHz	78.125 kHz
OFDM symbol duration	3.2 μs + 0.8/0.4 μs Cyclic Prefix (CP)	12.8 μs + 0.8/1.6/3.2 μs Cyclic Prefix (CP)
Highest modulation scheme	256-QAM	1024-QAM
MIMO streams	4	8 + 4 (eight spatial streams in the 5 GHz band, in conjunction with four spatial streams in the 2.4 GHz band, yielding up to 12 streams of Wi-Fi 6 connectivity)
Data rates	433 Mbps (80 MHz, 1 SS) 6933 Mbps (160 MHz, 8 SS)	600.4 Mbps (80 MHz, 1 SS) 9607.8 Mbps (160 MHz, 8 SS)
Guard intervals (smaller values allow an increase data rate; higher values make the transmission more resilient, for example in outdoor environments)	0.4 μs (400 ns), 0.8 μs (800 ns)	Three-guard intervals: 0.8 μs (800 ns), 1.6 μs and 3.2 μs.

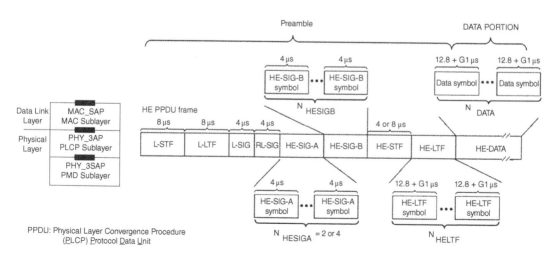

FIGURE 5.4   HE PPDU (data) frame (partially from [13]).

focus the RF energy toward each WN [7]. Beamforming is ideal at millimeter waves because of the high directionality of the signals, but there are other limitations as noted earlier.

The 802.11ax process has two modes of operation:

- *Single-user* (SU) operation: in this mode the wireless WNs send and receive data frames one at a time once they obtain access to the medium; the process results in sequential frame transmissions.
- *Multi-user* (MU) operation: this mode allows for simultaneous operation by multiple (non-AP) WNs.

For the multi-user mode, the specification further defines a *downlink multi-user* operation and an *uplink multi-user* operation. The downlink multi-user operation refers to an operation where the AP transmits data to multiple wireless WNs at the same time (this is similar to the operation in 802.11ac). The uplink multi-user process involves simultaneous transmission of data from multiple WNs to the AP. This functionality is new to 802.11ax.

In the MU mode of operation, the standard also specifies two multiplexing methods: (i) MU-MIMO (as used in 802.11ac implementation), and (ii) Multi-User Orthogonal Frequency Division Multiple Access (MU-OFDMA) (which has technological roots in 4G/Long-Term Evolution [LTE]). For both methods, the AP acts as the central controller of the operation (an 802.11ax AP can also combine these two methods) [8]. A key design goal of the 802.11ax was to be able to support 4X higher average per-user throughput in dense user environments. To achieve that, 802.11ax-ready WNs support the DL-MU-MIMO/UL-MU-MIMO operation, MU-OFDMA operation, or both.

802.11ax APs use beamforming techniques to simultaneously transmit packets to spatially diverse WNs: the AP calculates a channel matrix for each WN and steers simultaneous beams to different WNs, each beam containing specific packets for its target WN. Each MU-MIMO transmission could use a different MCSs and a different number of spatial streams. The specification allows up to eight MU-MIMO transmissions at a time. In the MU-MIMO uplink direction (UL-MU-MIMO), as a new feature, the AP can initiate a simultaneous uplink transmission from each of the WNs. This is accomplished using a TF (see Figure 5.5): when the multiple WNs

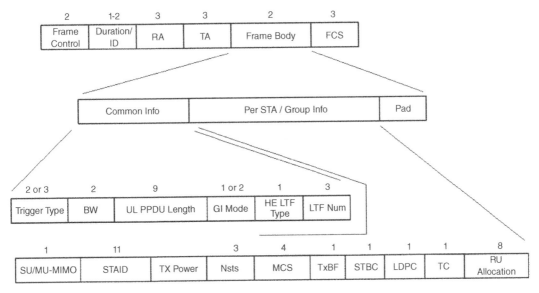

FIGURE 5.5   Trigger frame (partially from [2]).

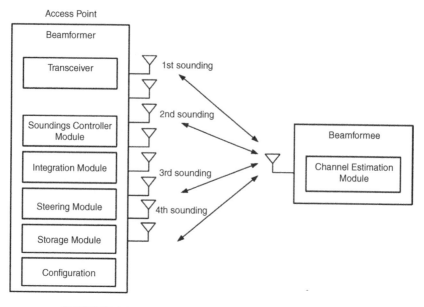

FIGURE 5.6   Access point supporting beamforming [19].

respond simultaneously with their own packets, the AP uses the channel matrix in the context of the received beams and separates the data that each uplink beam carries. The AP may also initiate uplink multi-user transmissions to receive beamforming feedback information from the active WNs. See Figure 5.6. To manage UL-MU-MIMO or UL-OFDMA transmissions, the AP broadcasts a TF to all users, identifying the number of spatial streams and/or the transmit frequency allocations of each WN (more on this below). Additionally, the frame also contains power control information, so that individual WNs can increase or reduce their transmitted power, in an effort to equalize the power that the AP receives from all uplink users and improve reception of frames from nodes farther away [7]. The AP also alerts all WNs as to when to start and stop transmitting using a multi-user uplink TF that indicates to the WNs the precise time at which they can start transmitting and the exact duration of their frame. Once the AP receives the frames from all users, it sends them back a block ACK to complete the operation.

OFDMA enables 802.11ax to multiplex more users in the same Channel Bandwidth (CBW) by extending the existing OFDM digital modulation scheme that 802.11ac already employs. OFDMA is a multi-user version of the OFDM digital modulation technology utilized in 802.11a/g/n/ac (in those systems, OFDM is employed to single-user transmissions at a given frequency). With this scheme the specification facilitates the assignment of specific sets of subcarriers to individual WNs. The 20-, 40-, 80-, and 160-MHz channels are divided into smaller subchannels with a predefined number of subcarriers (tones). By subdividing the channel, parallel transmissions of small frames to multiple users happen simultaneously. OFDMA affords frequency selectivity gains, where an AP can allocate resources to each STA/WN where those allocated resources offer highest frequency-gain for that STA [13]. Using acknowledgment procedures, the AP can obtain the information that is needed to harvest frequency selectivity gain for each STA/WN in the subsequent DL or UL OFDMA frames.

The smallest subchannel is called a Resource Unit (RU); a RU has a minimum of 26 subcarriers. RUs denote a group of 78.125 kHz bandwidth subcarriers utilized in the DL and UL transmissions; note that different transmit powers may be applied to different RUs. The 802.11ax system allows for a maximum of 9 RUs for a 20 MHz CBW, 18 in a 40 MHz channel, and more in 80 or 160 MHz channels. Currently, the following RUs are defined: 26-tone (26-subcarrier) RU, 52-tone RU, 106-tone RU, 242-tone RU, 484-tone RU, 996-tone RU, and 2 × 996-tone RU.

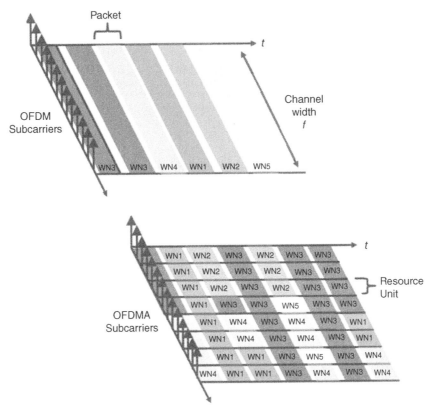

FIGURE 5.7   Concept comparison of a single user using the channel and OFDMA methods for multiplexing multiple users in the same channel.

Considering the instantaneous traffic needs of the multi-user environment under the control of a given AP, the AP decides how to allocate the channel by assigning any and all available RUs on the downlink: it may allocate the entire CBW to only one user at a given time – as is the case with the 802.11ac specification where only a single user can undertake an uplink transmission at a time using the entire allocated 20 MHz bandwidth, while all other WNs must wait their turn – or it may partition the channel with RU segments to support multiple users simultaneously (see Figure 5.7). The OFDMA mechanism is much more efficient use of the medium for smaller frames and it is particularly useful in dense user environments; in this environment, many users would normally have to wait and contend for their turn to use the channel but the new mechanism can support the WNs simultaneously making available a smaller, but temporarily bespoken, transmission subchannel.

The 802.11ax-based AP manages how many RUs are used within a given channel. For example, when subdividing a 20 MHz channel, the AP can designate 26, 52, 106, and 242 subcarrier RUs; these correspond approximately to 2, 4, 8, and 20 MHz (sub)channels, respectively. Within a channel, different combinations can be used: for example, an AP could simultaneously communicate with one 802.11ax client using 4 MHz of bandwidth while communicating with two other 802.11ax clients using 2 MHz subchannels. Table 5.3 depicts the number of WNs that can obtain frequency-multiplexed access when the 802.11ax AP and WNs utilize MU-OFDMA; for example, for standard 20 MHz bandwidth, nine 26-tone RUs can be utilized, but only four with 52-tone RUs. Applying this scheme, for example, in the case where an AP supports 52-tone RUs in its 20-MHz bandwidth, four users can transmit and/or receive simultaneously to the AP; in this setup each WN can use 52-tones; however, the total bandwidth for all users must be less

TABLE 5.3   Total Number of RUs by Channel Bandwidth

RU Type	Number of RUs in 20 MHz/ CBW20	Number of RUs in 40 MHz/ CBW40	Number of RUs in 80 MHz/ CBW80	Number of RUs in 160 MHz/ CBW160 and CBW80+80
26-subcarrier RU 24 data subcarriers and 2 pilot subcarriers	9	18	37	74
52-subcarrier RU 48 data subcarriers and 4 pilot subcarriers	4	8	16	32
106-subcarrier RU 102 data subcarriers and 4 pilot subcarriers	2	4	8	16
242-subcarrier RU 234 data subcarriers and 8 pilot subcarriers	1-SU/ MU-MIMO	2	4	8
484-subcarrier RU 468 data subcarriers and 16 pilot subcarriers	N/A	1-SU/ MU-MIMO	2	4
996-subcarrier RU 980 data subcarriers and 16 pilot subcarriers	N/A	N/A	1-SU/ MU-MIMO	2
2 × 996 subcarrier RU two 996-tone RUs, each located at each half of the PLCP (Physical Layer Convergence Protocol) Protocol Data Unit (PPDU) bandwidth for 160 MHz and 80+80 MHz HE (High Efficiency) PPDU formats: 1960 data subcarriers and 0 pilot subcarriers	N/A	N/A	N/A	1-SU/ MU-MIMO

than or equal the entire 20-MHz allocated bandwidth. When there is only one active WN, if needed the WN can still use the entire 20-MHz bandwidth by operating at 242-tone RU to sustain a higher data rate. Figure 5.8 illustrates conceptually how an 802.11ax system may multiplex the channel using different RU sizes (the actual positioning of the RUs within the frequency spectrum is not captured in this figure).

Figure 5.9 is a more detailed example of a RU allocation scheme in 802.11ax. The resource allocation scheme includes a plurality of allowed allocations. The respective resource unit allocation indications are provided to respective client stations, as is the case with the RU allocation subfield of the TF (per Figure 5.5). The[3] resource allocation scheme includes 137 allowed resource units in a 160 MHz bandwidths; the 137 allowed resource units include 74 26-tone resource units, 32 2 × 26-tone resource units, 16 4 × 26-tone resource units, eight resource units that occupy respective 20 MHz sub-bands of the 160 MHz bandwidth, four resource units that occupy respective 40 MHz sub-bands of the 160 MHz bandwidth, two resource units that occupy

---

[3]This discussion is based on reference [2].

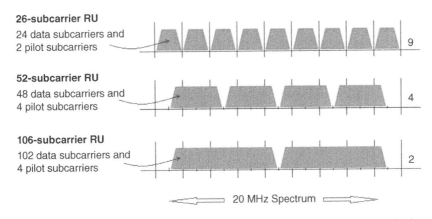

26-subcarrier RU

24 data subcarriers and
2 pilot subcarriers

52-subcarrier RU

48 data subcarriers and
4 pilot subcarriers

106-subcarrier RU

102 data subcarriers and
4 pilot subcarriers

20 MHz Spectrum

FIGURE 5.8   Illustrative 20 MHz spectrum allocation based on resource unit sizes.

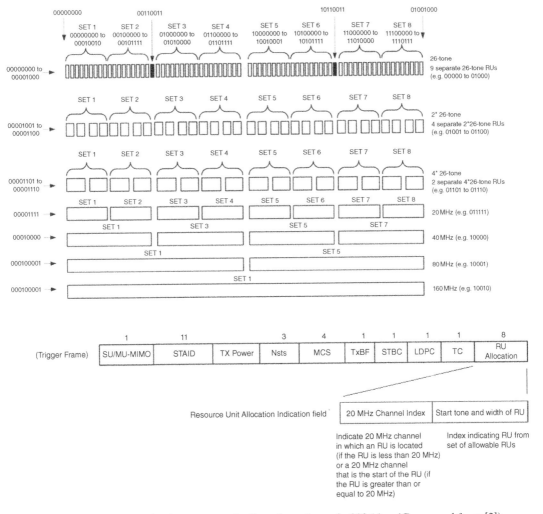

FIGURE 5.9   Example of resource unit allocation scheme in 802.11ax (Composed from [2]).

respective 80 MHz sub-bands of the 160 MHz bandwidth, and one resource unit that occupies the entire 160 MHz sub-band. The RU allocation indication for a client station, for example, included in the RU allocation subfield of the TF, indicates one of the allowed resource units as the resource unit allocated for transmission by the client station. The RU allocation indication includes eight bits: a first field used to indicate a 20 MHz channel that includes at least an initial basic resource unit block, and a second field to jointly indicate a location of the initial basic resource unit block of the resource unit and a width of the resource unit. In another implementations, the value of the eight-bit indication maps directly onto one of the allowed resource allocations, such as one of the allowed resource units.

The RU allocation indication field is used to indicate a resource allocation within an OFDMA transmission, the resource allocation corresponding to transmission by a client station. In some implementations, the RU allocation indication field is included in per-STA information fields. The RU allocation indication field includes a field to indicate a 20 MHz channel in which an RU is located (if the RU is less than 20 MHz) or a 20 MHz channel that is the start of the RU (if the RU is greater than or equal to 20 MHz). The RU allocation indication field also includes a field to indicate a width of the RU and/or a starting group of 26 OFDM tones within a 20 MHz channel. In some implementations, the field is set to an index value that indicates an RU from a set of allowable RUs for a given starting 20 MHz channel; the field is an index value indicating one of a plurality RUs, including RUs of different bandwidths and RUs having different positions within a 20 MHz communication channel, given the 20 MHz channel indicated by the field. In a typical implementation, the 20-MHz-Channel-Index field is three bits and the Start-tone-and-width-of-RU field is five bits – in some implementations, the 20-MHz-Channel-Index can be another suitable size (e.g. 2, 4, 5, 6, 8, 9, 10, etc., bits) and/or the Start-tone-and-width-of-RU field can be another suitable size (e.g. 3, 4, 6, 7, 8, 9, 10, etc., bits).

For example, the values of the five bits of the Start-tone-and-width-of-RU field are in the range of "00001" to "10010," with each value corresponding to a particular resource unit. In some cases, an additional value (e.g. 10100) in the field Start-tone-and-width-of-RU is used to indicate a 26-tone resource unit between two adjacent 40 MHz channels. When the Start-tone-and-width-of-RU field is set to indicate a 26-tone resource unit between two adjacent 40 MHz channels, the 20-MHz-Channel-Index field is set to indicate the 20 MHz channel that is on the left of the 26-tone resource unit. In another example (e.g. in Figure 5.9), the value of the eight-bit indication maps directly onto one of the allowed resource allocations, such as one of the allowed resource units – the figure illustrates example mapping between values of the eight-bit indication and allowed resource allocations, such as one of the allowed resource units. In this example, the allowed resource units are divided into a plurality of sets (e.g. eight sets) of resource units. A set to which each resource unit belongs is indicated above the resource unit in the figure. Consecutive logic values of the eight bits are assigned to consecutive resource units within each set of the resource units. Example consecutive values of the eight bits assigned to resource units in each set are indicated above the corresponding portion of the first row of resource units in Figure 5.9. The consecutive values are assigned to resource units in a particular set from left to right and from top to bottom in the particular set. For reference, example eight-bit values assigned to resource units in the first set (SET 1) are illustrated to the left of each row in the figure. Further, as illustrated in the figure, the value "00110011" is assigned to the 26-tone resource unit that is between the first 40 MHz and the second 40 MHz sub-band of the 160 MHz bandwidth, and the value "10110011" is assigned to the 26-tone resource unit that is between the third 40 MHz and the fourth 40 MHz sub-band of the 160 MHz bandwidth. The reader may consult reference [2] for more inclusive discussions of these issues.

First-generation 802.11ax radios are expected to support 1024-QAM, facilitating in principle higher data rates in some venues; however, a very high SNR threshold (e.g. 35 dB) is needed in order for 802.11ax radios to use this modulation scheme effectively. These environments – low noise floor and close proximity between an 802.11ax AP and 802.11ax client – are unlikely to exist in public campus venues (e.g. airports), especially public hotspos.

Note that OFDMA systems have a high susceptibility to frequency and clock offsets. Consequently, 802.11ax MU-OFDMA performance requires tight frequency synchronization and clock offset correction, especially at the higher MCS10 and MCS11 with 1024-QAM. Frequency synchronization ensures that all WNs operate within their allocated subchannels with minimal spectral leakage (4G LTE base stations can use GPS clocks to synchronize all associated devices; however, 802.11ax APs typically do not have the ability to use GPS, thus stable electronic circuitry is required).

To make efficient use of spectrum resources and to enhance the system performance in dense environments, the 802.11ax specification supports a spatial reuse technique described next.

WNs are able to identify signals from Overlapping Basic Service Sets (OBSS) and assess medium contention and, thus, undertake interference management based on this information. BSS Color (aka BSS Coloring) is a method for mitigating channel contention overhead due to OBSSs. When a WN that is actively monitoring the medium detects an 802.11ax frame, it checks the so-called BSS color bit in the MAC header: (i) if the BSS color in the detected PPDU is the same color as the one that its associated AP has already announced, then the WN considers that frame to be an intra-BSS frame; (ii) if the detected frame has a different BSS color than its own, then the WN considers that frame as an inter-BSS frame from an OBSS – the WN then treats the medium as being "busy" only during the time required by the WN to establish that the frame is from an inter-BSS, but not for a period longer than the time indicated as the length of the frame's payload. When WNs use the color code-based CCA rule, they are also permitted to adjust the OBSS signal detection threshold along with the transmit power control; additionally, WNs can adjust CCA parameters, such as the energy detection level and the signal detection level. Specifically, in an 802.11ax environment, an AP or a WN may increase a CCA threshold; in this manner, so that spatial simultaneous transmission can be performed on a link that simultaneous transmission could not be performed originally [20]. Clearly, this approach can improve the system's overall throughput.

Besides using CCA methods to determine if the medium is idle or busy for the current frame, the 802.11ac standard employs the NAV[4] – a timer mechanism that maintains a prediction of future traffic – for WNs to indicate the time required for the frames immediately following the current frame [11]. An 802.11ax AP can negotiate with the participating WNs the use of the Target Wake Time (TWT) function to define a specific time (or set of times) for individual stations to access the medium: the WNs and the AP exchange information that includes an expected activity duration, enabling the AP to control the level of contention and overlap among WNs. The WNs may use TWT to reduce energy consumption, entering a sleep state until their TWT arrives.

With MU-MIMO one can add WN devices to the localized network without reducing the data rates for the other clients in the same hotspot; however, the internal bus rate of the AP and the uplink from the AP to the core network has to be able to also sustain the high data rate.

- A single 2×2 WN device in the hotspot that is using MCS 11 with 80 MHz of CBW, can theoretically, achieve a data rate of 1.2 Gbps; adding additional WN devices lowers the individual data rate, which is shared among the set of WNs in the hotspot.

---

[4]In an 802.11ac system (also in 802.11n), the NAV setting is used as follows: after receiving a data packet, a node determines whether a receiver address of the data packet is that of that node or not; if the receiver address of the data packet is not the node, the node determines a relationship between a decoded duration field and a current NAV -- if the field is greater than a current NAV value, the node updates the NAV. Given the spatial reuse technology, if the CCA threshold is increased, it indicates that the system supports that the node performs spatial reuse. However, if the node determines that a current CCA is idle, according to an original NAV mechanism, the node needs to update the NAV according to the decoded duration field, and the node still cannot perform spatial reuse [20].

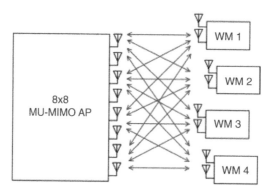

Multiple WMs use same time-
frequency resources simultaneously

FIGURE 5.10 Theoretical example of 8×8 MU-MIMO AP using differences in the spatial paths of the signals to different WNs.

- With a 4×4 MU-MIMO AP, two WN devices can share the same frequency and time resources, theoretically each achieving a 1.2 Gbps data rate. The total aggregate data rate for the AP is 2.4 Gbps.
- With an 8×8 MU-MIMO AP (e.g. see Figure 5.10), four WN devices can operate at the same time, and the total aggregate data rate for the AP is 4.8 Gbps.

An $8 \times 8 : 8$ AP (eight input [antennas], eight output [antennas] and supporting eight spatially separated streams on a given frequency channel/subchannel), can in theory modulate data on all eight radio chains to a single WN, which would then result in high data rates; however, it is very unlikely that there will be $8 \times 8 : 8$ WNs, due to the drain on battery life to manage all these streams. There are some APs on the market that operate as $8 \times 8 : 8$ in 5 GHz and $4 \times 4 : 4$ in 2.4 GHz (the battery life of an $8 \times 8 : 8$ WN client, such as a smartphone, is estimated to be about five minutes; most Wi-Fi WN devices such as smartphones use dual-frequency $2 \times 2 : 2$ radios because an $8 \times 8 : 8$ radio would drain battery life; in the future, one might see some $4 \times 4 : 4$ client radios in high-end laptops [9]).

802.11ax has some mechanisms that are useful in the IoT context. One such mechanism is the TWT, a capability first proposed under the 802.11h standardization activities. TWT uses negotiated policies based on expected traffic activity between 802.11ax clients, such as intelligent Wi-Fi-enabled sensors, and an 802.11ax AP, to specify a scheduled wake time for each client [9]. 802.1ax IoT clients can potentially sleep for relatively long periods (hours) and optimize battery life.

5.1.2 Key 802.11ax Processes

Figure 5.11 depicts a flow diagram for the trigger process implemented by the AP [2]. To start the process, one or more TFs are generated. The TFs include an indication of a trigger type and the frames are formatted according to the corresponding indicated trigger type. The frames are generated to trigger an uplink OFDMA transmission by multiple WNs. The TFs are transmitted to the multiple WNs. In some cases, a broadcast TF is transmitted to the multiple WNs; in other cases, unicast TFs generated are transmitted to the appropriate WN(s) in respective frequency portions of a downlink OFDMA transmission. In other environments, a downlink OFDMA transmission includes both (i) a broadcast TF is generated and transmitted in a first frequency portion of the downlink OFDMA transmission, and (ii) one or more unicast TFs are generated

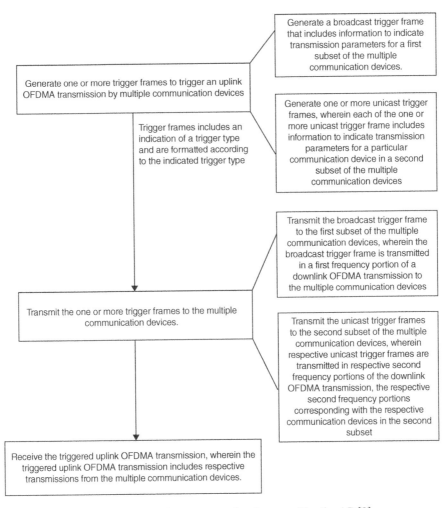

FIGURE 5.11   Trigger process implemented by the AP [2].

and transmitted in a respective second frequency portions of the downlink OFDMA transmission. The uplink OFDMA transmission triggered by the one or more TFs is received by the AP; the uplink OFDMA transmission includes respective transmissions from the multiple WNs; the respective transmissions from the multiple WNs are transmitted in respective frequency portions, and using respective transmission parameters, indicated, to respective (multiple) WNs, by one or more TFs.

Figure 5.12 is a flow diagram for the beamforming training process implemented by the AP [2]. To start, a beamforming training packet is transmitted to multiple WNs; the beamforming training packet includes one or more training fields that allow the multiple WNs to obtain measures of respective communication channels associated with the WNs. Next, a TF is generated; it is generated to trigger a beamforming feedback OFDMA transmission from at least some of the WNs. The TF includes information to indicate the respective frequency portions of the beamforming feedback OFDMA transmission, the respective frequency portions corresponding with respective ones of at least some of the multiple WNs. The TF is transmitted to the at least some of the WNs. The beamforming feedback OFDMA transmission is then received by the AP. The beamforming feedback OFDMA transmission includes respective beamforming training feedback packets from at least some of the multiple WNs. The respective beamforming

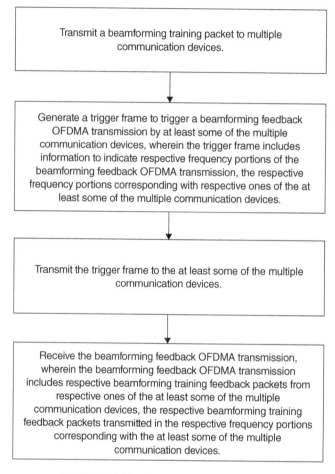

FIGURE 5.12 Beamforming process [2].

training feedback packets are transmitted in the respective frequency portions corresponding with at least some of the multiple WNs indicated by the TF.

Figure 5.13 captures a related view of the MU service negotiation and measurement exchange process.

5.1.3 Summary

In summary, MU-OFDMA allows for multiple-user access by subdividing a channel into subchannels. MU-MIMO allows for multiple-user access by using different spatial streams, where AP can send unique steams of data to multiple clients simultaneously (the 802.11ax specification allows for the combined use of MU-MIMO and MU-OFDMA but currently this combination is not expected to be broadly implemented by equipment vendors.) MU-OFDMA affords increased efficiency and reduced latency in environments where low bandwidth applications abound and the packets are relatively small. MU-MIMO affords increased capacity and higher WN speeds where high bandwidth applications predominate, and the packets are relatively large.

DL-MU-MIMO was introduced with Wave-2 802.11ac Aps; however, actual implementation of MU-MIMO for indoor environments is still rare. MU-MIMO requires transmit beamforming which requires sounding frames; these sounding frames add considerable overhead, especially

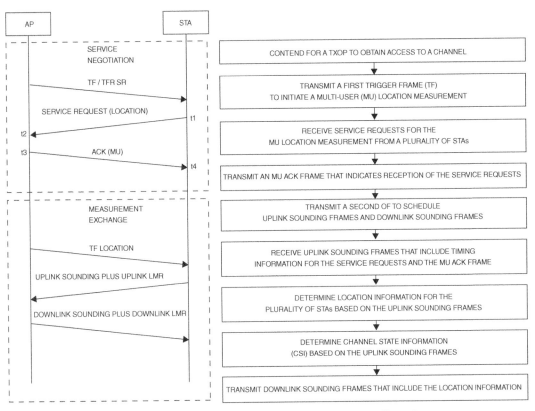

The AP contends for a transmission opportunity (TXOP) to obtain access to a channel. The AP transmits a Trigger Frame (TF) to initiate a Multi-User (MU) location measurement during the TXOP. The AP receives service requests for the MU location measurement from a plurality of STAs. The AP transmits an MU acknowledgement (ACK) frame that indicates reception of the service requests. The AP may receive, from the STAs, uplink sounding frames that include per-STA timing information for the service requests and the MU ACK frame. The STA may determine location measurement for the STAs based on the per-STA timing information included in the uplink sounding frames.

FIGURE 5.13   MU service negotiation and measurement exchange process (composed from [10]).

when the majority of the traffic data frames are small; MU-MIMO would only be a favorable option in very low density, high bandwidth environments. 802.11ax will continue to support DL-MU-MIMO and may also include UL-MU-MIMO, although support for UL-MU-MIMO is not expected to be included in any of the first generation of 802.11ax radios. Market adoption of MU-MIMO, in general, has yet to take place. Industry proponents posit that OFDMA will be the most relevant technology that 802.11ax offers. Downlink MU-MIMO was introduced with Wave-2 802.11ac APs; however, real-world implementation of MU-MIMO for indoor environments and enterprise networks is still rare and very few if any MU-MIMO capable clients exist in the current marketplace. MU-MIMO requires spatial diversity, therefore the physical distance between the clients is necessary; most existing enterprise deployments of Wi-Fi involve a high density of users that is not conducive for MU-MIMO conditions [9].

Additionally, 20 MHz channels are likely to continue to be prevalent. 40 MHz channels are more typical at 5 GHz (also considering the Dynamic Frequency Selection [DFS] channels discussed in Chapter 2). 80 MHz channels are best used when no legacy devices participate in the network (but this would be difficult to achieve in an environment such as an airport or a sporting venue, where many people enter and exit the environment).

Note that at least initially 802.11ax will not be using Citizens Broadband Radio Service (CBRS) or mmWave frequencies; in the context of CBRS there should not be technical limitations in using 802.11ax, but the appropriate RUs would need to be defined.

5.2 VOICE OVER Wi-Fi (VoWi-Fi)

We discussed the use of DASs in Chapter 3 as a way to support voice and Internet connectivity in buildings and large venues. Indeed, such an approach is valid; however, it depends on well-defined and well-paid cellular services. Both voice services and Internet services can be obtained over a venue's Wi-Fi infrastructure when using VoWi-Fi-enabled smartphones. In a number of situations, Wi-Fi can be considered "free" or available for a reduced cost – many public-facing businesses and professional services offer free guest Wi-Fi access in their premises. In a corporate setting, the intranet may serve that purpose; at a domicile, the Wi-Fi home network can serve that purpose; in a venue such as an airport terminal or a stadium, an appropriately provided network (say managed by a specialized operator or by a "venue" carrier, similar to the older concept of a Building Local Exchange Carrier [BLEC]) can serve that purpose.

Formally, Voice over Wi-Fi (VoWi-Fi, also known as VoWLAN) refers to the delivery of commercial telephony services using VoIP technologies from UEs connected to a Wi-Fi AP [21]. In definitional terms, VoWi-Fi stands for Voice over (EPC-integrated) Wi-Fi. VoWi-Fi is a complementary technology to VoLTE[5] and utilizes IP Multimedia Subsystem (IMS) technology to provide a packet voice service delivered over IP via a Wi-Fi network. Where possible, VoLTE calls may be seamlessly handed over between LTE and Wi-Fi and vice versa. Conversational video is also possible via Wi-Fi. VoWi-Fi is already used extensively within the home networks due to the ubiquity and low cost of Wi-Fi connectivity in domicile settings.

VoWi-Fi utilizes IMS to provide a packet voice service delivered to the user over IP links via a local Wi-Fi network. In particular, GSMA IR.51 defines a profile that identifies a minimum mandatory set of features that are defined in 3rd Generation Partnership Project (3GPP) specifications that a wireless device, the UE, and network are required to implement to guarantee interoperable, high-quality IMS-based telephony and conversational video services over Wi-Fi access. In the formal IMS construct, Wi-Fi access refers to WLAN-based access to the Evolved Packet Core (EPC) via untrusted access interface (S2b interface), as defined in 3GPP TS 23.402 [22]. The EPC is the core network of the LTE/4G cellular system. When designing the evolution of the 3G system, the 3GPP community decided to use Internet Protocol (IP) as the basic protocol to transport all services; it was therefore agreed that the EPC would not have a circuit-switched domain anymore and that the EPC should be an evolution to a packet-switched architecture. This decision had consequences on the architecture itself but also on the way that the services were provided [23]. It was also decided to separate the user data (also known as the user plane) and the signaling (also known as the control plane) to make the scaling independent.

A formal way of defining VoWi-Fi is to observe that it deals all core IMS services accessed from unlicensed spectrum and across untrusted access infrastructures, such as public Wi-Fi hotspot or even an intranet AP. More specifically, the S2b interface defines the connection between the Public Data Network (PDN) (i.e. the Internet) and the Evolved Packet Data Gateway (ePDG), as illustrated in Figure 5.14. The formal emphasis is on Mobile Network Operator (MNO)-provided services; therefore, while Over The Top (OTT) mobile communications

---

[5]Voice over LTE (VoLTE) refers to delivery of commercial telephony services (in packetized form) by a Mobile Network Operator (MNO) utilizing licensed spectrum in a 4G LTE.

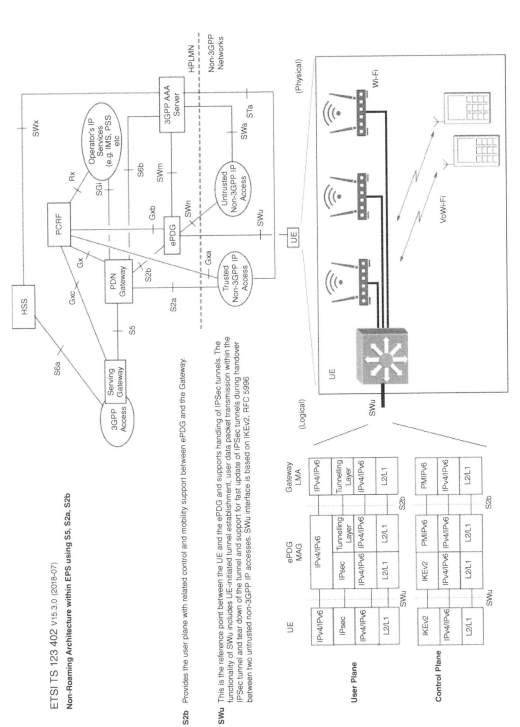

FIGURE 5.14 A formal view of VoWi-Fi (partially based on ETSI 123 402).

TABLE 5.4 **Key EPC Architecture Elements**

Architecture Elements	Basic Functionality
Evolved Packet Data Gateway (ePDG)	The functionality of ePDG includes the following: 1. Functionality for transportation of a remote IP address as an IP address specific to a Public Data Network (PDN); 2. Routing of packets from/to PDN Gateway to/from UE; the UE can maintain a Security Association (SA) with the PDN Gateway; 3. When security is used, an IPSec SA is established for the PDN connection, this includes routing of uplink packets based on the uplink packet filters rules; 4. Routing of downlink packets towards the IPsec SA associated to the PDN connection; 5. Decapsulation/encapsulation of packets for IPSec; 6. Mobile Access Gateway (MAG) according to the PMIPv6 specification, RFC 5213; 7. Tunnel authentication and authorization (termination of IKEv2 signaling and relay via AAA messages); 8. Transport level packet marking in the uplink; 9. Enforcement of Quality of Service (QoS) policies; and, 10. Accounting for inter-operator charging (if applicable)
Public Data Network (PDN) gateway	Handoff functionality (hardware) to IP-based voice/video services carrier services provided via the PDN such as VoIP (IMS), Packet-switched Streaming Service (PSS), and generic data services (Internet access)

services may also employ voice over Wi-Fi to limit the use of a subscriber's (licensed) data plan, this is not "officially" referred to as VoWi-Fi, since these services are not tightly integrated with a MNOs global service offering and infrastructure [24]; however, the underlying technology is similar if not identical. Making use of VoWi-Fi enables MNOs to efficiently extend their coverage while obviating the costs associated with building out additional Radio Access Network (RAN) infrastructure, acquiring licensed spectrum, or establishing roaming agreements with other providers.

Utilizing the 3GPP's TS 23.402 architectural model [22], connectivity from the UE to the EPC (the core packet-based voice infrastructure) can be established using the specifications of the S2b reference interface to connect to the Enhanced Packet Data Gateway (ePDG) function; the S2b specification defines not only the control plane and user plane protocol stack, but also authentication, policy control, handover between untrusted and trusted access infrastructures for both roaming and non-roaming scenarios, and billing. See Table 5.4 for a brief description of the (abstract) functional elements. The above-cited IR.51 extends the requirements, identified in IR.92 (Voice over LTE/VoLTE) and IR.94 (Video over LTE/ViLTE) to include untrusted Wi-Fi access, and identifies the requisite information a UE requires to connect to, and obtain services from, an IMS core; furthermore, defines the protocol stack across the access interface and it also specifies the basic Wi-Fi radio, and the underlying IMS and EPC capabilities. The specification also identifies supplementary services and the supported media types; dual-radio circuit switch call continuity and fallback are also described.

As noted, effectively, VoWi-Fi is VoIP media over a Wi-Fi network. While the formal mechanism described above can be used by a commercial wide-area operator, the MNO, the service can be supported by an institution in its intranet, by a residential user in their his/her own home, or by a local/campus/venue-based provider (for example, in an airport). Thus, the functionality can be supported as a service or simply as an application run by the user either with a traditional VoIP application such as Skype or by any of the videoconferencing applications such as Zoom,

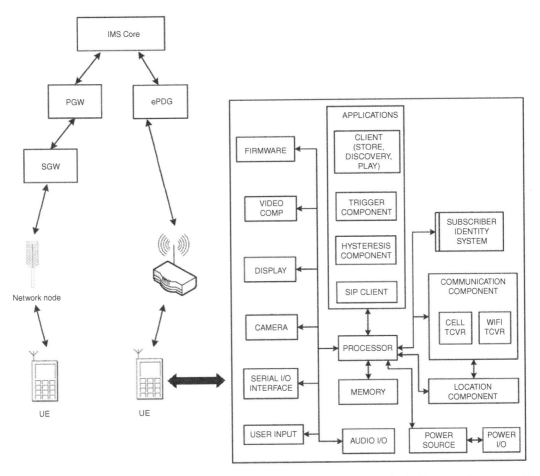

FIGURE 5.15 Simplified example Wi-Fi call through a Wi-Fi AP and block diagram of an example mobile handset [25].

Teams, Webex, and so on. VoWi-Fi can be supported over any Internet-connectable device (with a speaker and microphone), including a laptop, a tablet, or a smartphone. VoWi-Fi allows users to be connected to the Public Switched Telephone Network (PSTN) even without a cellular signal; this may obviate the need to install a DAS in some situations, making it possible for users to originate and receive calls, particularly when indoors, without having to install the DAS equipment or upgrade a smartphone to support new bands (e.g. the cellular version of CBRS).

Figure 5.15 is a simplification of Figure 5.14. As seen in the figure, instead of communicating through a mobile provider's network node, in a typical Wi-Fi call, a UE uses a Wi-Fi AP; the AP/ router is connected to the IMS core via a link such as a DSL or cable broadband network. With a typical Wi-Fi call, VoIP packets are forwarded via a Wi-Fi AP through the Internet to the ePDG and then routed to the IMS core. With a Wi-Fi call, the IP packets that are transmitted contain voice, but generally do not have a Quality of Service (QoS) guarantee (e.g. are "best effort" traffic) – although QoS mechanisms can indeed be engineered into the system. If a UE has Wi-Fi calling enabled, and if the device is connected via an AP, then the UE can make a call through the AP. This can offload some of the traffic from a mobile network to a land-connected Wi-Fi network; UEs can also be configured to make a Wi-Fi call first, before attempts to use the cellular operator's traditional LTE network, if Wi-Fi calling is enabled on the UE.

Figure 5.15 also depicts a block diagram of a UE capable of connecting to a Wi-Fi (or to a traditional cellular) network and supporting voice, data, and multimedia sessions, as described in [25]. The UE includes a processor for controlling and processing all onboard operations and functions. A communications component interfaces to the processor to facilitate wired/wireless communication with external systems, e.g. cellular networks, VoIP networks, and so on; the communications component can also include a suitable cellular transceiver (e.g. a global GSM transceiver) and/or an unlicensed transceiver (e.g. Wi-Fi 5, Wi-Fi 6, WiMax). The UE can process IP data traffic through the communications component to accommodate IP traffic from an IP network such as, the Internet, a corporate intranet, a home network, a person area network, and so on, through an ISP or broadband cable provider. Thus, VoIP traffic can be utilized by the UE and IP-based multimedia content can be received in either an encoded or decoded format. A video processing component (e.g. a camera) is typically provided for decoding encoded multimedia content, and the UE can also include a video component for processing video content received and, for recording and transmitting video content. A hysteresis component facilitates the analysis and processing of hysteresis data, which is utilized to determine when to associate with the access point. A software trigger component is provided to facilitate triggering of the hysteresis component when the Wi-Fi transceiver detects the beacon of the access point. A SIP client enables the UE to support SIP protocols and register the subscriber with the SIP registrar server.

When a UE is in WLAN/Wi-Fi coverage and supports VoWi-Fi, the UE can obtain an IP address from a PDN (i.e. a PDN address) and can use this PDN address to make VoWi-Fi calls.

Naturally, to use VoWi-Fi, a device must support the VoWi-Fi functionality as illustrated in Figure 5.15. VoWi-Fi is now typically implemented with native support embedded in the smartphones and working seamlessly through any Wi-Fi connection. VoWi-FI can leverage existing SIM-based security and authentication as is done in VoLTE: instead of relying on users' passwords, or on other weak cryptographic elements, VoWi-Fi uses existing SIM security mechanisms to authenticate and protect communications for subscribers. When using the SIM, the same security standards can be applied seamlessly across all connectivity technologies. VoWi-Fi is an inexpensive way for MNOs to augment services inside buildings, in place of deploying more complex DAS-based solutions; however, DAS-solutions are more robust and intrinsically provide QoS. When using the Wi-Fi service extension, the operator often (but not always) relies on enterprise-supplied Wi-Fi hotspots. From another perspective, an enterprise may determine that their service is poor is some parts of their building and they would therefore consider installing a privately owned DAS. The use of VoWi-Fi leverages an investment that, in most cases, has already been made by the enterprise. The service is universal, namely it is not specific to any mobile operator.

Naturally, a connection to the Internet is needed; furthermore, to use a VoWi-Fi-enabled device, the subscriber must register onto the network and establish the default data bearer to exchange data with the network or applications. The UE attaches to Wi-Fi AP and obtain IP connectivity, selecting the ePDG based on static IP configuration or DNS. The UE and the ePDG will perform mutual authentication during IPsec tunnel establishment using public key certificates, while the UE sends encapsulated EAP-AKA messages over IKEv2 to the ePDG including user identity (Network Access Identifier [NAI]) and potential Access Point Name (APN) (for IMS). The AAA (Authentication, Authorization, and Accounting) fetches Authentication and Key Agreement (AKA) vectors generated by the Home Subscriber Server (HSS) (if not available). AAA extract the International Mobile Subscriber Identity (IMSI) for the user. If successful, the AAA initiates the subscriber profile retrieval with the HSS to check if the user is authorized for the untrusted access and sends server registration. If successful, the AAA sends the final authentication answer including IMSI and Master Session Key. As is the case in VoLTE, once the bearer has been set, normal SIP/IMS procedures follow to enable making calls [26]. If the UE is moving out of the WLAN/Wi-Fi coverage and cannot discover another WLAN/Wi-Fi signal, the UE will try to establish a voice call over the LTE network (i.e. a VoLTE call). However, because no handover from WLAN/Wi-Fi to LTE network is available, the VoWi-Fi call is disconnected. Proposals have been made for facilitating VoWLAN call hand-over in either an inter-WLAN call handover scenario or a WLAN-LTE call handover scenario (e.g. [27, 28] among others).

The National Emergency Number Association (NENA) had advocated a few years back the deployment of femtocells and Universal Mobile Access (UMA), as one way to enhance cellular access in confined areas [29]; however, these specific constructs have seen limited field deploy-ment [30]. A UMA BS is a stand-alone unit that acts as a Wi-Fi "hot-spot" and is typically deployed in a building such as a home or small business. A femtocell (briefly discussed in Chapter 3) is a cell that operates in the geographic area of a carrier's licensed footprint. Femtocells communicate with UEs using the carrier's licensed spectrum of the carrier's net-work and interconnect to the carrier's network using a broadband connection. A femtocell may accept new call originations, process hand-offs between the femtocell and to the macro network, and process hand-offs between the macro network and the femtocell. A femtocell typically extends cellular communications service by providing BS capability in a small unit located at the small business premises (or customer's home), providing enhanced mobile coverage. In addition, the femtocell may provide an interface for landline wired phones, facilitate E9-1-1 call completion, and may provide service for Non-Service Initialized (NSI) phones, and allow both registered users of the associated service and/or registered users of the femtocell access to the wireless network.

UMAs and femtocells connect to the carrier's mobile network through a customer-supplied broadband connection. Because a femtocell uses a carrier's licensed frequencies, it allows the customer to use existing handsets to take advantage of the service. To use a UMA BS, a cus-tomer needs a handset that can operate both on a cellular network and over Wi-Fi, providing voice calls in the VoIP mode as well as over traditional cellular; however, UMA handsets may often also be able to operate outside the customer's premise, by accessing Wi-Fi "hotspots" in other locations. The UMA/femtocell initiatives have proven to be too expensive and lacking widespread support among phone manufacturers (also, the NMOs have sought to protect their cellular revenue, including revenue from roaming charges, thus deemphasizing OTT VoWi-Fi in the recent past.) Nonetheless, VoWi-Fi is now seeing more extensive deployment in homes, businesses, and campus venues.

5.3 5G TECHNOLOGIES

As noted elsewhere, 5G cellular networks are being deployed around the world. The 5G system[6,7] expands the 4G environment by adding new wireless capabilities but doing so in such a manner that LTE and New Radio (NR) can evolve in complementary ways. 5G NR is a 3GPP-defined new Radio Access Technology (RAT) for 5G mobile networks, effectively a global standard for the air interface. The possibility exists for using 5G services for voice, video, Internet access data, and IoT applications for wide-area, campus area, venue area, and building/local area environ-ments. Some applicable 5G capabilities are discussed in this section.

As it might be envisioned, a 5G system entails devices connected to a 5G access network, which in turn is connected to a 5G Core (5GC) network. The 5G access network may include 3GPP radio BSs and/or a non-3GPP access network. The 5GC core network offers major improvements compared with a 4G system in the area of network slicing and Service-Based Architectures (SBAs); in particular, the core is designed to support cloud implementation and

---

[6]Also called a "Beyond 4G Network" or a "Post LTE System" in the recent past.
[7]Portions of this section are based on reference [31].

the IoT. 5G systems subsume important 4G system concepts such as the energy-saving capabilities of Narrowband IoT (NB-IoT) radios, secure low latency small-data transmission for low-power devices. Network slicing allows service providers to deliver "Network as a Service (NaaS)" to large/institutional users affording them the flexibility to manage their own services and devices on the 5G provider's network. A goal of 5G networks is to be five times as fast as compared to the highest current speed of existing 4G networks, with download speeds as high as 5 Gbps – 4G offering only up to a maximum of 1 Gbps. Deployment of 5G networks started in 2018 in some advanced countries and is continuing at this time; naturally, the current 4G/LTE and 5G are expected to coexist for many years. 5G systems are expected to support:

- A large set of applications, initially (very likely) with focus on broadband applications for human use (e.g. high-quality OTT video over smartphones) and (distributed) IoT applications.
- A range of data rates, up to multiple Gbps, and tens of Mbps to facilitate existing and evolving applications, particularly broadband applications and IoT applications.
- Tight latency, availability, and reliability requirements to facilitate applications related to video delivery, surveillance and physical security, automotive locomotion, and mission-critical control, among others, particularly in an IoT context.
- Support for clustered users with very high data rate requirements as well as for a large number of distributed devices with low complexity and limited power resources, particularly in an IoT context.
- Network slicing.

5.3.1 Emerging Services

The ITU-R has assessed usage scenarios in three classes: Ultra-Reliable and Low-Latency Communications (URLLC), massive Machine-Type Communications (mMTC), and enhanced mobile broadband (eMBB). eMBB is probably the earliest class of services being broadly supported and implemented. Key performance indicators are identified for each of these classes, such as spectrum efficiency, area traffic capacity, connection density, user-experienced data rate, peak data rate, and latency, among others. The ability to efficiently handle device mobility is also critical. Some examples of eMBB use cases include Non-SIM devices, smart phones, home/enterprise/venues applications, UHD (4K and 8K) broadcast, and virtual reality/augmented reality. mMTC use cases include smart buildings, logistics, tracking and fleet management, and smart meters. URLLC cases include traffic safety and control, remote surgery, and industrial control. In some parallel work to the ITU-R, 3GPP/TR 22.891 has defined and/or described the following service groups: eMBB, Critical Communication, mMTC, Network Operations, and Enhancement of Vehicle-to-Everything (V2X). Next-Generation Mobile Networks (NGMN) Alliance has defined and/or described the following service groups: Broadband access in dense area, Indoor ultrahigh broadband access, Broadband access in a crowd, 50+ Mbps everywhere, Ultra low-cost broadband access for low ARPU areas, Mobile broadband in vehicles, Airplanes connectivity, Massive low-cost low long-range/low-power MTC, Broadband MTC, Ultralow latency, Resilience and traffic surge, Ultrahigh reliability and Ultralow latency, Ultrahigh availability and reliability, and Broadcast-like services.

Figure 5.16 depicts some of the key 5G services currently contemplated. Although some have associated smart cities with mMTC, likely the early applications will be more within the eMBB domain; also, one would expect eMBB to be deployed more broadly, driven by the commercial "appeal" of the video services it facilitates. All of these services are applicable to campus applications; additionally, IP-based video surveillance in campuses, venues, and smart cities that may be supported by IoT can operate rather well at the 0.384–2.5 Mbps bandwidth range. Figure 5.17

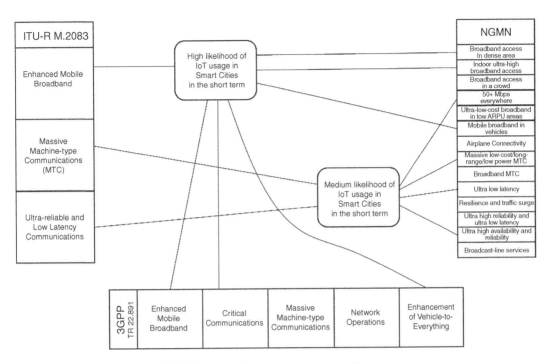

FIGURE 5.16   5G services under development.

highlights some technical features of 5G services in terms of data rates, latency, reliability, device density, and so on. 5G IoT overcomes the well-known limitation of unlicensed LPWAN technologies that utilize crowded license-free frequency bands, especially in large cities; therefore, 5G IoT is ideal for smart city for mission-critical and QoS-aware applications (for example, traffic management, smart grid, utility control.)

5.3.2   New Access and Core Elements

As alluded to above, to support 5G rollouts, 3GPP has specified new 5G RAT and new 5GC networks. Implementable standards for 5G have been incorporated in 3GPP Release 15 onward. 3GPP Rel 15 defines New 5G Radio and Packet Core technology. Specifically, 3GPP has defined a new 5GC network and a new RAT called 5G NR. The new 5GC architecture has several new capabilities built inherently into it as native capabilities: multi-Gbps support, ultralow latency, Network Slicing, Control and User Plane Separation (CUPS), and virtualization. To deploy the 5GC, new infrastructure will be needed. There is a firm goal to support for "forward compatibility." The 5G NR modulation technique and frame structure are designed to be compatible with LTE. As might be expected, however, it is possible to integrate into 5G elements of different generations and different access technologies – two modes are allowed: the SA (Stand-Alone) configuration and the NSA (Non-Stand-Alone) configuration (see Figure 5.18).

- 5G Stand-Alone (SA) Solution: in 5G SA an all-new 5G packet core is introduced. SA scenarios utilize only one RAT (5G NR or the evolved LTE radio cells); the core networks are operated independently.
- 5G Non-Stand-Alone Solution (NSA): in 5G, NSA Operators can leverage their existing EPC/LTE packet core to anchor the 5G NR using 3GPP Release 12 Dual Connectivity feature. This will enable operators to launch 5G more quickly and at a lower cost. This solution might suffice for some initial use cases. However, 5G NSA has a number of limitations,

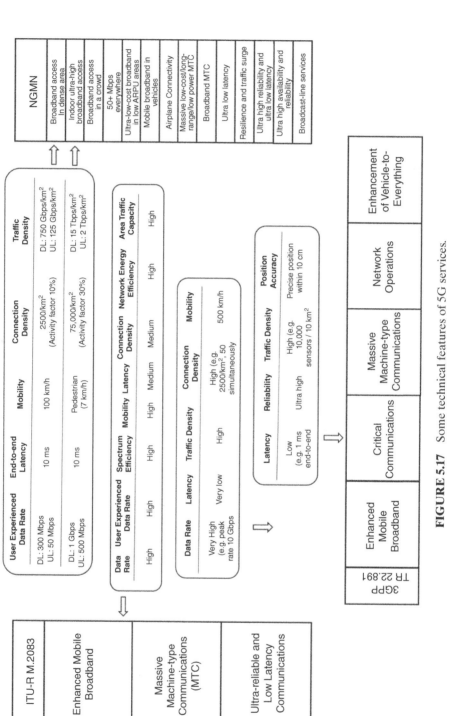

FIGURE 5.17 Some technical features of 5G services.

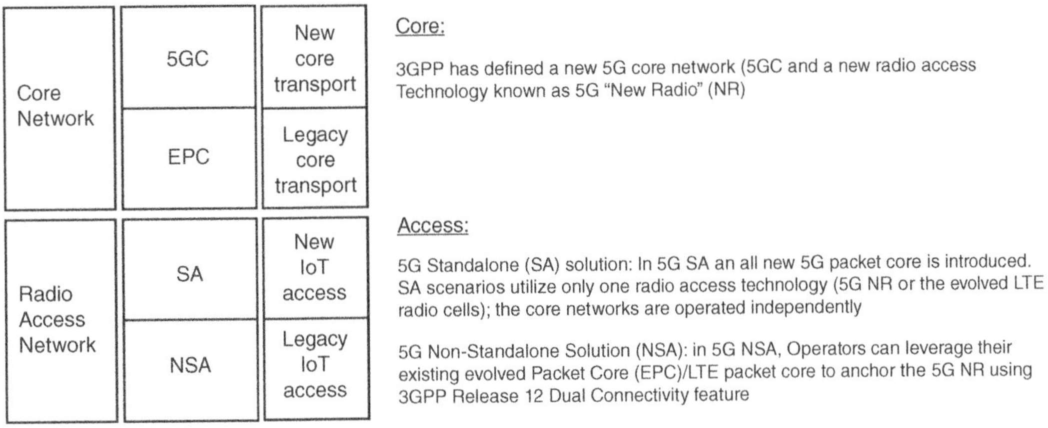

Core Network	5GC	New core transport
	EPC	Legacy core transport
Radio Access Network	SA	New IoT access
	NSA	Legacy IoT access

<u>Core:</u>

3GPP has defined a new 5G core network (5GC and a new radio access Technology known as 5G "New Radio" (NR)

<u>Access:</u>

5G Standalone (SA) solution: In 5G SA an all new 5G packet core is introduced. SA scenarios utilize only one radio access technology (5G NR or the evolved LTE radio cells); the core networks are operated independently

5G Non-Standalone Solution (NSA): in 5G NSA, Operators can leverage their existing evolved Packet Core (EPC)/LTE packet core to anchor the 5G NR using 3GPP Release 12 Dual Connectivity feature

FIGURE 5.18   5G transition options and IoT support.

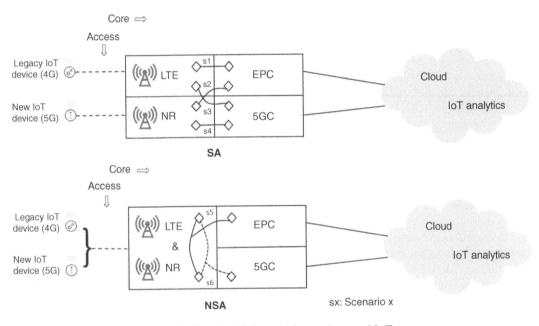

FIGURE 5.19   Detailed 5G transition options and IoT support.

thus these Operators will eventually be expected to migrate to 5G Stand-alone solution. NSA scenario combines NR radio cells and LTE radio cells using dual-connectivity to provide radio access and the core network may be either EPC or 5GC.

Multiple evolution/deployment paths may be employed by service providers (service providers of various services, including IoT services) to reach the final target configuration; this migration could well take a decade, and may also have different timetables in various parts of a country, e.g. top urban areas, top suburban areas, secondary urban areas, secondary suburban areas, exurbian areas, and rural areas. Figure 5.19 depicts the well-known migration paths.

The 5G NR duplex frequency configuration will allow 5G NR, NB-IoT, and LTE-M subcarrier grids to be aligned; this will enable the 5G NR UE to coexist with NB-IoT and LTE-M signals. The IoT implementer will need to be keenly aware of what 5G (5G IoT) services are available in a given area as an IoT implementation is contemplated. In Figure 5.19, Scenario 1 illustrates that the IoT Service provider will continue to use LTE and EPC to provide services (e.g. NB-IoT); here only legacy IoT devices can be supported. The provider only has a stand-alone radio technology, in this case LTE only. Scenario 2 illustrates an IoT Service provider has migrated completely to NR (again only providing a stand-alone radio technology), but will retain the existing core network, the EPC. (Only) new 5G IoT devices can be used. In scenarios 5 and 6, the service providers will support both the legacy LTE and the new NR (clearly in this non-stand-alone arrangement, both radio technologies are deployed.) Some of these providers retain the legacy core and some will deploy the new 5GC core. Both legacy and 5G IoT devices can be supported.

3GPP approved the 5G NSA standard at the end of 2017 and the 5G SA standard in early 2018 in the context of its Release 15. Release 15 also included the support eMBB, URLLC, and mMTC in a single network to facilitate the deployment of IoT services; Release 15 also supports 28 GHz millimeter-wave (mmWave) spectrum and multi-antenna technologies for access.

5.3.3   New 5GC Architecture

A new core network structure, the 5GC, has been defined by the 3GPP in order to evolve from the 4G LTE system to the 5G system. The 5GC for the 5G system supports new, differentiated functions, as discussed, for example in [32] (on which this subsection is based).

In particular, a network slice function is introduced. As requirements of the 5G system, the 5GC should support various terminal types and service types, for example, eMBB, URLLC, and mMTC. Such terminals and services have different requirements for the core network. For example, the eMBB service requires a high data rate, and the URLLC service requires high stability and a low latency. A network slice scheme is a technique proposed to satisfy these various service requirements. The network slice scheme is to virtualize a single physical network and thereby create multiple logical networks; the resulting Network Slice Instances (NSIs) may have different characteristics. This is made possible when each NSI has a Network Function (NF) adapted to the characteristics thereof. It is possible to effectively support various 5G services by allocating the NSIs suitable for the characteristics of a service requested by each terminal. This topic is revisited in Section 10.8.

It is possible to support the network virtualization paradigm through separation of mobility management and session management functions. In the 4G LTE, all terminals can perform services in the network through an exchange of signaling with a single core device called a Mobility Management Entity (MME) that performs all functions of registration, authentication, mobility management, and session management (these concepts were briefly discussed in Chapter 3). However, in the 5G system, given that the number of terminals is expected to increase significantly and also given that the mobility and traffic/session characteristics to be supported are varied depending on terminal types, supporting all functions at a single device such as the MME may lower scalability and impact the ability of adding an entity according to a required function. Therefore, in order to improve the scalability, various functions are being developed based on a structure for separating the mobility management and session management functions in terms of function/implementation complexity and signaling load of the core device responsible for a control plane.

Figure 5.20 depicts the emerging network architecture for the 5G system. As it can be seen, an Access and Mobility Management Function (AMF) for managing the mobility and network registration of a terminal (i.e. UE) and a Session Management Function (SMF) for managing an end-to-end session are separated from each other. The AMF and the SMF may exchange signaling with each other through the N11 interface.

Furthermore, a Service and Session Continuity (SSC) mode is introduced to support the requirements on various types of continuity for applications or services of the terminal, and the SSC mode may be designated and used for each PDU session. There are three SSC modes: SSC Mode 1, SSC Mode 2, and SSC Mode 3. SSC Mode 1 is a mode that does not change an anchor User Phone Function (UPF) (or a PDU session anchor [PSA]), which is a communication interface with an

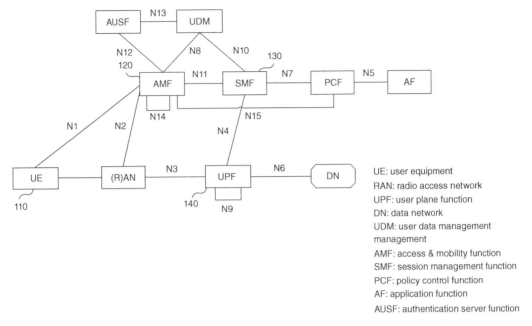

FIGURE 5.20   Network architecture and interface of a 5G cellular system [32].

external Data Network (DN), while a corresponding session is maintained, including a case where the UE moves. In this mode, because an IP address/prefix assigned to the session is not changed, the session continuity at the IP level can be achieved. SSC Modes 2 and 3 allow a change (i.e. relocation) of the above-mentioned anchor UPF. A difference between both modes is that SSC Mode 2 releases a connection with a current anchor UPF and immediately establishes a connection with a new anchor UPF, whereas SSC Mode 3 is able to maintain a connection with a current anchor UPF while establishing a connection with a new anchor UPF. Thus, the SSC Mode 3 session allows simultaneous data transmission through a plurality of anchor UPFs with respect to the same external DN (i.e. make-before-break scheme). However, the SSC Mode 2 session based on a break-before-make scheme may cause a service interruption when changing the anchor UPF at the time of transmitting terminal traffic, even though causing small overhead for inter-entity signaling and tunnel management in the core network.

Additionally, a Policy Control Function (PCF), which is a server for managing an operator policy for a terminal, can store policies for session request and selection with respect to respective terminals and provide the policy to each terminal in order for an operator to route terminal traffic. This policy is called a UE Route Selection Policy (URSP). In particular, the URSP may include a Network Slice Selection Policy (NSSP) for supporting network slicing technique, an SSC Mode Selection Policy (SSCMSP) for supporting the SSC mode, and a Data Network Name (DNN) selection policy for selecting a DNN corresponding to an APN used in the EPC. The URSP may be managed in conjunction with a traffic filter to indicate a rule for specific traffic. In order to deliver UE-specific URSP to a terminal, the PCF may first send it to the AMF via a standard interface (e.g. N15), and then the AMF may deliver it to the terminal via a standard interface (e.g. N1) by means of Non-Access Stratum (NAS) signaling.

5.3.4   Frequency Spectrum and Propagation Challenges

As discussed in Chapter 3, there are a number of spectrum bands that can be used in 5G; these bands can be grouped into three macro categories: sub-1 GHz, 1–6 GHz, and above 6 GHz. The more advanced features, especially higher data rates require the use of the millimeter wave spectrum. New mobile generations are typically assigned new frequency bands and wider spectral bandwidth per frequency channel (1G up to 30 kHz, 2G up to 200 kHz, 3G up to 5 MHz, and 4G up to 20 MHz). Up to now, cellular networks have used frequencies below 6 GHz. Generally, without advanced MIMO antenna technologies one can obtain about 10 bits-per-Hertz-of-channel bandwidth. But the integration of new radio concepts such as Massive MIMO, Ultra-Dense Networks, Device-to-Device, and mMTC will allow 5G to support the expected increase in the data volume in mobile environments and facilitate new IoT applications.

As discussed in Chapter 3, the millimeter wave spectrum, also known as Extremely High Frequency (EHF), or more colloquially mmWave, is the band of electromagnetic spectrum running between 30 and 300 GHz. Bands within this spectrum are being considered by the ITU and the Federal Communications Commission in the United States as a mechanism to facilitate 5G by supporting higher bandwidth. The use of a 3.5 GHz frequency to support 5G networks is also gaining some popularity, but the higher speeds networks will use other frequency bands, including millimeter-wave frequencies (these bands ranging from 28 to 73 GHz, specifically the bands 28, 37, 39, 60, and 72–73 GHz bands). In the United States, recently the FCC approved spectrum for 5G, including millimeter-wave frequencies in the 28, 37, and 39 GHz bands, although these targeted cellular frequencies may nominally overlap with other pre-existing users of the spectrum, for example point-to-point microwave paths, Direct Broadcast satellite TV, and High Throughput Satellite (HTS) systems (Ka-band transmissions). Initially, 5G will, in many cases, use the 28 GHz band, but higher bands will very likely be utilized later on; initial implementations, will support a maximum speed of 1 Gbps. Lower frequencies (at the so-called C band) are less subject to weather impairments, can travel longer distances, and penetrate building walls more easily. Waves at higher frequencies (Ku, Ka, and E/V bands) do not naturally travel as far or penetrate walls or objects as easily. However, a lot more CBW is available in millimeter-wave bands. Furthermore, developers see the need for "an innovative utilization of spectrum"; "small cell" approaches are required to address the scarcity of the spectrum, but at the same time covering the geography. V band spectrum covers 57–71 GHz, which in many countries is an "unlicensed" band, and E band spectrum covers 71–76, 81–86, and 92–95 GHz.

Due to RF propagation phenomena that are more pronounced at the higher frequencies, such as multipath propagation due to outdoor and indoor obstacles, free space path loss, atmospheric attenuation due to rain, fog, and air composition (e.g. oxygen), small cells will almost invariably be needed in 5G environments, especially in dense urban environments. Additionally, Line-of-Sight (LOS) will typically be required. ITU-R P series of recommendations has useful information on radio wave propagation, including ITU-R P.838-3, 2005; ITU-R P.840-3, 2013; ITU-R P.676-10, 2013; and ITU-R P.525-2, 1994. The following figures highlight the issues at the higher frequencies, including the millimeter-wave frequencies. Figure 5.21 depicts the path loss as a function of distance and frequency. Figure 5.22 shows the attenuation a function of precipitation and frequency. Figure 5.23 illustrates the attenuation a function of fog density and frequency. Figure 5.24 depicts the attenuation a function of atmospheric gasses and frequency (notice high attenuation around 60 GHz).

5.3.5 Resource Management

In an OFDM-based communication system, a resource element can be defined by a subcarrier during on OFDM symbol duration. In the time domain, a Transmission Time Interval (TTI) subframe can be defined which is composed of multiple OFDM symbols; in the frequency domain, a Resource Block (RB) can be defined that is composed of multiple OFDM subcarriers. The resources can be divided into TTIs in time domain and RBs in frequency domain. Typically, an RB can be a baseline resource unit for scheduling in the frequency domain, and a TTI can be a baseline resource unit for scheduling the time domain (however, depending on different service features and system requirements, there can be other options). See

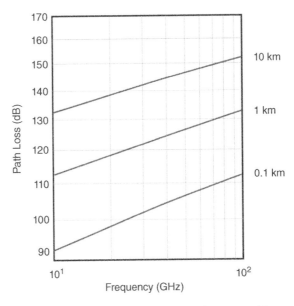

FIGURE 5.21   Path loss as a function of distance and frequency.

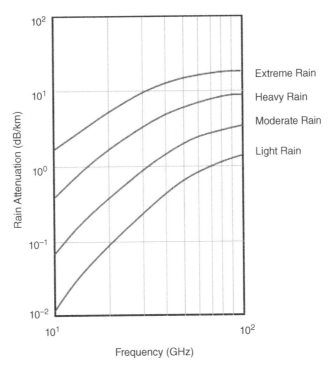

FIGURE 5.22   Attenuation as a function of precipitation and frequency.

Figure 5.25 [33]. To support multiplexing of different services, a BS of the next-generation radio network (gNodeB (NB)) or a NR can preconfigure semi-statically some resources for different services: based on the performance requirement and traffic feature of a certain service, the BS (or gNB) decides how to preconfigure the resources in an efficient and flexible manner. The resource configurations can be signaled in the system information of a cell.

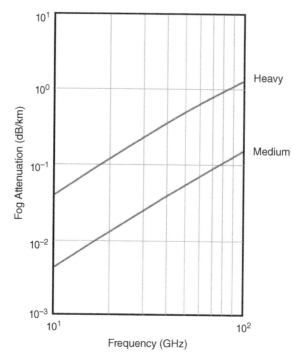

FIGURE 5.23   Attenuation as a function of fog density and frequency.

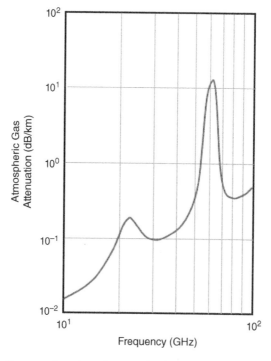

FIGURE 5.24   Attenuation as a function of atmospheric gasses and frequency (notice high attenuation around 60 GHz).

Traditionally, in 4G/LTE, the spectral resources assigned for downlink and uplink data transmission are a number of Physical Resource Blocks (PRBs) pairs as a baseline, that occupies one subframe in the time domain and several contiguous or noncontiguous PRBs in the frequency

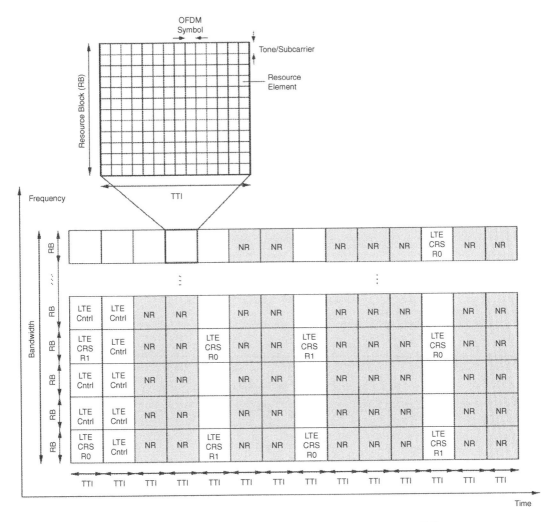

FIGURE 5.25   OFDM resources and LTE legacy support [33].

domain. When the NR network is expected to coexist with other networks, such as LTE, the reserved resource for LTE can be static where partial symbols in a TTI can be reserved in a periodic manner. For example, if a TTI has 14 symbols which is same as LTE, some symbols in a TTI can be reserved for LTE control region and Cell-Specific Reference Signal (CRS) symbols, since these channels/signals are always transmitted in LTE. As shown in the example of Figure 5.25, during one TTI, 5 symbols can be reserved for LTE, including the first 2 symbols for LTE control region and 3 other symbols for CRS symbols; the other nine symbols in one TTI can be used for NR.

Synchronization is needed by UEs to undertake cell search and obtain system information to establish connections. As is the case in LTE, two types of synchronization signals are defined for NR: The Primary Synchronization Signal (PSS) and the Secondary Synchronization Signal (SSS). The Synchronization Signal/PBCH block (also known as SSB) consists of PSS, SSS, and the Physical Broadcast Channel (PBCH): the combination of synchronization signals and the PBCH comprises the SS/PBCH block.

The UE uses the synchronization signals and PBCH to derive the necessary information required to access the cell. For a stand-alone NR system deployed independently from LTE, SSB is essential for a device to carry out cell search and acquire system information to establish an initial connection, initiate potential handover, or perform beam management [34]. The UE

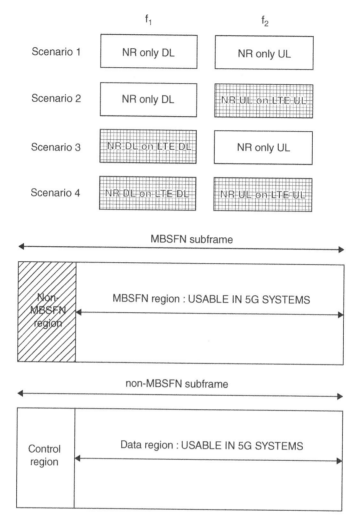

FIGURE 5.26 Scenarios of providing 5G services sorted according to usage bands [35].

needs to first decode PBCH/MIB (Master Information Block) in order for it to receive other system information transmitted on the Physical Downlink Shared Channel (PDSCH) that carries user-specific data. PBCH payload size including 24-bit CRC is 56-bits; among other fields it includes the 10-bit System Frame Number (SFN).

A 5G communication system, however, may be deployed to operate in the same frequency band as a 4G communication and, in this case, it may be necessary for the 5G communication system to keep certain resources vacant for Sounding Reference Signal (SRS) transmission in the 4G communication system.

Figure 5.26 depicts a number of scenarios [35]. Scenario 1 depicts the case where 5G downlink and uplink carriers are located in frequency bands separated from legacy LTE frequency bands. Scenario 2 illustrates the case where the 5G downlink carrier is located in a frequency band separated from the legacy LTE frequency band while the 5G uplink carrier is located in a frequency band overlapped with a legacy LTE frequency band. Scenario 3 shows the case where the 5G downlink carrier is located in a frequency band overlapped with a legacy LTE frequency band while the 5G uplink carrier is located in a frequency band separated from legacy LTE frequency bands. Finally, Scenario 4 illustrates the case where 5G downlink and uplink carriers are located in frequency bands overlapped with legacy LTE frequency bands. It has been agreed in the industry to provide 5G communication services with Multicast Broadcast Single Frequency

Network (MBSFN) subframes of LTE/4G in the case where the 5G downlink carriers coexist with the LTE downlink carriers as shown in scenarios 3 and 4. In the LTE standard, each network can designate MBSFN subframes among ten subframes constituting a radio frame according to the duplexing mode (FDD or TDD). The bottom part of Figure 5.26 shows a MBSFN subframe of LTE; the subframe includes a non-MBSFN region occupying one or two OFDM symbols for transmitting physical control channels such as a Physical Downlink Control Channel (PDCCH) and a Physical Control Format Indicator Channel (PCFICH) – the length of the non-MB SFN region is notified to an LTE terminal by means of a Control Format Indicator (CFI) being transmitted through the PCFICH in the corresponding subframe; the remaining OFDM symbols (except for the OFDM symbols allocated for the non-MBSFN region) is referred to as MBSFN region with which the 5G communication services are provided. In this manner, the 5G communication services are provided in the frequency band having the LTE downlink carriers. The last portion of Figure 5.26 illustrates a resource in a non-MBSFN subframe for providing a 5G communication service; unlike the MBSFN subframe, the non-MBSFN subframe consists of a control region carrying physical control channel and a data region carrying data, which can be used for providing the 5G communication services.

5.3.6   Requirements for Small Cells

High bandwidth will typically require a wide spectrum. Millimeter wave frequencies (signals with wavelength ranging from 1 to 10 mm) support a wide usable spectrum. The millimeter wave spectrum includes licensed, lightly licensed, and unlicensed portions. Bandwidth demand and goals are the main driver for the need to use the millimeter wave spectrum, particularly for eMBB-based applications, allowing users to receive 100 Mbps as a bare minimum and 20 Gbps as a theoretical maximum. The use of millimeter wave frequencies, however, will imply the use of a much smaller tessellation of cells and supportive towers or rooftop transmitters due, as noted, to transmission characteristics, such as high attenuation and directionality. This is an important design consideration for 5G, especially in dense city/urban environments. The aggregation of these towers will by itself require a significant backbone network, whether a mesh based on some point-to-point microwave links, an optical fiber network, or a set of "wireless fiber" links. Millimeter wave systems utilize smaller antennas compared to systems operating at lower frequencies: the higher frequencies, in conjunction with MIMO techniques, can achieve sensible antenna size and cost. The millimeter wave technology can be utilized both for indoors and outdoors high-capacity fixed or mobile communication applications. The term "densification" is also used to describe the massive deployment of small cells in the near future.

mmWave products used for backhauling typically operate at 60 GHz (V Band) and 70/80 GHz (E Band) and offer solutions in both Point to Point (PTP) and Point to Multipoint (PtMP) configurations providing end-to-end multi-gigabit wireless networks, for example, 1 Gbps up to 10 Gbps symmetric performance. Very small directional antennas, typically less than a half-square foot in area, are used to transmit and/or receive signals which are highly focused beams. These stationary radio systems are often installed on rooftops or towers. mmWave products are now appearing on the market targeting high-capacity smart city applications, 5G Fixed, Gigabit Wireless Access solutions, and Business Broadband. Urban canyons, however, may limit the utility of this technology to very short LOS paths. Mobile applications of mmWave technology are more challenging. On the other hand, one advantage of this technology is that short transmission paths (high propagation losses) and high directionality allow for spectrum reuse by limiting the amount of interference between transmitters and/or adjacent cells. Near LOS (NLOS) applications may be possible in some cases (especially for short distances).

mmWave transmission performance will drive the requirement for small cells. Small(er) cells have been used for years to increase area spectral efficiency – the reduced number of users per cell provides more usable spectrum to each user. However, the smaller cells in 5G are also dictated by the propagation characteristics. In the *5G context*, UMi typically have radii of 5–120 m

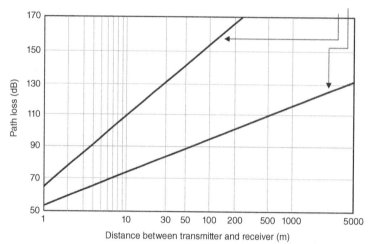

Path Loss results as obtained by
5GCM, 3GPP, METIS simulations
under various conditions at 28 GHz
fall between these two boundary lines

Distance between transmitter and receiver (m)

FIGURE 5.27 Path loss simulations for 5G by various entities.

for LOS and 20–270 m in NLOS; UMa typically have radii of 60–1000 m for LOS and 50–1500 m for NLOS [36]. Given their size, 5G/mmWave UMi cells will be able to support high bandwidth enabling eMBB services over small areas of high traffic demand. At the mmWave operation, user-device proximity with the antenna will enable higher signal quality, lower latency, and by definition, high data rates and throughput. Also, to be noted, mmWave frequencies make the size of multielement antenna arrays practical, enabling large MU-MIMO solutions.

Diffraction Loss (DL) and Frequency Drop (FD) are just two of the path quality issues to be addressed. Although greater gain antennas will likely be used to overcome path loss, diffuse scattering from various surfaces may introduce large signal variations over travel distances of just a few centimeters, with fade depths of up to 20 dB as a receiver is moved by a few centimeters. These large variations of the channel must be taken into consideration for reliable design of channel performance, including beamforming/tracking algorithms, link adaptation schemes, and state feedback algorithms. Furthermore, multipath interference from coincident signals can give rise to critical small-scale variations in the channel frequency response. In particular, wave reflection from rough surfaces will cause high depolarization. For the LOS environment, Rician fading of multipath components, exponential decaying trends and quick decorrelation in the range of 2.5 wavelengths have been demonstrated. Furthermore, received power of wideband mmWave signals has a stationary value for slight receiver movements but average power can change by 25 dB as the mobile transitions around a building corner from NLOS to LOS in an UMi setting. Additionally, human body blockage causes more than 40 dB of fading at the mmWave frequencies. Figure 5.27 depicts the path loss according to various simulations for 5G by various stakeholder entities.

The main parameter of the radio propagation model is the Path Loss Exponent (PLE), which is an attenuation exponent for the received signal. PLE has a significant impact on the quality of the transmission links. In the far-field region of the transmitter, if $PL(d_0)$ is the path loss measured in dB at a distance d_0 from the transmitter, then the loss in signal power expected when moving from distance d_0 to d $(d \rightarrow d_0)$ is [37–39]

$$PL_{d_0 \rightarrow d}\{dB\} = PL\{d_0\} + 10n\log_{10}\left(\frac{d}{d_0}\right) + \chi \quad d_f < d_0 < d$$

where

$PL(d_0)$ = Path Loss in dB at a distance d_0

$PL_{d \to d_0}$ = Path Loss in dB at an arbitrary distance d

n = PLE

χ = A zero-mean Gaussian distributed random variable with standard deviation σ. (This is utilized only when there is a shadowing effect; if there is no shadowing effect, then this random variable is taken to be zero.)

See Figure 5.28. Usually, PLE is considered to be known upfront but in most instances PLE needs to be assessed for the case at hand. It is advisable to estimate the PLE as accurately as possible for the given environment. PLE estimation is achieved by comparing the observed values over a sample of measurements to the theoretical values. Obstacles absorb signals, thus treating the PLE as a constant is not an accurate representation of the real environments, both indoors and outdoors (for example, treating PLE as a constant may cause serious positioning errors in complicated indoor environments [37]). Usually, to model real environments the shadowing effects cannot be overlooked, by taking the PLE as a constant (a straight-line slope). To capture a shadowing effect, a zero-mean Gaussian random variable with standard deviation σ is added to the equation. Here the PLE (slope) and the standard deviation of the random variable should be known precisely for better modeling.

As pragmatic working parameters, one has the following [40]:

- PLE values are in the 1.9 and 2.2 range for LOS and at the 28 and 60 GHz bands; PLE is approximately 4.5 and 4.2 range for NLOS in the 28 and 60 GHz bands.
- Rain attenuation of 2–20 dB/km can be anticipated for rain events ranging from light rain (12.5 mm/hr) to downpours (50 mm/hr) at 60 GHz (higher for tropical events). For a 200-m cells, the attenuation will be around 0.2 db for 5 mm/hr rain at 28 GHz and 0.9 dB for 25 mm/hr rain at 28 GHz. The attenuation will be around 0.5 db for 5 mm/hr rain at 60 GHz and 2 dB for 25 mm/hr rain at 60 GHz.
- Atmospheric absorption of 1–10 dB/km occurs at the mmWave frequencies. For 200-m cells, the absorption will be 0.04 dB at 28 GHz and 3.2 dB at 60 GHz.

Penetration into buildings is an issue for mmWave communication, this being a lesser concern for contemporary sub 1 GHz systems and even systems operating up to 6 GHz. O2I (Outdoor-to-Indoor) losses must be taken into account. Actual measurements (e.g. at 38 GHz) demonstrated a penetration loss of 40 dB for brick pillars, 37 dB for a glass door, and 25 dB for

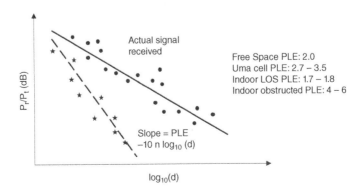

FIGURE 5.28   PLE.

a tinted glass window (indoor clear glass and drywall only had 3.6 and 6.8 dB of loss) [41]. This is why DASs are expected to be important for 5G in general and 5G IoT in particular.

Signal penetration indoors may represent a challenge, just as is the case even at present with 3G/4G LTE, even for traditional voice and internet access and data services. This has driven the need for DAS systems, especially in densely constructed downtown districts. Free space attenuation at the higher frequency, power budgets, directionality requirements, and weather, all impact 5G and 5G IoT. Outdoor small cells and building-resident DAS systems utilize high-speed fiber-optic lines or "wireless fiber" to interconnect the sites to the backbone and the Internet cloud.

5.3.7 Comparison to Wi-Fi 6

5G NR and Wi-Fi 6 are both emerging at this juncture and both aiming at providing higher bandwidth, more robust connectivity, and support for high-density environments. Wi-Fi 6 is primarily a local access technology that requires separate backhaul into the cloud; 5G offers the backhaul as part of the service. For some use cases (but certainly not for all) either solution might be applicable, but by and large, the technologies are complementary and 5G NR cannot replace Wi-Fi 6 and vice versa (especially in the case of real-time geographic mobility): Wi-Fi is targeted to and is ideal for indoor coverage, while 5G NR will be best for outdoor coverage.

In theory, if there was a strong, reliable 5G NR indoors signal in a given environment, some applications could utilize this technology in lieu of Wi-Fi, particularly since the range might be somewhat higher for the 5G NR solution; these applications might relate to IoT sensors in some open but confined environment such as a stadium or a college campus.

For indoor applications, it will be unlikely that the signal is actually from cell sites on nearby buildings, rooftops, or hills; most likely, the signals are coming from the indoor operator antennas as part of a DAS system. While with a DAS, 5G NR antenna can provide signal coverage with a range very similar to that of a typical Wi-Fi AP, building out that indoor 5G NR coverage is more expensive than building a brand-new Wi-Fi infrastructure (although the use of licensed 5G bands provide better reliability and grade of service, and the technology offers seamless hand-off even in a high speed): fiber optics, Base Band Units (BBU), Remote Radio Unit (RRU), RRU hub (RHUB), and pico RRU (pRRU) are telecommunications-grade devices that are more expensive than the Wi-Fi AP and controllers; while sharing such an investment with the service provider will help the landlord to enable 5G NR on their properties, having a Wi-Fi infrastructure will give customers full control over the network at a much lower cost [42].

5.4 IoT

The use of 5G for IoT was already highlighted in Chapter 4. Reference [32] makes note of the fact that various attempts have been made to apply 5G communication systems to IoT networks; for example, technologies such as WSN, MTC, and Machine-to-Machine (M2M) communication may be implemented by beamforming, MIMO, and array antennas. Application of a cloud RAN with Big Data processing technology may also be considered to be as an example of convergence between the 5G technology and the IoT technology [32].

As discussed elsewhere, IoT sensors that are found in large geographic areas such as in smart city environments can use Low Power Wide Area Network (LPWAN) technologies such as LoRa; Sigfox; or cellular services, especially 5G-focused solutions. As noted, one of the design goals of 5G is to support the IoT environment in a reliable, efficient, and cost-effective manner. Open-air campus/venue applications (e.g. stadiums) can use the same technologies just listed for smart cities, or can make use of local systems such as Wi-Fi 6, Bluetooth Low Power (BLE), or ZigBee, perhaps with more localized clusters interconnected via some kind of shared or dedicated intranet facility. Smart Building applications or enclosed venues (such as airports, train stations, hospitals) can in principle use all of the just-cited technologies; however, local-oriented solutions may be preferred (at least in the short term); 5G cellular IoT solutions have the building-penetration challenges, and would, perhaps, have to depend on DAS-based technologies.

Although Wi-Fi-enabled sensors are still generally somewhat more expensive than other sensor solutions, the overall simplifications in ecosystem management and operations may make such solution relatively attractive in a number of situations, especially when the sensors are fed with line power or Power Over Ethernet (PoE). Sensors such as security cameras, function-control sensors (such as when attached to escalators, baggage carousels, weather monitoring stations, and so on), and sensors in devices such as refrigerators, ovens, HVAC (Heating, Ventilation, and Air Conditioning) typically have available line power (but converted to DC). A partial compromise to the use of end-point Wi-Fi-ready sensors is to use BLE or ZigBee sensors but utilize Wi-Fi-enabled gateway nodes to integrate the sensors into the Wi-Fi-based intranet.

5.5   5G DAS SOLUTIONS

As discussed above, transmission issues will require small-cell designs for 5G, especially when using the mmWave frequencies. These "small cells/small BSs" systems could be deployed by a service operator that is interested in providing services to a given high-rise office building or venue (e.g. stadiums, airports, convention centers) via some deployment/billing/revenue-share arrangement with the venue landlord, or be deployed by the landlord, or specialized BLEC. As of press time, just a handful of 5G-ready DAS systems were available, but more systems will likely emerge in the near future.

5.6   INTEGRATED SOLUTIONS

This chapter has discussed several solutions that are often deployed as stand-alone systems. In some cases, the use of multiple solutions from multiple providers is useful, practical, and even inevitable. Such an approach, however, requires the separate administration, engineering, provisioning, operations, managing, sparing, vendor-stewardship, and billing of the various systems and service providers.

The movement toward an integrated solution that is consistent both at the technology level as well at the administration level (including engineering, provisioning, operations, managing, sparing, vendor-stewardship, and billing) is both achievable and desirable. Furthermore, we discussed in Chapter 1 a more organic, all-inclusive evolution to what we called a "super-integrated" network, which is further explored in Chapter 10.

While 5G solutions may bring a degree of conversion at the wide-area level, it is likely that *bona fide* LAN technologies (such as 802.11ax) will continue to serve the needs of institutions at the local level for the foreseeable future.

A major advancement toward integrated solutions took place when a large number of siloed technologies, including voice, entertainment video, OTT-video, OTT-voice, videoconferencing, video security, situational awareness, and all web-based activity has migrated to TCP/IP.

The question to ponder now is if all Layer 2 technology will migrate to Ethernet, both at the local level, for wired and wireless access, as well as in the metro and regional level (e.g. see [43]). Can all services, such as voice, entertainment video, OTT-video, OTT-voice, videoconferencing, video security, situational awareness, and ubiquitous IoT-sensing (or a large majority thereof) be delivered using Wi-Fi?

Integration can occur at the technology level as well at the administration and service provisioning/service monitoring level, where one provider bundles all requite systems by doing the

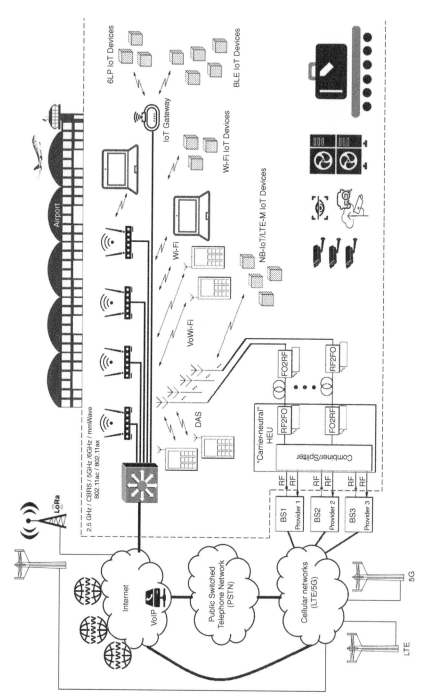

FIGURE 5.29 A state-of-the-art integrated system.

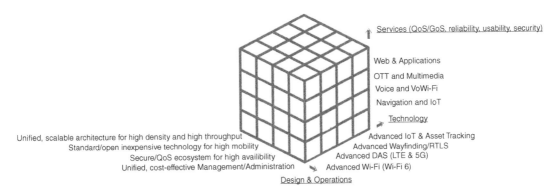

Services (QoS/GoS, reliability, usability, security)

Web & Applications
OTT and Multimedia
Voice and VoWi-Fi
Navigation and IoT

Technology

Advanced IoT & Asset Tracking
Advanced Wayfinding/RTLS
Advanced DAS (LTE & 5G)
Advanced Wi-Fi (Wi-Fi 6)

Unified, scalable architecture for high density and high throughput
Standard/open inexpensive technology for high mobility
Secure/QoS ecosystem for high availibility
Unified, cost-effective Management/Administration

Design & Operations

FIGURE 5.30   Conceptual view of an advanced high-density high-impact network.

design, the feature selection, the deployment, the grade-of-service monitoring, the intrinsic multi-vendor management, the technology tracking and refreshment, the overall security tracking, and the cost and billing support. For public venues such as airports, stadiums, convention centers, and so on, benefit from installing such integrated solutions, for example as depicted in Figure 5.29. Chapters 9 and 10 of this text describe one such deployment undertaken by these authors.

An advanced network in this context typically has a multidimensional formulation in terms of services, technologies, and design and operations, (e.g. see Figure 5.30).

- Services that are QoS-rich, reliable, usable, and secure span at least these classes: (i) Web and apps access; (ii) OTT and multimedia; (iii) voice and VoWi-Fi; and (iv) navigation and broad IoT support.
- Technologies spans at least the following domains: (i) advanced Wi-Fi services; (ii) advanced DAS (including LTE, private LTE, 5G, and 5G slicing); (iii) advanced wayfinding/RTLS; and, (iv) advanced IoT sensing and asset tracking.
- Design and Operations encompass at least these features: (i) unified, scalable architecture for high density and high throughput; (ii) standard/open inexpensive technology for high mobility; (iii) secure and QoS-enabled ecosystem for high service availability; and (iv) unified, cost-effective management/administration.

In conclusion, the reader should note that a next-generation Wi-Fi, Wi-Fi 7, was already being planned at press time, with standardization development being done under the IEEE P802.11be Working Group. The goal of Wi-Fi 7 is to expand the availability and cost-effectiveness of Wi-Fi to transport most of the wireless traffic in enterprise, public and residential environments and further improve the efficiency of the spectrum resources, while maintain interoperability with previous generations of Wi-Fi. The new amendment will define Extreme High Throughput (EHT) PHY and MAC layers capable of supporting a maximum throughput of at least 30 Gbps. The amendment will also aim at reducing worst case latency and jitter to improve support for time sensitive applications such as virtual reality, augmented reality, gaming, and so on. The early expectation was that Wi-Fi 7 equipment will be offered by 2025. With the new PHY/MAC, OFDMA is planned be more flexible and efficient, with the added option of 4096-QAM modulation; additionally, MU-MIMO will support 16 spatial streams, up from 8 in Wi-Fi 6. Wi-Fi 7 has a doubled maximum channel band (namely, 320 MHz) making the most out of the new unlicensed bands being added.

REFERENCES

1. IEEE (2020). *IEEE P802.11ax - IEEE Draft Standard for Information technology-- Telecommunications and Information Exchange Between Systems Local And Metropolitan Area Networks--Specific requirements Part 11: Wireless LAN Medium Access Control (MAC) and Physical Layer (PHY) Specifications Amendment 1: Enhancements for High Efficiency WLAN*, P802.11ax/D8.0 - Unapproved Draft, https://standards.ieee.org/project/802_11ax.html.

2. Chu, L., Sun, Y., Zhang, H. et al. (2019). Trigger frame format for Orthogonal Frequency Division Multiple Access (OFDMA) communication. US Patent 10,764,874; September 1, 220, filed 21 June 2019. Uncopyrighted.

3. Khorov, E., Kiryanov, A., Lyakhov, A. et al. (2019). A tutorial on IEEE 802.11ax high efficiency WLANs. *IEEE Communications Surveys & Tutorials* 21 (1): 197–216, First Quarter. https://doi.org/10.1109/COMST.2018.2871099.

4. Woojin, A.H.N., Kim, Y., Son, J. et al. (2020). Wireless communication method for simultaneous data transmission, and wireless communication terminal using same. US Patent 10,813,139, filed 1 April 2019. Uncopyrighted.

5. Porat, R. (2015). Clear Channel Assessment (CCA) levels within wireless communications. US Patent 9,204,451; filed 19 February 2014. Uncopyrighted.

6. Oteri, O., Xia, P., Wang, X. et al. (2017). Clear Channel Assessment (CCA) threshold adaptation method. US Patent 9,807,699; filed 12 September 2014. Uncopyrighted.

7. Zhou, Y., Pramod, A., Cherian, G. et al. (2020). Distributed MIMO communication scheduling in an access point cluster. US Patent 10,820,333; filed 8 March 2018. Uncopyrighted.

8. Naik, G., Bhattarai, S., and Park, J. (2018). Performance analysis of uplink multi-user OFDMA in IEEE 802.11ax. *2018 IEEE International Conference on Communications (ICC)*, Kansas City, MO, pp. 1–6, doi: https://doi.org/10.1109/ICC.2018.8422692.

9. Coleman, D., Correll, P., and Vucajnk, G. (2020). What is 802.11ax (Wi-Fi6)?. https://www.extremenetworks.com/wifi6/what-is-80211ax/ (accessed 20 December 2020).

10. Segev, J., Alpert, Y., Ghosh, C. et al. (2019) Access Point (AP), Station (STA) and method of Multi-User (MU) location measurement. US Patent 10,512,047, filed 29 March 2017. Uncopyrighted.

11. Staff (2020). Introduction to 802.11ax High-Efficiency Wireless, 11500 N Mopac Expwy, Austin, TX 78759-3504, https://www.ni.com.

12. Taylor, C. (2019). The Ultimate Wi-Fi Access Point: Which Wi-Fi 6 Features Define the New Premium Tier?, Strategy Analytics. https://www.qualcomm.com/media/documents/files/the-ultimate-wi-fi-access-point-which-wi-fi-6-features-define-the-new-premium-tier.pdf (accessed 20 December 2020).

13. Hedayat, A.R. (2020) Aggregation methods and systems for multi-user MIMO or OFDMA operation. US Patent 10,790,884; filed 29 May 2018. Uncopyrighted.

14. Ito, T., Murakami, K., and Ishihara, S. (2015). Improving wireless LAN throughput by using concurrent transmissions from multiple access points based on location of mobile hosts. *The 12th IEEE International Workshop on Managing Ubiquitous Communications and Services*, pp. 99–104.

15. Miyamoto, S., Hayata, N., Sampei, S. et al. (2014) Wide-area centralized radio resource management for DCF-based multi-hop Ad hoc wireless networks. *IEEE International Conference on Computing, Networking and Communications, Wireless Networks Symposium*, Kauai, Hawaii, pp. 710–715.

16. Ishihara, K., Murakami, T., Asai Y. et al. (2013). Selective beamforming for inter-cell interference mitigation in coordinated wireless LANs. *2013 16th International Symposium on Wireless Personal Multimedia Communications (WPMC), NICT*, New Brunswick, New Jersey, USA, pp. 1–5, XP032493950, ISSN: 1347-6890.

17. Michaloliakos, A., Rogalin, R., Balan, V. et al. (2013). Efficient MAC for distributed multiuser MIMO systems. *IEEE 10th Annual Conference on Wireless On-Demand Network Systems and Services (WONS)*, Banff, AB, Canada, pp. 52–59.

18. Syed, A.U. and Trajković, L. (2015). Improving VHT MU-MIMO communications by concatenating long data streams in consecutive groups. *IEEE Wireless Communications and Networking Conference, Workshop, Next Generation WiFi Technology*, New Orleans, LA, USA, pp. 107–112.

19. Porat, R., Puducheri, S., Nassiri Toussi, K. et al. (2020). Frame formats for distributed MIMO. US Patent 10,797,928, filed 26 August 2019. Uncopyrighted.

20. Luo, J., Ma, C., Pang, J. et al. (2020). NAV setting method In wireless communications system and related device. US Patent 10,820,304, filed 17 January 2018. Uncopyrighted.

21. GSMA (2016). IR.51: IMS profile for voice, video and SMS over Wi-Fi, Version 4.0. https://www.gsma.com/newsroom/wp-content/uploads/IR.51-v4.0.pdf (accessed 20 December 2020).

22. 3GPP(2018). 3GPP TS 23.402 architecture enhancements for non-3GPP accesses, Version 15.3.0 Release. https://www.etsi.org/deliver/etsi_ts/123400_123499/123402/15.03.00_60/ts_123402v150300p.pdf (accessed 20 December 2020).

23. Firmin, F. (2018). The evolved pPacket core. 3GPP White Paper. https://www.3gpp.org/technologies/keywords-acronyms/100-the-evolved-packet-core.

24. Metaswitch Staff (2020). What is voice over Wi-Fi (VoWiFi)?. https://www.metaswitch.com/knowledge-center/reference/what-is-vowifi-voice-over-wifi.

25. Jardon, M. (2020). VoLTE roaming using general purpose packet data access. US Patent 10,694,457, filed 21 February 2019. Uncopyrighted.

26. Infradata Staff (2020). What is VoWiFi - how voice over WiFi works. https://www.infradata.com/resources/what-is-vowifi/ (accessed 20 December 2020).

27. Zhang, H., Liu, T., and Dong, L. (2018). VoWLAN call handover method, UE and core network node. US Patent 10,057,829, filed 30 June 2016. Uncopyrighted.

28. Zhang, J., Payyappilly, A.T., Santhanam, A.V. et al. (2019). Efficient transition between a trusted WLAN and a WWAN. US Patent 10,448,443, filed 22 July 2016. Uncopyrighted.

29. National Emergency Number Association (NENA) Network Technical Committee Working Group (2011). NENA Femtocell and UMA, Technical Information Document, NENA 03-509, Version 1, January 27, 2011 Issue 1.

30. Aptilo Staff (2020). What is Wi-Fi calling?. https://www.aptilo.com/solutions/wifi-calling/what-is-wifi-calling-vowifi/ (accessed 20 December 2020).

31. Minoli, D. (2019. Hindawi & John Wiley & Sons (https://doi.org/10.1155/2019/5710834). Practical aspects for the integration of 5G networks and IoT applications in smart cities environments, Special Issue titled "Integration of 5G Networks and Internet of Things for Future Smart City,". *Wireless Communications and Mobile Computing* 2019: 5710834, 30 pages. (With B. Occhiogrosso).

32. Lee, J., Lee, J., Park, J., et al. (2020) Method and apparatus for supporting session continuity for 5G cellular network. US Patent 10,708,824, filed 8 May 2018. Uncopyrighted.

33. Peng, X., Hyunseok, R., Agiwal, A. et al. (2020) Method and apparatus of data transmission in next generation cellular networks. US Patent 10,602,516, filed 11 August 2017. Uncopyrighted.

34. Lin, Z., Li, J., Zheng, Y. et al. (2018). SS/PBCH block design in 5G New Radio (NR). *2018 IEEE Globecom Workshops (GC Wkshps)*, Abu Dhabi, United Arab Emirates, pp. 1–6, doi: https://doi.org/10.1109/GLOCOMW.2018.8644466.

35. Yoon, S., Euichang, J.U.N.G., Park, S., et al. (2020). Method and apparatus for facilitating coexistence of 4th and 5th generation communication systems. US Patent 10,681,743, filed 8 January 2018. Uncopyrighted.

36. Sun, S., Rappaport, T.S., Rangan, S. et al. (2016). Propagation path loss models for 5G urban micro and macro-cellular scenarios. *2016 IEEE 83rd Vehicular Technology Conference (VTC2016-Spring)*, Nanjing, China.

37. Lu, Y.-S., Lai, C.-F., Hu, C.C. et al. (2010). Path loss exponent estimation for indoor wireless sensor positioning. *KSII Transactions on Internet and Information Systems* 4 (3): 243 ff.

38. Srinivasan, S. and Haenggi, M. (2009). Path loss exponent estimation in large wireless networks. *Information Theory and Applications Workshop*, La Jolla, CA, USA, pp. 124–129.

39. Viswanathan, M. (2013. Log distance path loss or log normal shadowing model, Gaussianwaves. https://www.gaussianwaves.com/2013/09/log-distance-path-loss-or-log-normal-shadowing-model/ (accessed 20 December 2020).

40. 3GPP 38.901-e20 (2017). 3rd Generation Partnership Project; Technical Specification Group Radio Access Network; Study on Channel Model For Frequencies From 0.5 to 100 GHz (Release 14).

41. Rappaport, T., Xing, Y., MacCartney, G.R. et al. (2017). Overview of millimeter wave communications for fifth-generation (5G) wireless networks-with a focus on propagation models. *IEEE Transactions on Antennas and Propagation* 99 https://doi.org/10.1109/TAP.2017.2734243.

42. Electronics360 (2020). 5G NR vs Wi-Fi 6. https://electronics360.globalspec.com/article/15771/5g-nr-vs-wi-fi-6 (accessed 20 December 2020).

43. Minoli, D. (2002). Ethernet-Based Metro Area Networks – Planning and Designing the Provider Network (co-authored). McGraw-Hill.

6 De-densification of Spaces and Work Environments

The first part of this book dealt with High-Density Communication (HDC) in campus environments such as airports, stadiums, convention centers, shopping malls, cruise ships, train stations and subway stations, boardwalks, and other venues. However, as the second decade of the twenty-first century rolled along, a new requirement presented itself due to the worldwide pandemic: almost the orthogonal requirement to HDC, namely physical/desk distancing. Fortunately, the wireless technologies discussed in earlier chapters can and have been harvested to address and manage these pressing issues. Real-Time Location systems (RTLSs) have been employed for several years to automatically identify and track the location of objects or people in real time within a building or in other constrained locations; RTLSs are now seeing renewed interest and applications.

In the COVID-19 era, social[1] distancing in general, at the workplace in particular, has become important; additionally, there is a need to deploy a plethora of other infection containment and control measures. As depicted in Figure 6.1, various technologies can assist stakeholders in accomplishing these tasks. Available solutions range from simple to more advanced, with underlying cost implications. Emerging Internet of Things (IoT) and Artificial Intelligence of Things (AIoT) approaches, as well as Blockchain Mechanisms (BCMs), can play a role in these solutions.

This chapter focuses on technologies that can be utilized for office physical/desk distancing in support of Office Social Distancing (OSD) and Office Dynamic Cluster Monitoring and Analysis (ODCMA). The chapter provides an overview of the topic, while the chapters that follow furnish more technical details.

6.1 OVERVIEW

Smart buildings can make use of IoT and AIoT sensors to monitor a variety of building functions, including but not limited to environmental conditions, wellness parameters, Heating, Ventilation, and Air Conditioning (HVAC) operation, Building Management Systems (BMSs) status, people presence and density, people/face recognition, physical security, and physical cleaning protocols [1].

In response to the COVID-19 pandemic, many organizations have recently been planning to deploy effective OSD policies, documenting and managing key factors, such as:

- Real-time people count and density in each work zone or shared space.
- Distance between work zone occupants.
- Location tracking; determining locations of groups or individual employees.

---

[1]Some (more properly) use or prefer the term "spatial distancing."

High-Density and De-Densified Smart Campus Communications: Technologies, Integration, Implementation, and Applications, First Edition. Daniel Minoli and Jo-Anne Dressendofer.
© 2022 John Wiley & Sons, Inc. Published 2022 by John Wiley & Sons, Inc.

Sophistication/cost

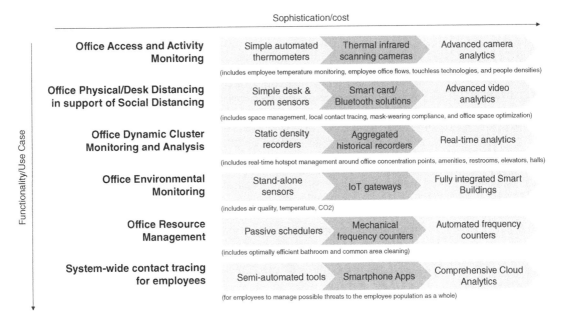

FIGURE 6.1 Social distancing other infection containment and control measures in the age of COVID-19.

- Proximity alerting; generating alarms and alerts if employees are not complying with social-distancing guidelines.
- Movement between work zones.
- Contact tracing; determining the contact and proximity history of individual employees.
- Implementation of (more) rigorous hygiene measures.

Important aspects of occupancy, many of which are critical for OSD/ODCMA are shown in Table 6.1.

While some of these policies and precautions may be relaxed when vaccines become widely available and widely administered, the lessons learned by managing this virus can be utilized to address the "next pandemic," which medical observers say "will come in the future" (with observers noting that "...*this is not the first pandemic and it does not seem to be the last one*" [2]), with episodes occurring at faster rates given the global economy, the global supply chain, the desire or need for global travel, the long-established progression toward concentration and migration of populations to cities, the deplorable hygiene conditions in third world countries in several continents, the peculiar and idiosyncratic eating habits of some societies, the overall global population growth, the aging population, and the ever-increasing cases of societal pre-existing medical conditions (e.g. obesity). Some of these policies and precautions should be built and incorporated into a well-developed, well-maintained, and well-tested Business Continuity Plan (BCP) that the organization should always have at ready access.

Advanced wireless networks, cloud services, and traditional or AI-based analytics can support the multitude of in-building sensors and personal electronic devices (such devices including smartphones, wearables, Body Area Network nodes, fitness monitors, and medical devices [3–7]), thus enabling the realization of intelligent monitoring and response in smart building environments. RTLSs are systems that can be used to automatically identify and track location of objects or people in real time within a building, or in other constrained locales; these systems have been around for several years (the term itself was coined in the late 1990s). Some

TABLE 6.1 Occupancy Aspects that are Important to OSD/ODCMA Policies

Occupancy Aspects	Definition/Scope
Presence (aka "detection")	An assessment that provides information about "when" individuals or occupants are present in a particular zone, room, office, or building. Answers the question "Is there at least a person present?"
Location	Information that relates to occupants "coordinates" within a particular zone, room, office, or building in which individuals or occupants are situated. Answers the question "Where is a person who is present?"
Count (aka "estimation")	An assessment that provides information as to the "number" of individuals or occupants are present in a particular zone, room, office, or building. Answers the question "How many people are present?"
Activity	A determination that provides information on "what" activity is being carried out by occupants in a particular zone, room, office, or building. Answers the question "What is the person doing?"
Identity	An assessment that relates to information as to "who" is in a particular zone, room, office, or building. Answers the question 'Who is the person?
Track/Tracking	An assessment that provides information about a particular occupant's movement history across different zones, rooms, offices, or buildings. Answers the question "Where was this person before?'

refer to these systems as Indoor Positioning Systems (IPSs). An RTLS can be thought of as a type of "indoor GPS". GPS is generally not suitable for determining indoor locations given that radio waves at the 1.1–1.6 GHz range are attenuated and scattered by walls and other internal objects, with multiple reflections from internal surfaces causing multipath propagation and ensuing errors. In addition, a receiver would need to receive signals from four satellites, which is unlikely to occur inside a building. RTLSs utilize a network of signal-receiving nodes to locate objects or people inside structures such as high-rise buildings, parking garages, airports, or underground locations when the GPS service is unreliable or unavailable. To date, the majority of RTLS applications have been to track warehouse objects and for monitoring patients in clinical settings.

Occupancy and utilization monitoring typically utilize space-deployed motion sensors; monitoring can also use more advanced anonymous/cluster image sensors, where people are not identified personally but abstractly. At the other end of the spectrum, one finds tag-based or other technology-based identity recognition systems, and even facial/trait recognition using advanced video analytics. Facial recognition systems, also occasionally used by Law Enforcement operatives, have become somewhat controversial of late in the Western World. Movement through door openings, for example to a semi-secure room or office, can be monitored using simple contact sensors.

RTLSs can be used for tracking people (personnel, clients, and customers) and assets (mechanical parts, packages, and equipment) [8]. RTLSs are employed in numerous industries, such as in logistics, airports, amusement parks, mining, the food industry, the automotive industry, the aerospace and defense sectors, and the retail sector. This chapter focuses on healthcare applications; RTLSs have been used extensively in healthcare applications at least since the early late 2000s/early 2010s. In fact, COVID-19 management, both in the hospital and nursing home context, as well in the OSD/ODCMA context, can be viewed as a "healthcare-related" issue [9–11]. RTLSs have been used to monitor the adherence to hand hygiene protocols to reduce infection risk [12]; in this case, personnel are monitored or alerted whenever they forget to wash their hands at specified moments. Furthermore, RTLSs can be used to monitor whether personnel have been in contact with infectious patients: as early as 2012, the University Health

Network (UHN) in Toronto was testing an RTLS to prevent the transmission and spread of new infections, as well as control any existing infections, by tracking equipment, patients, and employees [13].

People counting/presence methods, thus, include (among others):

- (Stationary) motion/presence sensors.
- User-worn smart tracking tags.
- Embedded systems in user-worn clothing.
- Non-wearable methods including smart floors and generic Ultra-Wideband (UWB) systems.
- Computer vision with anonymous analytics.
- Full video with face recognition.

In the context of user-worn (smart) tracking tags, RTLSs utilize physical tags that can be connected to mobile objects or to people and make use of sensors and digital middleware (i.e. software that acts as a bridge between an operating system or database and applications, especially over a network) to process information in order to locate people or objects in real time. The signals that the sensors send to and receive from the tags are processed by location-engine software, which converts the signals into readable information. RTLSs can be utilized both indoors and outdoors and cover a limited area, depending on the size and signal strength of the system. By placing location sensors throughout a building in strategic spots, the location of the tags (and, thus, the person or object) can be determined and displayed in real time, as well as stored for additional offline processing [8]. Indoor navigation by people, as well as the tracking of people indoors, are also important for safety and security, for example, evacuation and rescue operations. Figure 6.2 depicts pictorially how some of the methods can be realized; in the figure, the sensors can be specific to the method in question (e.g. motion sensors, tag-tracking transponders, computer vision cameras, and so on).

Radio Frequency Identification (RFID) and Wi-Fi are at the basis of most commercial RTLS applications [14, 15]; other solutions utilize UWB, Bluetooth, ZigBee, infrared (IR), and ultrasound. In practice, a combination of technologies is often applied. Some of these technologies require the installation of new and separate hardware in a building, which affects the overall costs. Every technology has its set of advantages and disadvantages. Wi-Fi-based tracking is relatively inexpensive and easy to install (typically as an overlay to an existing infrastructure), but the signal can penetrate walls and cannot, therefore, be used to locate people at the room level; IR is suitable for room-level location but requires a clear line-of-sight from the tag to the sensor [8]. An environment such as an office, a warehouse, a hospital, or an airport must be

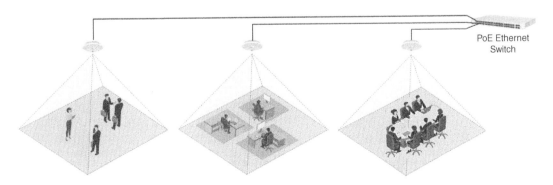

FIGURE 6.2   Monitoring of people presence and/or density – general concept.

carefully assessed to determine which technology (or combinations) works best for the requirements at hand.

Use cases for these tracking technologies include the following [16]:

- Managing COVID-19 OSD: real-time people count and density in each space.
- Dynamic workspace space utilization: on-demand assignment of desks and rooms to address employees' needs throughout the workday.
- Smart facility management using people-counting sensors: enabling and automating core facility management services.
- Managing employee collaboration: capturing employees' interactions to increase their productivity and wellbeing.
- Employee-centric services: providing employees with real-time information on corporate services, such as bathroom and common area cleaning.
- Traditional office utilization and planning: flexible allocation and design of building space according to historical and actual utilization data to maximize real-estate use and improve employee productivity.

Some implementations utilize static sensors that are physically attached to office assets such as desks, chairs, printers, and the like; these sensors are not worn by people. Sensors can transmit simple status data traceable to the physical assets in question and its use or occupancy. Figures 6.3 and 6.4 depict one example of results achievable with this kind of technology. Figure 6.3 depicts instantaneous occupancy at a desk, based on the desk sensor state; this data can be captured as

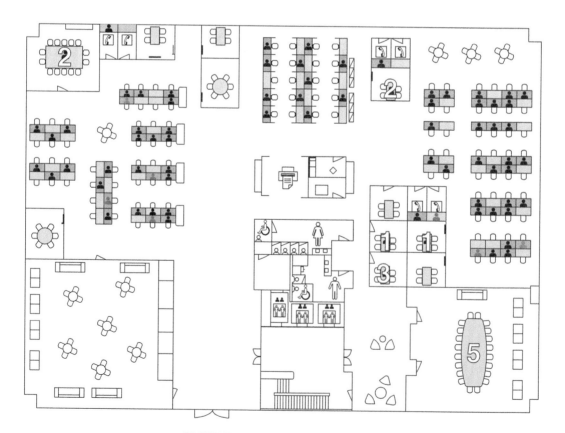

FIGURE 6.3   Seat/desk-specific tags.

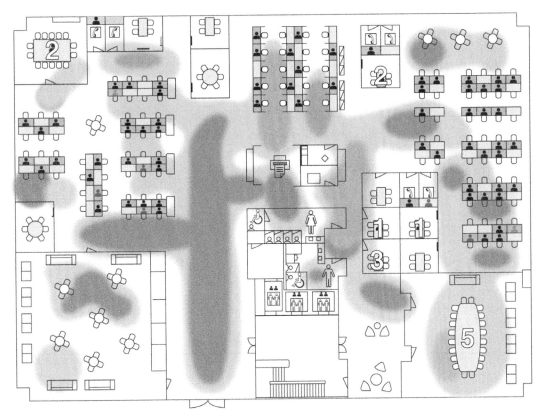

FIGURE 6.4   Heat maps generated over a time window.

a function of time and analyzed. Unless there is a fixed association between a desk and an individual, these data do not provide exact mapping between specific individuals and presence at a defined location. These seat-attached sensors can also be used in conference rooms, but again the data are aggregated, not individualized. Figure 6.4 depicts a "heat map" of presence activity at various points on the floor, accumulated over a specified time window. A pre-COVID-19 system such as implied in Figure 6.3 might have been used for a "hoteling" office setup where the green seats depict availability. In a post-COVID-19 area, the green areas depict seats to be kept empty for OSD.

6.2   BASIC APPROACHES

Table 6.1 identified key concepts in the building occupancy estimation and detection arena. The following aspects are of specific interest in the context of OSD/ODCMA:

- Presence: methods to establish how many people (or objects) are present in a space. Addresses occupancy estimation and detection. Tends to be an aggregate assessment of the number, often an approximation. Typically, does not establish the presence of a specific individual entity. Typically, sensor-based, e.g. Pyroelectric ("Passive") InfraRed (PIR) sensors can be used to detect the presence and absence of occupants in an office; cameras, can be used to obtain the number of occupants in a room (computer vision and/or full video); CO_2 sensors can be used to obtain an estimation of occupancy [17–20]. Typical applications focus on energy conservation, including HVAC management and lighting control.

- Motion detection/tracking: methods to establish and/or track the movement of people or objects in a space, for example, moving close to other employees or from one building zone to another zone.
- Identification: methods to establish the identity of people or objects present in a space. Can be sensor-based (e.g. RFID) or camera-based.
- Distancing: method to establish the real-time distance between two or more people (or objects) present in a space. Can be sensor-based (e.g. RFID), camera-based, or use other methods. Generally used in conjunction with identification methods.

Presence and absence of occupants clearly impacts the operation time of HVAC systems in buildings since the number of occupants in a thermal zone will indicate the cooling or heating load of this area. Some occupancy-based building climate control systems have been proposed in recent years to optimize usage energy while providing a comfortable office environment; savings in the 5–15% have been documented [21–24]. Occupancy-based lighting has been around for a number of years, whether with stand-alone sensors or in the context of connected lighting [25, 26]; savings in the 30–60% or better have been documented. This chapter does not focus on energy management[2] but on OSD/ODCMA issues.

For people-specific presence detection and tracking, two commonly used approaches are wearable tags and video cameras; a third, less common approach is nonintrusive external/passive means.

In the smart-tag-based monitoring arrangement, each user is required to wear a tag that advertises its unique identity (ID), which is then captured by in-building nodes utilizing various wireless communication technologies. With smart tags, identification becomes straightforward, given the administrative knowledge of the linkage between the tag ID and a specific individual carrying it. The direct tracking of people is mainly conducted in the United States and not in Europe due to privacy legislation on the continent [27, 28]; however, anonymized tracking for COVID-19-driven OSD/ODCMA may perhaps prevail.

IPSs/RTLSs can utilize a grid approach or a broad range approach, as follows:

- The grid approach makes use of a relatively dense set of interconnected low-range (typically low-cost) receivers that are deployed in a regular grid pattern in the space of interest; the tags are identified just by a few receivers in the proximity of the tag. The IPS system maps the topology of the grid to the received signals from a handful of the receivers in the zone of interest to determine the approximate location of the tag.
- A broad-range sensor approach typically uses continuous physical measurement (such as angle and distance or distance only), derived from electromagnetic signals, in conjunction with the tag's ID data in one combined signal to determine the tag's location. The reach by these IPS sensors often covers an entire floor or some appropriate subzone.

There are different types of tags used in RTLS, namely, passive, semi-passive, and active tags. The difference between these tags lies in their ability to transmit signals. Passive and semi-passive tags are not able to directly transmit signals and can only be detected by location sensors by returning an echo signal. The return signal typically has very low power; therefore, the applicability of this technology is specific to a well-defined environment. Active tags are battery powered and actively send information about their location; this gives them a longer detection range than passive and semi-passive tags. Active tags can have Wi-Fi transmitters [29], Bluetooth

---

[2]There is an extensive body of literature and standards, in particular from the U.S. Department of Energy (U.S. DoE) and from American Society of Heating, Refrigerating and Air-Conditioning Engineers (ASHRAE), related to energy management in the context of occupancy and occupancy density. This topic is not further addressed in this text.

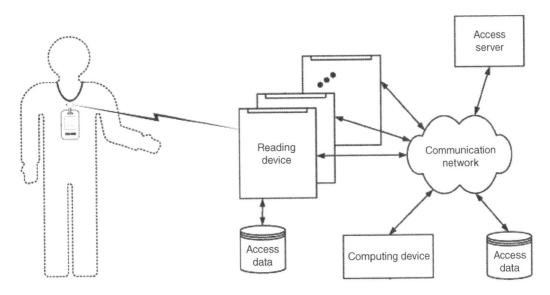

FIGURE 6.5   Pictorial view of the Tag-Based Approach.

transmitters [30], active RFID transmitters [31], and UWB transmitters [32], among others. Figure 6.5 provides a pictorial view of the tag-based approach [33]. Semi-passive tags also use batteries, but mainly for secondary functionalities, such as emergency pushbuttons, temperature sensors, or accelerometers.

Information from these sensors is included in the data the stationary sensor nodes receive and provide additional contextual information; for example, whether a device is in use. The ability of a tag to transmit signals or collect additional data influences its price. For example, battery-powered tags require the replacement of batteries on a regular basis, which adds to maintenance costs. There are tags that only transmit a signal when the tag is being moved, which conserves battery power. For some applications, the location data do not have to be available 24-7, but only upon request. For example, some physical assets are only moved once or a few times a day, or even less frequently. Moreover, tags can be located with different levels of accuracy, from just knowing whether a tag is or is not present in a certain area to knowing the exact location of the tag. Different options for locating the tags are as follows: locating the tag at choke points (knowing when a tag passes a door), locating the tag by association (knowing which tags are close to each other), locating the tag at room level (knowing in which room a tag is), locating the tag precisely (knowing the exact spatial coordinates), and locating the tag at sub-room level (knowing the location of a tag in a smaller area within a room). Strategic placement of the location sensor nodes can optimize the network coverage inside a building. To determine the position of a tag, various localization methods are used to determine the distance between the tag and the location sensor(s); thereafter, estimation algorithms calculate the position of the tag [8].

Methods based on smart tags, however, face the problem that the people may be reluctant to wear them, although this may be made a condition of employment or entering a space. Furthermore, because of the portability, the tag may be stolen or misused by others. In addition, the cost of smart tag-based system can be high. In some "advanced" cases, people may receive implanted transponders, but this is a relatively rare solution in commercial applications [34–36]. Another approach is to use signals generated by smartphones to support in-building tracking (comparable in a way to GPS tracking of phones while in open space); however, privacy concerns may exist.

Due to the inconvenience and relative inflexibility of wearable devices, and the potential intrusiveness of full-video cameras, the adoption rate of these approaches has remained relatively low prior to the COVID-19 pandemic. Non-wearable (but non-video) identification methods address these issues. Some of the non-wearable, nonintrusive external/passive approaches discussed in the literature include, but are not limited to, the following [37]:

- The SensFloor method [5], where user tracking and localization is obtained based on an array of commercial capacitive sensors placed under the floor.
- The floor vibration sensor method [38–41], an indoor identification system where identity can be determined by extracting the user's gait pattern from the footstep induced structural vibration.
- The Wi-Fi tracking method [6, 42–44], where user identification from a small group of people can be achieved using the Channel State Information (CSI) to recognize the user's walking steps and walking gait (other Wi-Fi methods can also be employed, but they entail the use of tags).
- The UWB non-tag direct method [45] that can operate without tags, where an identification system for office or factory employees is created with ambient nonintrusive UWB sensors installed at an entrance area. Since each individual has a different body figure and walking gait, they are "seen" differently by UWB sensor; the system generates a unique UWB signature for each individual, which will be used for their identification.
- The wireless pyroelectric IR method [46] is where a distributed and wireless PIR sensor network is constructed to detect and track human targets. In general, a PIR sensor typically has two slots, and made of materials sensitive to IR radiation. When the sensor is idle, both slots detect the same amount of ambient IR energy radiated from the room or walls, but when a warm body passes by the sensing area, it causes a positive differential change between the two halves of the sensor; when the warm body leaves the sensing area, the reverse happens, with the sensor generating a negative differential change; these change pulses are detected by the electronics in the sensor and transmitted downstream. In conjunction with Empirical Mode Decomposition (EMD) and Hilbert–Huang Transform (HHT), features of targets' (e.g. humans) can be extracted in both the time domain and the frequency domain for identification, enabling identification. EMD is a method of breaking down a signal without leaving the time domain; this method is useful for analyzing nonlinear and nonstationary signals. HHT is a technique to decompose a signal into Intrinsic Mode Functions (IMF) coupled with a trend, and, thus, enables one to obtain instantaneous frequency data. Since PIR-based sensors can detect the motion of occupants, indicating the presence of occupants, some occupancy detection systems have been developed using PIR sensors; however, counts, proximity, and identification are generally impractical using these methods. Furthermore, although PIR sensors are of low cost and are easy to be deployed, they can only detect the motion of moving occupants, which means that the static occupants will be missing, limiting even the energy management applicability (except for lighting applications) [20].
- Another approach for nonintrusive sensing is to use transducing mechanisms. This entails embedding sensors in floors to acquire the sensory information from human walking, including indoor position, activity status, individual identity, as well as the number of people entering and/or leaving a room; this also enables automation of air conditioning/lighting and security monitoring. Transducing mechanisms include resistive, capacitive, piezoelectric, and triboelectric mechanisms. With the self-generated electrical signals in response to the mechanical stimuli, piezoelectric and triboelectric mechanisms have advantages of self-powering elements of the system. In particular, triboelectric-based deep-learning-enabled smart mats (DLES-mats) have been described in the literature as

being smart floor monitoring systems that are able to achieve person-identification-tracking functionality at low cost and high scalability [37]. Triboelectric nanogenerators are able to convert basic mechanical energy into usable electric power: mechanical energy from rain, wind, body motion such as hand touching or walking can be harvested as a mechanical energy source for triboelectrification; an electrical potential is generated between the nanogenerator contact surfaces and the resulting alternating potential generated from the dynamic mechanical motions can be utilized for powering electric devices and sensors (or stored in a storage unit). In this specific application, the triboelectric floor mat sensors can produce self-generated electrical signals under pressure from footsteps, which, in turn, eliminates the power requirement for sensors. The self-generated electrical signal can also be used as a wake-up signal to trigger the operation of the entire system that can be in a "sleep" mode when no one is walking to reduce overall power consumption. The smart floor monitoring system enables personal identification and position sensing without the camera-based privacy concern or the need to carry tags or devices [37]. This smart floor monitoring system is a convenient detection method, having the advantage that the personal identity (i.e. dynamic walking gait) cannot be misused by others, unlike the tags/cards (even fingerprints in sophisticated infractions) can be stolen and/or borrowed.

- Environmental sensors such as CO_2, temperature, humidity, light, and pressure are widely available in modern lighting and HVAC systems; given that occupants directly influence indoor environments, environmental sensor readings – particularly CO_2 sensors, since indoor CO_2 concentrations are related to the number of occupants and the ventilation level – can be used as indicators for building occupancy estimation and detection [20]; these approaches, however, only provide a coarse occupancy view, not adequate for OSD and ODCMA applications.

- Several other sensor types have also been used for occupancy estimation and detection, including chair sensors, vibration sensors [47], LEDs [48], and ultrasonic [49] sensors. In particular, [50] describes a fine-grained occupancy estimation system in a conference room (or open desk spaces or private offices, for that matter) using chair sensors, where a simple thresholding method is utilized to detect the occupied/unoccupied states of a chair: the total occupancy can be derived from the number of chairs to be occupied. Since each sensor has some unique properties and limitations for occupancy estimation and detection, some have proposed the fusion of multiple sensor types compensating limitations of each sensor (e.g. CO_2 + camera [51]; temperature + humidity + CO_2 + light + sound + PIR + door status [52]; and many other combinations [20, 50, 53–56]).

Although in some cases just the generic concentration, or heat map, of people in a confined space is sufficient, in other cases – especially in the context of COVID-19 OSD/ODCMA and contact tracing – the specific identity of the individuals involved in that confined area is needed. In general, camera-based surveillance for monitoring and face recognition is used in city areas, especially in the Far East. Basic video monitoring of common areas is almost ubiquitous in commercial Class A or Class B buildings; however, such methods are less common within most office environments per se. Facial recognition in commercial building settings in the United States is currently rare. Video technology, especially when coupled with facial recognition, raises privacy concerns. To mitigate concerns about full video monitoring, some other technologies have been advanced, including optical approaches such as nonintrusive computer vision sensing and laser beam scanning (in addition to the systems discussed just above). With laser beam scanning, the acquired sensory information is limited and the laser beam is easily blocked by other objects, resulting in information loss and inaccurate sensing; additionally, the implementation and operation of such a system is usually expensive.

Computer vision technologies support nonintrusive sensing in office spaces (typically 24×24 ft areas) using anonymous computer vision mechanisms. With these technologies, images are processed for body counts (but not IDs) in the sensor utilizing deep-learning techniques (but not stored anywhere); only the analyzed counts are forwarded to the cloud, but not the video. These nonintrusive and anonymous Power over Ethernet (PoE)/Wi-Fi-based intelligent workspace occupancy sensors are typically installed in the ceiling and monitor predefined "Areas of Interest." The sensors detect people's presence, locations, and numbers – but not IDs or actual distinguishable individual pictures – and send analytics data to a cloud management system. With these types of sensors, all the image processing is done on the sensor itself; the sensor does not store or output images and has no interface to extract images, therefore privacy is protected. Because the sensor tracks peoples' movement rather than personal devices (e.g. mobile phones), the system and the aggregate results are completely anonymous; the sensors just report how people use the workspace (e.g. count, location) [16]. When needed, multiple counting/presence areas can be defined per sensor (e.g. multiple 8×8 ft subareas).

Security cameras or bespoken cameras can also be used for office or building occupancy estimation and detection [57–67]. Besides using "brute-force" unprocessed full-resolution video that captures each person in a recognizable manner, computer vision methods can be used. There are a number of vision-based occupancy estimation and detection systems discussed in the literature. Documented counting techniques include, among others, location estimation; background subtraction, tracking, and recognition; background subtraction and manual counting; Bayesian network approach; convolution neural network approaches; and probabilistic state-space models [20]. At the far end of the monitoring spectrum, full video and associated video analytics can be used not only for face recognition but also compliance for mask-wearing protocols and office distancing protocols. Generally, video camera-based indoor occupancy counting systems can be developed which enjoy high accuracy results; however, these video systems have high computational complexity and also face the issue of illumination conditions; these technologies tend to be expensive if they operate in a real-time automatic mode, especially if the search database is large. Additionally, these methods and technologies also give rise to privacy concerns.

6.3   RTLS METHODOLOGIES AND TECHNOLOGIES

A taxonomy of occupancy detection systems can be based on (i) *method*: terminal-based (e.g. use of a terminal, e.g. smartphone or tag), *or* nonterminal-based detection systems (e.g. use passive sensors, such as PIR or carbon-dioxide [CO_2]); (ii) *function*: individualized systems (ability to detect, identify, and track individual building occupants) or non-individualized systems (with only the ability to provide aggregate occupancy without knowledge of user identities or exact building locations); and (iii) *requisite infrastructure*: occupancy systems requires a bespoken infrastructure (detection systems installed for the sole purpose of measuring building occupancy detection), or systems that provide building occupancy information as a secondary overlay function (for example, occupancy information can also be inferred from the use pattern of building appliances such as computers and/or aggregate Wi-Fi or Bluetooth energy) [50]. See Table 6.2. Obviously, there would be eight classes of occupancy detection systems, e.g. M1-F1-I1 (e.g. RFID systems), M1-F1-I2 (e.g. using the Wi-Fi signal from a smartphone over an existing corporate intranet), and so on.

RTLSs can be seen as being a subtype of the general occupancy detection category. As noted earlier, RTLSs are systems employed to automatically identify and then track the "precise" location of objects or people in real time within a building or in other constrained locations, including airports (particularly for tracking and processing baggage and other airport assets); thus, in most instances, they are "individualized, terminal-based" systems in the taxonomy

TABLE 6.2   Taxonomy of Occupancy Detection Systems, with Focus on RTLSs

Taxonomy Element	Branch 1	Branch 2
Method	M1: terminal-based: RTLSs using smartphones, tags, etc.	M2: nonterminal based: RTLSs using PIR, smart mats, video cameras, computer vision, etc.
Function	F1: individualized systems: RTLSs using smartphones, tags, etc.	F2: non-individualized systems: RTLS using CO_2 or other environmental sensors
Requisite infrastructure	I1: bespoke infrastructure: RTLSs using tags	I2: overlay secondary function: RTLSs using Wi-Fi, Bluetooth WLANs/WPANs

above, which may or may not require a dedicated infrastructure. RTLSs typically entail tags, badges, readers and exciters, wireless link technologies, and other elements (servers, middleware, and mapping software). RTLS tags are attached to objects or worn by people, and fixed reference monitors receive signals from tags to establish their location. Typically, wireless RTLS tags are carried by people or adjoined to objects; stationary reference nodes at known physical locations in the area of interest receive wireless signals from tags, from which the system can the determine their location. Tags and stationary reference nodes can be receivers, transmitters, or both.

There are two types of specific technologies traditionally used to support RTLS: (i) passive RFID (operating in the UHF band below 1 GHz), that utilizes inexpensive tags and somewhat expensive readers (also known as interrogators), providing coarse tracking; and (ii) active tags, that utilize RFID, UWB or a combination of technologies such as Wi-Fi/Bluetooth and IR (operating at much higher frequencies); these solutions are generally more expensive, but also yield more precise location determinations [33]. RFID systems utilize Radio Frequency (RF) signals (in some cases, other signals) to support communication between transponders (tags) and readers; RFID tags in their simplest form comprise a tag ID that is modulating a carrier signal, such as an electromagnetic wave, where the modulated signal is then propagated by the tag. RFID systems have become very popular in a large number of applications; however, a limitation of existing RFID systems is that they have limited capability with respect to longer-distance ranging or locating, that is, determining the range or location of a tag, object, or wireless device that is located at some distance from the reader – while the range of local devices may be determined with existing RFID systems, problems remain when endeavoring to determine how many devices are local and their relative locations with respect to other devices [68].

In many applications, such as but not limited to warehousing applications, industrial process control for robotics, and vehicular, traffic location accuracy is important [69]. Accuracy, in terms of true location compared to estimated location, is typically rated for a given distance, for example, 90% accurate for a 30-ft range. While some applications require very high accuracy (1–2 cm), many indoor applications can operate at the decimeter level. OSD applications require some level of accuracy but not at the centimeter level of accuracy. For comparison, technologies for outdoor applications for managing vehicle separation, lane positioning, time to collision, autonomous vehicular systems, Unmanned Aerial Vehicles (UAV) or land surveying, and so on, have a 2-centimeter precision accuracy, when using Real-time Kinematic (RTK) as an enhancement technique for GPS.

IPS systems entail stationary hardware, mobile device hardware, and software elements. Several methods and entities are utilized to provide indoor positioning with RTLS. Any number of wireless technologies can be used for locating entities or objects. Some systems avail themselves of *in situ* wireless infrastructure for indoor positioning. Wireless technologies include radio beacons/Wireless Local Area Networks and Personal Area Network/BLE; specifically,

Wi-Fi-based Positioning System (WPS) including Wi-Fi Simultaneous Localization and Mapping (SLAM), Bluetooth, and UWB. There are several other RTLS wireless technologies including Ultrasound Identification (US-ID)/Ultrasonic ranging (US-RTLS); Active RFID; Active RFID-Infrared hybrid (Active RFID-IR); Semi-active RFID; Passive RFID with phased array antennas; low-frequency signpost identification (e.g. 433.92 MHz); Visible Light Communication (VLC)/Li-Fi; and acoustic signals. Multimode RTLSs that use a combination of wireless technologies are also available. The focus of this chapter is on electromagnetic waves. Signal-emanating/receiving devices include smartphones, Wi-Fi-enabled devices, and Bluetooth systems, and bespoken solutions; in all instances, the design goal is to deploy relays and beacons throughout the space of interest. More specifically, RTLS systems typically rely on distance measurement to nearby anchor nodes – signal-emanating devices with known stationary in-office positions, for example, Wi-Fi Access Points (WAPs), Bluetooth beacons, Li-Fi access points, or UWB beacons. The anchor nodes actively locate mobile devices or tags; alternatively, they provide ambient location for entities to be sensed.

Figure 6.6 top illustrates how a mobile user inside an indoor environment, e.g. an office complex, a stadium, an airport, or a shopping mall, etc., can use a mobile receiver to calculate its location with respect to tagged landmarks ("tags"). The tags are attached to the storefronts,

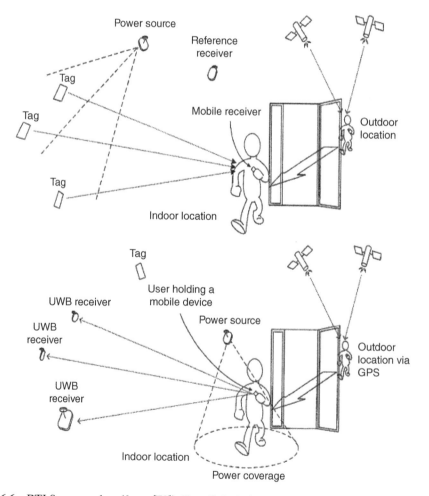

FIGURE 6.6 RTLS approaches (from [70]). Top: Calculation of location with respect to tagged landmarks. Bottom: Determining the location of the mobile device carried by the user.

doors, staircases, ticketing counters, gates, restrooms, etc. Tags can be powered by strategically placed exciter sources that typically emit electromagnetic power signals; alternatively or additionally, tags can be powered by scavenged energy. The tags that are in a reception range of the user's receiver device are considered a "constellation" of tags for the receiver device. Both the physical properties of the signal, such as strength, time and difference of Time of Arrival, angle of arrival, phase of the signal at the receiving antenna, frequency of the signal, its polarization, and so on, and the data carried by the signal may be used to make the receiving device aware of the location, identity, codes, messages, and any other information broadcast by the transmitter. This information is compared to the similarly and often simultaneously received signals from other broadcasting tags in the receiving range. This comparison provides the receiver with the information about how far it is located from each transmitting tag and thereby where it is located. In another arrangement, a reference receiver, positioned at a known location, can compensate for lack of synchronicity in the network. In this embodiment, the deployment of the landmark tags is easy as the tags can be implemented as stickers that are attached to, for example, the front door or facade of the premises. The receiver can be associated with a hosting mobile device, connected as a wired or wireless attachment to the mobile device, or act as a stand-alone device. In this example, each tag transmits its location and optionally additional information such as its identity and such information that is informative to the user. For example, the business hours of a store, promotional coupons, discounts, the temperature of the surroundings, the latest news, menus, etc., can be pushed to mobile visitors as they walk through a shopping mall and as they enter the transmission range of the tags mounted at the stores [70]. (The mobile device of users can be located by GPS satellites when the mobile user is outdoors and then, as they enter an indoor environment, the mobile device detects the presence of tags and switches to indoor location determination. The transition from outdoors GPS navigation to indoors navigation is preferably implemented seamlessly in the mapping presentation of the mobile device such that the users do not detect any interruptions or jitters as they enter the indoor environment.)

Figure 6.6 bottom illustrates an inverse example of Figure 6.6 top. Here, the user is holding a mobile device that is equipped with a transmitter tag. This transmitter broadcasts signals or impulses that are received by appropriate receivers mounted at the landmarks of interest, e.g. storefronts. When the user is in the range of the receivers, then the constellation of the receivers that receive impulses from the mobile device tag will be capable of figuring out the location of the mobile device carried by the user. In this case, the deployment of the transmitter in the mobile device can be as simple placing a passive transmitter sticker on the mobile device, but the receiver infrastructure is more elaborate than in example of Figure 6.6 top. The receiver infrastructure can communicate with the mobile device though the transmitter tag (if configured as a transceiver or a transmitter and receiver) or through other modalities, such as Bluetooth, Wi-Fi, a cellular network, the internet, etc., to provide mobile device identity and other information to the mobile device. For examples where a tag passive, it can be powered by a power source, as the case was in the example of Figure 6.6 top; alternatively, the tag can be an integral part of the mobile device or scavenge its power from the mobile device, e.g. by means of power coupling to the power circuitry of the mobile device or use the power available in the mobile device through a direct connection to its power source – these solutions can also be integrated into the mobile device [70]. Furthermore, a reference tag is also used to avoid the need for synchronicity in the system.

See Table 6.3 for a comparison of key RTLS solutions (also see [71]). Figure 6.7 depicts an illustrative example of an RFID setup.

In general, positioning accuracy can be increased by deploying a denser site-based wireless infrastructure. IPSs may or may not need to individually report the ID of the various targets under consideration; when individual IDs are required, the sensor network must be able to establish from which tag it has received a given signal and then undertake an appropriate

TABLE 6.3 RTLS Comparison

	RFID-passive	RFID-active	Bluetooth	Wi-Fi (WPS)	UWB	IR	Gen2IR™	Cellular
Deployment/usage	Very common	Very common	Broadly deployed technology	Broadly deployed technology	Currently more common in manufacturing/warehouses	Well-established technology	Newer	Ubiquitous
Approach	Passive tags	Tags carry a small battery to boost signal strength; receiving sensors determine tag location by calculating signal strength	Similar to Wi-Fi, but functioning on a shorter range	Multiple WAPs measure the relative signal strength of entities to approximate their position inside a facility	Relatively complex tags	Infrared system	Infrared system	Cellular: GSM, CDMA, 4G/LTE, 5G
Power usage	None	Low-to-Medium	Medium	High	High	Low-to-Medium	Low-to-Medium	High
Coverage area	Low	Medium	Medium	Medium-to-High	Medium	Low	Medium	Medium-to-High
Hardware costs	Tags: Low; Readers: Medium	Medium	Medium	Low if using existing wireless access points	Medium-to-High	Low	Medium-to-High	High if purpose-built
Advantages	No battery required in tag, low tag cost	Reliable RTLS system; Small sensors; Low reader cost; Works well in crowded environments; No Line-of-Sight (LOS) needed.	Can "piggyback" on existing infrastructure; No LOS needed; signals can pass through clothing and other light materials; Ubiquitous: most laptops, tablets, smartphones, and other electronics already support	Can 'piggyback' on existing wireless infrastructure with minimal additional hardware; No LOS required	Highly precise; Resilient	Well-established cost-effective technology	Highly precise; No LOS is necessary; Readers can filter out ambient radiation; Virtual "walls" can be set up to fit specific floorplans	No Interference, cellular networks operate in regulated bands, so interference is not an issue; Wide Range

Disadvantages	Short range (needs multiple readers)	Tag requires battery, but can last 1–3 years	Power consumption is relatively high (battery life typically a few weeks); does not scale since it is short range system; interference possible since standard is widely used on many devices	Bleed-through: relatively low accuracy, high power consumption (battery life typically a few days); not ideal for precision positioning	Relatively expensive; batteries in tags need routine recharging every few days	LOS required; light-based: what blocks or disrupts light also affects IR signals; ideal for smaller, controlled indoor spaces with appropriate lighting	New sensor infrastructure networks are required; difficult to "piggyback" on existing wireless infrastructure.	Difficult to use indoors; signal repeaters or DASs may be necessary for cellular RTLS to function indoors. lower relative accuracy (lower than with WPS)

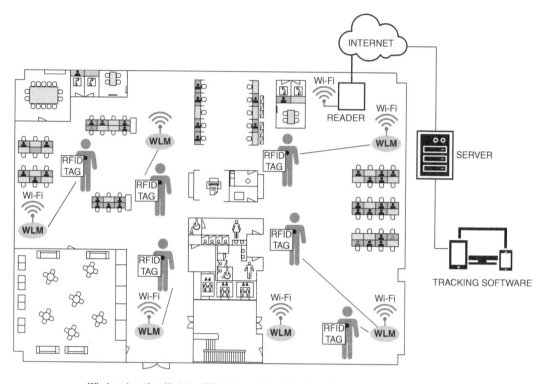

Wireless location Modules (WLMs) are strategically located within and around the office. The read/range-zone of each WLM is often adjustable to detect tags and badges within that specific zone or area. The read range zone of each WLM can be adjusted to read badges or tags between 2 feet and up to 250 feet. The WLMs transmit radio signals to the readers to report the exact location of the badge or tag in real time.

RFID readers receive radio signals from all WLMs within their read zone. The readers can be tuned to read all the badges or tags between 2 and 600 feet zones. RFID readers transmit the location of personnel and assets to the software application. The software application monitors the real time location of all personnel and assets and enforces the individual OSD or other rules.

FIGURE 6.7 Illustrative example of an RFID RTLS.

mapping. Typically, IPSs can detect the location of an object but they cannot be used to detect the direction of motion or the orientation of an object. Tracking, however, may be possible by establishing a sequence of locations, thus forming a trajectory from an initial location to the current location. The most-widely used techniques entail distance/angle estimation combined with position and localization computation algorithms.

The "M1 (method)" taxonomy for location tracking and positioning systems discussed earlier can be further expanded by classifying the measurement techniques used to determine mobile device location, given that approaches that can be utilized differ in terms of the specific technique employed to sense and measure the position of the mobile terminal in the zone of interest. Thus, the "M1 (method)" can be extended by grouping RTLS systems into four basic categories of systems that determine position based on the following: (i) M11: micro-zone or cell of origin (nearest cell); (ii) M12: distance (lateration, often based on power assessments); (iii) M13: angle (angulation); and (iv) M14: location patterning (pattern recognition). In practice, an RTLS system can implement several of these techniques. For example, some approaches attempt to optimize performance in two or more environments with different propagation characteristics, as is the case for dual-scope indoor and outdoor RTLSs where two different techniques might be appropriate.

Most IPSs utilize a continuous physical measurement (such distance, or angle and distance) in conjunction with the identification information in one combined signal. Angle of Arrival (AoA), Time of Arrival (ToA, aka Time of Flight), Received Signal Strength Indicator (RSSI), and RF Fingerprinting (RFF) methods are often used.

- AoA is the angle from which a signal arrives at a receiver. AoA is typically used with triangulation and a known baseline to establish the location relative to two anchor nodes. AoA can be determined by measuring the time difference of the signal arrival between multiple antennas in a tag, for example in a smartphone. In other cases, AoA can be determined by an array of highly directional sensors.
- ToA/Time (Difference) of Arrival (TDoA), is the propagation time – Time of Flight – a signal incurs from the transmitter to the receiver. The travel time of the signal is used to calculate distance; trilateration, multilateration, and successive measurements can be combined to determine the object's location, by processing signals from multiple tags. The signals that a tag transmits may include the location of the transmitting tag. Alternatively, the tag may transmit a reference, e.g. an ID to the outside world that can be used to look up its location. The user's mobile receiver can observe the Time of Arrival for signals from multiple tags and by calculating this difference in Time of Arrival, it can figure out its own location based on the locations of the tags. It should be noted that TDoA normally refers to the time difference of arrival from the same source to different receivers, but in certain implementations it also refers to the time difference of arrival from different sources to the same receiver. A common time reference in a TDoA network is necessary; however, since each tag is completely unaware of its network, it cannot be synchronized. In RTLS systems, a reference transmitter is normally used as a way to provide a time reference. For example, a reference receiver whose location is known and that can tell the mobile receiver about when the reference receiver captures a particular tag signal is provided so that mobile devices including the mobile receivers can calculate their position using that reference information. Once the mobile device acquires its own location, e.g. inside an airport or shopping mall, it can calculate the relative locations of registered businesses and services, what specials they offer and how to get there [70]. Multipath reflection and RF diffraction can deteriorate the measurements.
- RSSI is a measurement of the power level received by the sensor in the tag. Electromagnetic waves propagate in free space according to the inverse-square law; namely, when there are no losses caused by absorption or scattering, the power per unit area perpendicular to the direction of propagation, of a spherical wavefront varies inversely as the square of the distance from the source. At short ranges typical of office environments, distance can be calculated in an approximate manner based on a function between transmitted and received signal strength. Buildings internals have reflecting and absorbing elements, not the least being walls, doors, furniture, and even people; therefore, signal strength measurements can be noisy and position accuracy may be degraded.
- RFF combines the simplicity of an RSSI-based lateration approach with calibration capabilities to achieve improved indoor performance. RF Fingerprinting enhances RSSI-based lateration using RF propagation models developed from data gathered in the target environment (or environments very similar to it). RFF enables one to calibrate an RF model to a particular environment in a fashion similar to (but more efficient than) that of location patterning; unlike location patterning; however, a unique custom site calibration is not always required, especially in situations where multiple floors of similar construction, contents, and layout are deployed [72].

Once the readers have derived the distance between themselves and the tag, the next step is to compute the position of the tag within in the area. Several computation methods exist [73]:

- Trilateration/Multi(Tri)lateration: after the tag distance from each reader has been calculated using one or a mix of the methods discussed above, a circle is logically drawn from each reader with the radius equal to the distance of the tag from the respective reader. If there are three readers, then the three logical circles drawn meet at one or more points; the locus, where the most intersections exist, is taken to provide the location of the tag. For many basic positioning applications, trilateration using the RSSI algorithm provides sufficient accuracy and achieves the lowest Total Cost of Ownership (TCO).
- Triangulation, with the use of AoA: each reader in the network knows the angle between the reader and the tag. The distance between readers is predefined. Thus, using trigonometric identities, the other readers reference their angles to the respective readers, to calculate the position of the tag.
- Probabilistic approaches with trilateration: these methods (which have a high computational requirement) are based on the use of readers and reference tags to create a matrix of possible locations which is refined as more data are acquired from other readers about the tag.

The Appendix at the end of the chapter provides a glossary of some basic terms and concepts relative to positioning technologies for RTLS (some terms based on references [70, 74–76, 77] among others). Basic office/site topological designs include the following:

- Locating tag-wearers at choke points: this approach employs short-range tag signals from a moving tag which are received by a single fixed (but networked) reader – networking allows the signals to be relayed, typically using a backbone wireless channel, to a location-processing system.
- Locating tag-wearers in relative coordinates: in this approach, tag signals from a moving or stationary tag are received by a multiplicity of known-position readers, and a tag's position is estimated using locating mechanisms, such as triangulation, trilateration, or multilateration – again, networking allows the signals to be relayed, typically using a backbone wireless channel, to a location-processing system.

A high-level description of some of the key wireless systems in use follow. Additional details are provided in other chapters: position/distance tracking using UWB is discussed in more details in Chapter 7 and position/distance tracking using Wi-Fi, Bluetooth, and Cellular Technologies is discussed in Chapter 8.

6.3.1 RFID Systems

As noted earlier, RFID systems can be passive, semi-active, or active. "Passive" refers to tags that acquire their electrical power primarily by scavenging incident or ambient sources of energy, such as electromagnetic (e.g. RF, visible light, etc.), mechanical, and thermal energy. Currently, passive RFID tags are widely used for RTLS in tracking goods, assets, and people.

The RFID system includes a reader, a tag, and an application system. When the tag appears in the operational range of the reader, it starts receiving both energy and data via its antenna from the reader via its transmitter/receiver and antenna. A rectifier circuit in the tag collects and stores the energy for powering the other circuits (e.g. control/modulator) in the tag. After collecting enough energy, the tag may operate and send back prestored data to the reader. The reader then passes the received response data via a communications interface to the server system/database of the application system for system applications [68]. Radio communication between an RFID transponder tag and an interrogator/reader can be carried out in two different approaches [78]:

- The first approach involves use of a circuitry in the tag. When exposed to the electromagnetic or acoustic field generated by the reader, the tag antenna comes into oscillation or similarly can couple with the reader field. The tag can use this coupling effect, which manifests itself as an alteration of the original field generated by the reader to present its ID or data. This coupling can be used to link the reader and the tag together. When RF is used, this coupling can be magnetic coupling (near-field electromagnetic coupling) or backscattering (far-field electromagnetic coupling). The electromagnetic field generated by a reader's antenna induces a current in the tag whose receiver is tuned to the frequency of the field. When the wavelength of the frequency range used greatly exceeds the distance between the reader's antenna and the tag, the electromagnetic field may be treated as an alternating magnetic field and be considered as a transformer with one coil (antenna) located on the reader and the other coil (antenna) located on the tag. Magnetic coupling is commonly deployed in LF (Low Frequency) and HF (High Frequency) bands. The most popular frequencies for magnetic coupling are 135 kHz and 13.56 MHz. When the far-field electromagnetic coupling is deployed, the tag modulates its data back on to the electromagnetic field of the reader by changing the impedance of its own receiving antenna. This change of impedance causes the tag antenna to effectively act as a reflector. Changes in the antenna impedance effectively reflect some of the electromagnetic energy back to the reader; the reader can then perceive the pattern of the modulation in the reflection; this phenomenon is called backscattering. Under these circumstances, the reader can sense the presence of the tag, transmit data, and receive the response back from the tag by demodulating the data that the tag has modulated into the field pattern caused by magnetic coupling or backscattering. There are variations of the first approach that use backscattering in a band whose center frequency is an integer multiple or fraction of the center frequency of the received signal, but this flexibility is limited to similar simple techniques.
- The second approach is to have a setup similar to the one in conventional RF communication. The readers transmit signals that are received by the tags and the tags transmit signals, by means of a transmitter stage, that can be detected and decoded by the readers. With this approach, the structure of the signal transmitted by the tag is inherently independent of the signal received by it. The tag can receive information from the reader in one band and transmit it in a completely unrelated band and with a different signal structure and technology.

In the first approach, the tag can be a completely passive element: the tag can be powered up by rectifying the incident signal and since it is merely reflecting back the incident continuous wave (CW), it does not need to take on the power-consuming task of generating a CW as a signal carrier for transmission. The passive tag responds by presenting its ID or other data through manipulating the incident signal that is in turn sensed by the reader monitoring the frequency band in which the particular modulation is expected. Backscattering relies on small signal reflections that only offer a limited range and a low bandwidth for data exchange between the tag and the reader. As described in [68], backscattering is typically used in microwave band RFID systems. The antenna of the reader couples energy to the tag and by modulating the reflection coefficient of the tag's antenna, data may be transmitted between the tag and the reader (also as seen in Figure 6.8). Power P_{in} is emitted from the reader's antenna. A small proportion of P_{in} is received by the tag's antenna and is rectified to charge the storing capacitor in the tag for serving as a power supply. After gathering enough energy, the tag begins operating. A portion of the incoming power P_{in} is reflected by the tag's antenna and returned as power P_{return}.

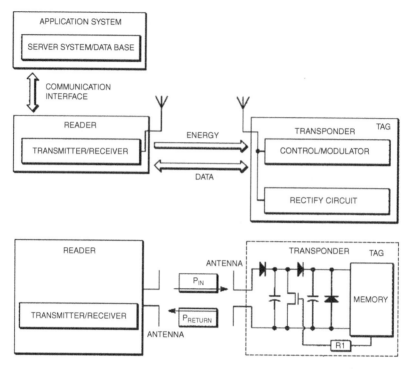

FIGURE 6.8   RFID environment (top) and backscatter (bottom) [68].

The reflection characteristics may be influenced by altering the load connected to the antenna. To transmit data from the tag to the reader, for example, a transistor may be switched on and off in time with the transmitted data stream. The magnitude of the reflected power P_{return} may thus be modulated and picked up by the reader's antenna.

Passive tags are inexpensive to manufacture. At short range, they can supply their needed power by capturing electromagnetic energy through simple and affordable power rectification circuitry located on the tag [78]. A problem with passive tags is that the energy scavenged is generally quite small. Such small amounts of energy are normally not enough to generate powerful and high-performance radio carrier waves for data transmission, and, therefore, a system based on these tags tends to be limited in range and data capacity [70].

In the second approach, transmitting the data back to the reader requires power just like any other RF transmission because a carrier needs to be generated and depends in part on the frequency of the carrier, the complexity of the modulation scheme, and the required power output. The amount of required power can easily fall outside what can be recovered from the incident signal. Therefore, the second approach is often only applicable to the category of active or semi-active transponders.

In addition to the mechanism needed to modulate and propagate the response of the tag to the reader, other functional units in the tag require power; one such unit is the logic engine that processes and transports the stored ID or data. Power can be provided by a source of energy that is integrated with the tag, e.g. a capacitor, a battery, or an accumulator of some kind. It can also be generated by other means, e.g. by capturing the electromagnetic energy propagated by the reader or similar sources of energy-carrying signals. The former category of tags is called active and the latter is called passive; the hybrids constitute the category of semi-active tags. In the case of RF signals, the process of power recovery from the incident signal requires a circuitry that can convert electromagnetic energy to such current and voltage levels that can satisfy the power needs of the tag.

Regardless of whether the tag acts as an active transmitter or backscatters passively, communication between a tag and a reader is performed in specific, regulated frequency bands. The amount of output power in each band is regulated to protect other devices and bands against interference and saturation [78]. These bands are normally narrow bands in LF (0.03–0.3 MHz), HF (3–30 MHz), UHF (Ultra High Frequency (300–3000 MHz)) and Microwave portions of the RF spectrum.

The approach of location indexing and presence reporting for tagged objects makes use of known sensor identification only. This approach does not report the signal strengths and various distances of single tags; furthermore, they do not update the location coordinates of the sensor or tags. The reach of these sensors typically covers a floor, a corridor, or a single room; thus, they require a defined entry/passage locus for the tags to be "read." Alternatively, a relatively dense network of low-range receivers may be laid out in a grid pattern permeating the space to be observed. The tagged entity will be identified by only a few neighboring receivers allowing a coarse approximation of the tag's location.

The passive tags transmit signals that are captured by a mobile receiver of a passerby. The mobile receiver can simultaneously capture signals from multiple transmitters and, by correlating the condition of the received signals, can calculate a geometric relationship between the transmitting tags and the receiving handheld device. This relationship can be mapped onto an absolute frame of reference.

An RTLS typically outputs the location of each sensor or RFID tag with two or three coordinates relative to a fixed, known position of the RFID/sensor reader. In a fixed-reader infrastructure in indoor environments such as warehouses, distribution centers, or retail stores, the coordinates of each reader, excitation node, and associated antennas are typically manually measured and individually input to the system by a human operator when initially configuring the system. RFID systems can also include mobile RFID readers. Mobile RFID readers can be handheld units operated by humans, robots, or drones equipped with an embedded RFID reader. In mobile applications, the position of the reader varies when the reader is moved for the purpose of covering different regions in a physical area. Figure 6.9 depicts (conceptually) some of the methods at the block-processing level, illustrating different systems for generating location estimates from location information sources [79].

For comparison, contactless smart cards (SCs) are more sophisticated than RFID tags, being that they contain a microprocessor that enables (i) onboard computing, (ii) two-way communication including encryption, and (iii) storage of predefined and newly acquired information. Because of their more restricted capabilities, RFID tags are typically less expensive than SCs.

6.3.2 Wi-Fi-based Positioning System (WPS)

In this approach, WAPs operate in conjunction with Wi-Fi-based tags. WPSs typically utilize WAPs to measure the received signal strength (RSS) of transmitted signals from mobile stations and to determine the RSSI. WPS also uses the methodology of "fingerprinting." The SSID and the MAC address of the WAP or of the Wi-Fi hotspot are utilized to associate the mobile device with the anchor node.

Typically, location systems will use either (i) one of WAPs for RSSI processing or (ii) TDoA processing with various WAPs and tag transmitters that conform with the ISO 24730 RTLS standards (see Section 6.4). With TDoA methods, a device receives a transmitted signal at multiple physical locations corresponding to a receiver to locate a tag transmitter. Time of Arrival measurements made with respect to independent (randomly related) clocks contained in non-synchronized receivers can be accurately related to each other if the receivers observe signals from sources at known locations. A WLAN signal source can be placed in a known fixed location that propagates directly to all participating receivers. Each receiver can measure the Time of Arrival for transmissions from the reference source and communicate them to a central clock

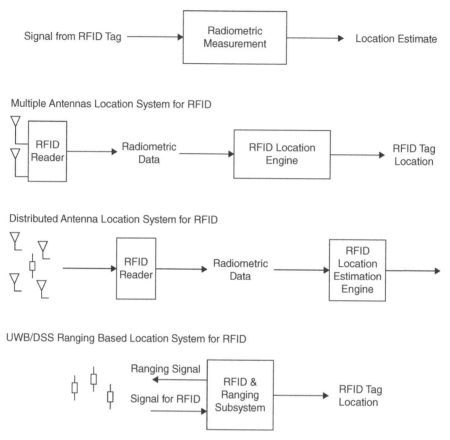

FIGURE 6.9 Various systems for generating location estimates from location information sources (from [79]).

tracking service, which may be implemented in hardware or software. Knowing how much time is required for the reference source's signal to propagate to each receiver, the tracking service can evaluate what time each receiver clock indicated when the source is transmitted. By observing reference signals sufficiently often, the tracking service can determine a continuous relationship between the independent clocks. This relationship can then be used to accurately relate Time of Arrival measurements for signals arriving from unknown locations [80].

SLAM is a Wi-Fi IPS system that enables a smartphone to pinpoint its location in real time to 10-ft accuracy utilizing ambient Wi-Fi (the developing firm WiFiSLAM was acquired by Apple in 2013). The SLAM system collects location information by recording "trajectories" from smartphone sensors such as accelerometers, gyroscopes, and magnetometers; these data are based on relative Wi-Fi signal strength in order to assess where the smartphone user is in a building. When a sufficient number of people walk through a building with Wi-Fi activated on their smartphones, a map of the building interior begins to form: raw trajectories from the inertial sensors instantiated by several smartphones, combined with Wi-Fi data, will help to create a map of the inside of a building [81].

6.3.3 Bluetooth

In the context of this application, originally, Bluetooth™ was aimed at proximity (geofencing), allowing it to be employed as an indoor proximity methodology, not an IPS *per se*. More recently, indoor mapping has been achievable in Bluetooth, particularly Bluetooth Low Energy (BLE).

The mechanism of iBeacons has been utilized to that end. Bluetooth 5.1 allows the angle to the antenna to be measured, thus enabling position accuracy to the decimeter level.

6.3.4 UWB

A UWB tag-based RTLS provides the best possible accuracy for localized environments. UWB allows one to pinpoint location within millimeters at a range of up to 30 m (~100 ft), facilitating the tracking of goods and assets within a centimeter or millimeter – depending on the application. However, the power usage of the UWB tag may be fairly high, reducing battery life. Micro-location and presence detection systems are useful to a large set of businesses; for example, UWB beacons support logistics operations, improving tracking and location of assets in large warehouses to track and locate assets at the site.

6.3.5 Automatic Vehicle Location (AVL)

On larger geography, AVLs are systems for automatically establishing and relaying the location of a vehicle. Signpost systems are used to track and locate vehicles along fixed routes, for example, on transit routes, tunnels, rail lines, subways, tramways, where the tracked vehicles continually travel on the same, well-defined route.

6.4 STANDARDS

A number of standards have emerged for RTLS, but they only cover a subset of the available technologies. Furthermore, the methodology for computing locations or measuring distances are not often specified, if at all. Vendor interoperability is not necessarily a major stakeholder concern at this juncture.

ISO/IEC 19762: 2016, entitled Information Technology - Automatic Identification and Data Capture (AIDC) Techniques - Harmonized Vocabulary, defines general terms and definitions in the field of automatic identification techniques and data entry. In particular, Part 5, ISO/IEC 19762-5: 2016 provides terms and definitions unique to locating systems in the area of automatic identification and data capture techniques.

The major family of standards is ISO/IEC 24730; this family defines air interface protocols and an Application Program Interface (API) for RTLSs used in asset management. ISO/IEC 24730: 2014 consists of the following parts [82]:

- Part 1: Application Programming Interface (API)
- Part 2: Direct Sequence Spread Spectrum (DSSS) 2.4 GHz air interface protocol
- Part 21: Direct Sequence Spread Spectrum (DSSS) 2.4 GHz air interface protocol: Transmitters operating with a single spread code and employing a DBPSK data encoding and BPSK spreading scheme
- Part 22: Direct Sequence Spread Spectrum (DSSS) 2.4 GHz air interface protocol: Transmitters operating with multiple spread codes and employing a QPSK data encoding and Walsh offset QPSK (WOQPSK) spreading scheme
- Part 5: Chirp Spread Spectrum (CSS) at 2.4 GHz air interface
- Part 61: Low-rate pulse repetition frequency Ultra-Wideband (UWB) air interface
- Part 62: High-rate pulse repetition frequency Ultra-Wideband (UWB) air interface

An API is a boundary across which application software uses facilities of programming languages to invoke services. These facilities may include procedures or operations, shared data objects, and resolution of identifiers. A wide range of services may be required at an API to support applications. Different methods may be appropriate for documenting API specifications for different types of services. The information flow across the API boundary is defined by the syntax and semantics of a particular programming language, such that the user of that language may access the services provided by the application platform on the other side of the boundary. This API describes the RTLS service and its access methods to enable client applications to interface with the RTLS system. This RTLS service is the minimum service that needs to be provided by an RTLS to be API compatible with this standard.

There are a number of standards for RFIDs; some of the key standards include the following:

- The ISO 14443 standard describes components operating at 13.56 MHz frequency that embed a CPU; power consumption is about 10 mW; data throughput is about 100 Kbps and the maximum working distance (from the reader) is around 10 cm.
- The ISO 15693 standard describes components operating at 13.56 MHz frequency, but it enables working distances as high as 1 m, with a data throughput of a few Kbps.
- The ISO 18000 standard defines parameters for air interface communications associated with frequency such as 135 KHz, 13.56 MHz, 433 MHz, 860–960 MHz, 2.45 GHz, and 5.8 GHz. The ISO 18000-6 standard uses the 860–960 MHz range and is the basis for the Class-1 Generation-2 UHF RFID, introduced by the EPCglobal Consortium.

The following is a more detailed listing of key specifications supporting basic RFID operations:

- EPCglobal™: *EPC™ Tag Data Standards*
- EPCglobal (2004): *FMCG RFID Physical Requirements Document*
- EPCglobal (2004): *Class-1 Generation-2 UHF RFID Implementation Reference*
- EPCglobal (2005): *Radio-Frequency Identity Protocols, Class-1 Generation-2 UHF RFID, Protocol for Communications at 860–960 MHz*
- European Telecommunications Standards Institute (ETSI), EN 302208: *Electromagnetic compatibility and radio spectrum matters (ERM) – Radio-frequency identification equipment operating in the band 865–868 MHz with power levels up to 2 W, Part 1 – Technical characteristics and test methods*
- European Telecommunications Standards Institute (ETSI), EN 302208: *Electromagnetic compatibility and radio spectrum matters (ERM) – Radio-frequency identification equipment operating in the band 865–868 MHz with power levels up to 2 W, Part 2 – Harmonized EN under article 3.2 of the R&TTE directive*
- ISO/IEC Directives, Part 2: *Rules for the structure and drafting of International Standards*
- ISO/IEC 3309: *Information technology – Telecommunications and information exchange between systems – High-level data link control (HDLC) procedures – Frame structure*
- ISO/IEC 15961: *Information technology, Automatic identification and data capture – Radio frequency identification (RFID) for item management – Data protocol: application interface*
- ISO/IEC 15962: *Information technology, Automatic identification and data capture techniques – Radio frequency identification (RFID) for item management – Data protocol: data encoding rules and logical memory functions*
- ISO/IEC 15963: *Information technology – Radio frequency identification for item management – Unique identification for RF tags*
- ISO/IEC 18000-1: *Information technology – Radio frequency identification for item management – Part 1: Reference architecture and definition of parameters to be standardized*

- ISO/IEC 18000-6: *Information technology automatic identification and data capture techniques – Radio frequency identification for item management air interface – Part 6: Parameters for air interface communications at 860–960 MHz*
- ISO/IEC 19762: *Information technology AIDC techniques – Harmonized vocabulary – Part 3: radio-frequency identification (RFID)*
- U.S. Code of Federal Regulations (CFR), Title 47, chapter I, Part 15: *Radio-frequency devices, U.S. Federal Communications Commission*

EPCglobal is a neutral, not-for-profit standards organization consisting of manufacturers, technology solution providers, and retailers. Many industries participate in the EPCglobal standards development process, such as aerospace, apparel, chemical, consumer electronics, consumer goods, healthcare and life sciences, and transportation and logistics.

Additional RTLS standards are identified in Chapter 7.

6.5 APPLICATIONS

Usage of IPSs has gained popularity in recent years, and RTLSs are widely used. RTLSs are especially popular in the healthcare industry for a variety of applications ranging from tracking assets and people (e.g. patients and staff), sensing patient vital signs (e.g. temperature), hygiene compliance, elopement (i.e. unauthorized patient disappearance), theft prevention, and so forth [83]. Additionally, technologies for indoor location tracking have been used for location-aware services in public buildings such as museums, transit stations, airports, or malls. RTLS applications have also included retail, inventory tracking industries, and healthcare applications, with OSD now being contemplated as the latest application.

The social distancing feature of RTLSs helps employees maintain a separation (e.g. 6 feet) with an alert from a wearable tag. Typically, tags automatically identify each other when they come near. Contact tracing mechanisms document the history of these contacts in a database. In some environments, the tags are able to relay their contacts to an institutional server or the cloud software platform over existing wireless networks. Tags can be utilized to only support alerting for OSD management or can be utilized to also support contact tracing. Contact tracing requires the use of an ancillary software platform, which among other functions it typically provides reports to monitor trends and evaluate the effectiveness of established or newly modified workflows and instituted OSD policies.

Among many other applications, RTLSs have also been proposed to track time and attendance of an individual at the workplace. Such a location tracking system detects the presence of a portable electronic device carried by the individual and includes a time clock system that records clock-out registration time of the individual at the workplace. The system includes a main controller. If the main controller receives an electronic communication indicating that the portable electronic device is located at a predefined break or nonwork area, and if the main controller does not detect recording of the individual clock-out registration time at the time clock system, the controller automatically causes an alert at the portable electronic device prompting the individual to transmit a feedback signal acknowledging need to record the clock-out registration time [84]. Such a system could be adapted to support OSD/ODCMA requirements.

Within the domain of healthcare, RTLSs are predominantly used in hospitals and to a lesser extent in nursing homes [33]. The healthcare industry routinely deals with issues concerning the safety of the facility, the quality of care provided to patients, and the cost of operations and associated workflows. RTLSs are used to provide real-time or near-real-time tracking and management of medical equipment, specimens, expensive medications, staff, and patients. Hospital-based RTLS solutions typically include battery-based location sensors with unique IDs that are attached to key assets. Typical RTLS solutions allow tracking to the granularity of a hospital's unit or floor, whereas UWB-based systems are able to achieve room, bed, bay, and shelf-level tracking. The hierarchy of location accuracy within a healthcare facility often falls into the following scheme [85]: locating at entry or exit points, presence-based locating, room-level locating, sub-room level locating, precise pinpoint locating. Along this continuum, wireless technologies such as Wi-Fi, UHF Active RFID, and BLE are utilized for the coarse-grain applications (e.g. 3-m accuracy is possible with BLE), while technology such as UWB can be used for the fine-grain applications. Within nursing homes, RTLSs are mainly used to track residents, for instance, in the case of wandering behavior. The localization functionality of such systems is often combined with other monitoring technologies, such as fall detection [8]. RTLSs are used for two primary purposes in nursing homes: to ensure the safety of residents and to support personnel in (efficiently) caring for the residents. Risky wandering behavior is managed by tracking the location of residents and preventing them from wandering off too far, exiting, or accessing restricted areas [8]. Tags for people with dementia can be designed as wearables, for example, in shoes, incorporated in wristbands, or worn on the body; smartphones are gaining popularity when it comes to the location of older people [86–89].

Instead of needing separate tags, future medical assets may have integrated tags; currently, active and semi-active tags still need battery power to be operational, and wireless charging possibilities may offer a solution to this problem. Furthermore, in the near future, unobtrusive, inexpensive, and simple IPSs will no longer be based on tags that have to be worn by people; instead, the building itself and its infrastructures will monitor the people inside [88, 89]. The use of smartphone applications in relation to real-time monitoring of people and objects (say with BN-IoT technology [90]) is expected to become more dominant. Such approaches could potentially enhance the ease of use, as smartphones are familiar and frequently used devices [8].

Maintaining the safety and health of workers is a major concern across many industries. Various rules and regulations have been developed to aid in addressing this concern, which provide sets of requirements to ensure proper administration of personnel health and safety procedures. To help maintain worker safety and health, some individuals may be required to don, wear, carry, or otherwise use Personal Protective Equipment (PPE) articles if the individuals enter or remain in work environments that have hazardous or potentially hazardous conditions. Known types of the PPE articles include respiratory protection article (RPE), e.g. for normal condition use or emergency response; protective eyewear, such as visors, goggles, filters, or shields; protective headwear, such as hard hats, hoods, or helmets; hearing protection; protective shoes; protective gloves; other protective clothing, such as coveralls and aprons; protective articles, such as sensors, safety tools, detectors, global positioning devices, mining cap lamps, and any other suitable gear. Respirator rules or regulations may define that certain classes of filtering-facepieces be used with particular contaminant or virus and/or for certain exposure times based on contaminant concentration levels. Compliance with the predetermined criteria may become an issue in work environments involving a relatively large number of workers and/or respirators because of the relative difficulty in tracking worker habits and diligence. The PPE article tracking compliance systems have been proposed, as described in reference [74]. Such system may include an information retrieval system networked to a computer or system; such system supports processes for determining a condition of one or more PPE articles; the PPE articles may be coupled to a passive or active RFID tag in an RTLS that can determine the position of the smart tag in a two-dimensional or three-dimensional space, as seen in Figure 6.10. As noted, the data from the smart tag may be acquired by data-acquiring devices, such as readers, readers/writers, scanners, or receivers, such as wireless receivers, as well as other suitable devices. The readers may be linked to a remote system that enables tracking usage of the PPE articles against at least a predetermined criterion. The readers may be programmable electronic systems themselves. These functionalities may include but are not limited to certifications regarding using, servicing, repairing, cleaning, maintaining, decontaminating, or other processing of the

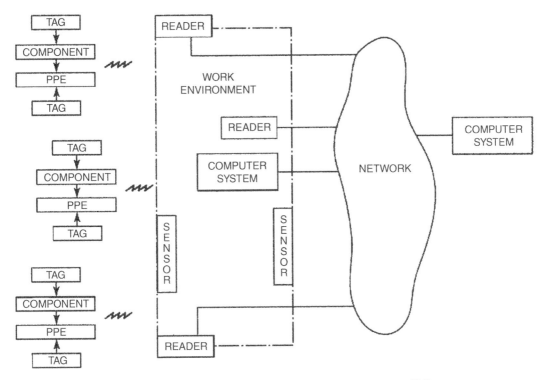

FIGURE 6.10   PPE article tracking compliance systems [74].

PPE articles. For example, if the cumulative exposure time of the RPE article in the workplace exceeds a certification value(s); the concentration level(s) of particular contaminants exceed certification value(s); the presence of unexpected contaminants in the working environment; persons with particular profiles should not be exposed to various contaminants; particular kinds of PPE articles should or should not be used when certain contaminants are present. The reader may be stationed at the entrance of the work environment and acquires relevant data of the wearer; component, and the PPE article, such as one or both of: at the start of the workday or shift and at the end of the workday or shift. Alternatively, or additionally, one or more readers may be located within the actual work environment so as to provide opportunities for wearers to obtain readings in the work environment.

While the principal application of RTLS is in confined space, they can also be used in smart city applications as noted in Table 6.4 – in particular, smart buildings, smart hospitals, smart services are all part of the fabric that constitutes a smart city [91]. Besides other amenities and opportunities, people opt to live in major cities due to access to quality healthcare institutions and hospitals; however, the recent pandemic may perhaps modulate certain in-progress trends related to urbanization, at least for the short-to-medium term.

RTLSs have also been used in the context of Location-Based Services (LBS) to identify and track devices or people and, thus, obtain and use location data to control features for devices. In many daily business and social contexts, the position of mobile users can guide them toward premises, services, and businesses. For example, knowing one's position in a shopping mall, a department store, an office complex, an airport, etc., can simplify the task of navigating through the place and finding the locations of interest. These locations of interest may comprise retailers, restrooms in the locations of events and activities [70]. For example, stores in a downtown street or in an airport that are interested in attracting customers and guiding them to their locations attach passive tags to their locations, e.g. their storefronts. These passive tags scavenge, for

TABLE 6.4   Possible RTLS Applications in Smart Cities

Smart City Applications	Possible RTLS Applications
Intelligent Transportation Systems (including Smart Mobility, Vehicular automation, and Traffic control)	Autonomous Cars Navigation; Proximity determination; Presence determination; Real-Time Location systems (RTLSs); Toll reading; Goods tracking; Asset tracking; People Tracking; Vehicular Safety; Fleet Management; Automatic Vehicle Location (AVL); Rail Signpost systems; Tunnel Signpost systems; Onboard sensors and cameras for train positioning; Tags for sidewalk navigation of blind people
Smart Building	Office/conference room occupancy management
Smart Institutions	RTLS; Equipment tracking; Health Care; Patient tracking; Doctor tracking; Security/Access control; Student tracking; Smart living
Sensing (including Crowdsensing, Smart Environments, and Drones)	Asset tracking; Robotics; Haptic devices; Environmental monitoring with sensors on city vehicles to monitor environmental parameters; Crowdsensing, where the citizenry at large utilizes wearables, smartphones, and car-based sensors to collect and forward data for aggregation for a variety of visual, signal, and environmental information
Surveillance/Intelligence	Security/Access control; Proximity determination; Presence determination; RTLS
Smart Public Services	Asset tracking; Near Field Communication (NFC) in IoT-Based Payment/Banking; Intelligent Bus Service
Goods and products Logistics (including Smart Manufacturing)	Asset/device tracking in warehouse/distribution; Micro-location and presence detection; Precision Robotics; Monitoring safety and supply chain-based data
Smart Grids	Inventory management
Lighting Management	Proximity determination; Presence determination
Waste Management	Fleet Management
Water Management	Asset tracking

example, power from ambient light, vibrations, or RF signals. A store or service site that identifies itself by an associated tag will help users to identify it electronically from a distance. Moreover, a constellation of several tags from multiple installations can cooperatively provide customers carrying a mobile receiver the ability to locate navigate effectively.

RFID providers include but are not limited to the following as of press time: Zebra Technologies, AeroScout Industrial, Alien Technology, Avery Dennison, Impinj, Roper Technologies, Inc., FleetCor Technologies, Inc., Omnicell, Inc., Gaming Partners International Corporation, SuperCom Ltd., Checkpoint Systems, Nedap, Tyco Retail Solutions, Smartrac, Alliance Tech, Radiant RFID, GlobeRanger, OrbComm, Mojix, ThinkMagic, Gao RFID, Caen RFID, NXP Semiconductors, Broadcom, and Omni-ID.

REFERENCES

1. Minoli, D., Sohraby, K., and Occhiogrosso, B. (2017). IoT considerations, requirements, and architectures for smart buildings – energy optimization and next generation building management systems. IEEE Internet of Things Journal 4 (1): 269–283. https://doi.org/10.1109/JIOT.2017.2647881.
2. Hsu, J. (2020). Can AI make bluetooth contact tracing better? Machine learning shows some promise in boosting contact tracing technologies meant for detecting nearby phones. IEEE Spectrum (8 September 2020).
3. Tian, X., Lee, P.M., Tan, Y.J. et al. (2019). Wireless body sensor networks based on metamaterial textiles. Nature Electronics 2 (6): 243–251.

4. Corchia, L., Monti, G., De Benedetto, E. et al. (2020). Fully-textile, wearable chipless tags for identification and tracking applications. Sensors 20: 429.

5. Sousa, M., Techmer, A., Steinhage, A. et al. (2013). Human tracking and identification using a sensitive floor and wearable accelerometers. 2013 IEEE International Conference on Pervasive Computing and Communications (PerCom 2013) (March 2013), pp. 166–171.

6. Zeng, Y., Pathak, P.H., and Mohapatra, P. (2016). WiWho: WiFi-based person identification in smart spaces. 2016 15th ACM/IEEE International Conference on Information Processing in Sensor Networks, IPSN 2016 - Proceedings.

7. Niu, S., Matsuhisa, N., Beker, L. et al. (2019). A wireless body area sensor network based on stretchable passive tags. Nature Electronics 2 (8): 361–368.

8. Oude Weernink, C.E., Felix, E., Verkuijlen, P.J.E.M. et al. (2018). Real-time location systems in nursing homes: state of the art and future applications. Journal of Enabling Technologies 12 (2): 45–56. ISSN: 2398-6263. https://doi.org/10.1108/JET-11-2017-0046. Also at https://www.emerald.com/insight/content/doi/10.1108/JET-11-2017-0046/full/html#ref033. License: Published by Emerald Publishing Limited. This article is published under the Creative Commons Attribution (CC BY 4.0) license. Anyone may reproduce, distribute, translate and create derivative works of this article (for both commercial & non-commercial purposes), subject to full attribution to the original publication and authors.

9. Madrid, C., Korsvold, T., Rochat, A., and Abarca, M. (2012). Radio frequency identification (RFID) of dentures in long-term care facilities. The Journal of Prosthetic Dentistry 107 (3): 199–202.

10. Lai, H.M., Lin, I.C., and Tseng, L.-T. (2014). High-level managers' considerations for RFID adoption in hospitals: an empirical study in Taiwan. Journal of Medical Systems 38 (2): 3.

11. Krohn, R. (2008). The optimal RTLS solution for hospitals: breaking through a complex environment. Journal of Healthcare Information Management 22 (4): 14–15.

12. Baslyman, M., Rezaee, R., Amyot, D. et al. (2015). Real-time and location-based hand hygiene monitoring and notification: proof-of-concept system and experimentation. Personal and Ubiquitous Computing 19 (3–4): 667–688.

13. Swedberg, C. (2012). Toronto general hospital uses RTLS to reduce infection transmission (28 February 2012). https://www.rfidjournal.com/. Available online on August 25, 2020 at https://www.rfidjournal.com/toronto-general-hospital-uses-rtls-to-reduce-infection-transmission.

14. Wang, B. and Toobaei, M. (2013). Evaluation of RFID and Wi-Fi technologies for RTLS applications in healthcare centers. *Proceedings of PICMET '13 Technology Management in the IT-Driven Services*, pp. 2690–2703.

15. Kirov, D.A., Passerone, R., and Ozhiganov, A.A. (2015). A methodology for design space exploration of real-time location systems. Scientific and Technical Journal of Information Technologies, Mechanics and Optics 15 (4): 551–567.

16. Staff. Use cases. https://www.pointgrab.com/use-cases/ (accessed 10 August 2020).

17. Dodier, R.H., Henze, G.P., Tiller, D., and Guo, X. (2006). Building occupancy detection through sensor belief networks. Energy and Buildings 38 (9): 1033–1043.

18. Zou, J., Zhao, Q., Yang, W., and Wang, F. (2017). Occupancy detection in the office by analyzing surveillance videos and its application to building energy conservation. Energy and Buildings 152: 385–398.

19. Jiang, C., Masood, M.K., Soh, Y.C., and Li, H. (2016). Indoor occupancy estimation from carbon dioxide concentration. Energy and Buildings 131: 132–141.

20. Chen, Z., Jiang, C., and Xie, L. (2018). Building occupancy estimation and detection: a review. Energy and Buildings 169: 260–270.

21. Oldewurtel, F., Sturzenegger, D., and Morari, M. (2013). Importance of occupancy information for building climate control. Applied Energy 101: 521–532.

22. Agarwal, Y., Balaji, B., Gupta, R. et al. (2010). Occupancy-driven energy management for smart building automation. ACM Workshop on Embedded Sensing Systems for Energy-Efficiency in Building, pp. 1–6.

23. Yang, Z. and Becerik-Gerber, B. (2014). The coupled effects of personalized occupancy profile-based HVAC schedules and room reassignment on building energy use. Energy and Buildings 78: 113–122.

24. Wang, F., Feng, Q., Chen, Z. et al. (2017). Predictive control of indoor environment using occupant number detected by video data and CO_2 concentration. Energy and Buildings 145: 155–162.

25. Candanedo, L.M. and Feldheim, V. (2016). Accurate occupancy detection of an office room from light, temperature, humidity and CO_2 measurements using statistical learning models. Energy and Buildings 112: 28–39.

26. Minoli, D. and Occhiogrosso, B. (2020). IoT-driven advances in commercial and industrial building lighting and in street lighting. In: Industrial IoT: Challenges, Design Principles, Applications, and Security (ed. I. Butun). Springer. ISBN: ISBN 978-3-030-42500-5.

27. Ebbers, C.W.J.M., van Hoof, J., and Oude Weernink, C.E. (2017). Privacyaspecten van track en tracet-echnologie in de zorg. Privacy & Informatie 20 (1): 24–32.

28. Ebbers, C.W.J.M., van Hoof, J., and Oude Weernink, C.E. (2017). De toepassing van track en tracetech-nologie in de zorg (2). Privacy & Informatie 20 (6): 256–263.

29. Zhang, J., Wei, B., Hu, W., and Kanhere, S.S. (2016). WiFi-ID: human identification using WiFi signal. *Proceedings of 12th Annual International Conference on Distributed Computing in Sensor Systems*, DCOSS 2016, pp. 75–82.

30. Alhamoud, A., Nair, A.A., Gottron, C. et al. (2014). Presence detection, identification and tracking in smart homes utilizing bluetooth enabled smartphones. *Proceedings of Conference on Local Computing Networks*, LCN (November), pp. 784–789.

31. Helfenbein, T., Király, R., Töröcsik, M. et al. (2017). Extension of RFID based indoor localization systems with smart tags. Infocommunications Journal 9 (3): 25–31.

32. Großwindhager, B., Stocker, M., Rath, M. et al. (2019). SnapLoc: an ultra-fast UWB-based indoor localization system for an unlimited number of tags. 2019 18th ACM/IEEE International Conference on Information Processing in Sensor Networks (IPSN), Montreal, QC, Canada, pp. 61–72.

33. Hoyer, P., Lovelock, J.E., and Robinton, M. (2020). Tracking for badge carrier. US Patent 10,659,917, 19 May 2020, filed 22 January 2018. Uncopyrighted.

34. Dong, Z., Li, Z., Yang, F. et al. (2019). Sensitive readout of implantable microsensors using a wireless system locked to an exceptional point. Nature Electronics 2 (8): 335–342.

35. Halamka, J., Juels, A., Stubblefield, A., and Westhues, J. (2006). The security implications of verichip cloning. Journal of the American Medical Informatics Association: JAMIA 13 (6): 601–607. https://doi.org/10.1197/jamia.M2143.

36. Voas, J. and Kshetri, N. (2017). Human tagging. Computer 50 (10): 78–85. https://doi.org/10.1109/MC.2017.3641646.

37. x,y,z (2020). Deep Learning Enabled Smart Mats as a Scalable Floor Monitoring System. Nature Communications.

38. Middleton, L., Buss, A.A., Bazin, A., and Nixon, M.S. (2005). A floor sensor system for gait recognition. *Proceedings Fourth IEEE Work Automatic Identification Advanced Technologies*, AUTO ID 2005, vol. 2005, pp. 171–180.

39. Li, Y., Gao, Z., He, Z. et al. (2018). Multi-sensor multi-floor 3D localization with robust floor detection. IEEE Access 6: 76689–76699.

40. He, C., Zhu, W., Chen, B. et al. (2017). Smart floor with integrated triboelectric nanogenerator as energy harvester and motion sensor. ACS Applied Materials & Interfaces 9 (31): 26126–26133.

41. Pan, S., Wang, N., Qian, Y. et al. (2015). Indoor person identification through footstep induced structural vibration. HotMobile 2015 - 16th Int Work Mob Comput Syst Appl, pp. 81–86.

42. Sapiezynski, P., Stopczynski, A., Gatej, R., and Lehmann, S. (2015). Tracking human mobility using WiFi signals. PLoS One 10 (7): e0130824. Published 2015 July 1. https://doi.org/10.1371/journal.pone.0130824.

43. Adib, F., Hsu, C.Y., Mao, H. et al. (2020). RF-capture: capturing the human figure through a wall. Computer Science & Artificial Intelligence Laboratory, Massachusetts Institute of Technology. http://rfcapture.csail.mit.edu.

44. Miyaki, T., Yamasaki, T., and Aizawa, K. (2007). Visual tracking of pedestrians jointly using Wi-Fi location system on distributed camera network. 2007 IEEE International Conference on Multimedia and Expo, Beijing, pp. 1762–1765. https://doi.org/10.1109/ICME.2007.4285012.

45. Mokhtari, G., Zhang, Q., Hargrave, C., and Ralston, J.C. (2017). Non-wearable UWB sensor for human identification in smart home. IEEE Sensors Journal 17 (11): 3332–3340.

46. Xiong, J., Li, F., Zhao, N., and Jiang, N. (2014). Tracking and recognition of multiple human targets moving in a wireless pyroelectric infrared sensor network. Sensors 14 (4): 7209–7228.

47. Pan, S., Bonde, A., Jing, J. et al. (2014). Boes: building occupancy estimation system using sparse ambient vibration monitoring. Sensors and Smart Structures Technologies for Civil, Mechanical, and Aerospace Systems.

48. Yang, Y., Hao, J., Luo, J., and Pan, S.J. (2017). Ceiling see: device-free occupancy inference through lighting infrastructure based LED sensing. IEEE International Conference on Pervasive Computing and Communications (Per-Com), pp. 247–256.

49. Shih, O. and Rowe, A. (2015). Occupancy estimation using ultrasonic chirps. ACM/IEEE International Conference on Cyber-Physical Systems, pp. 149–158.

50. Labeodan, T., Zeiler, W., Boxem, G., and Zhao, Y. (2015). Occupancy measurement in commercial office buildings for demand-driven control applications: a survey and detection system evaluation. Energy and Buildings 93: 303–314. https://doi.org/10.1016/j.enbuild.2015.02.028.

51. Wang, F., Feng, Q., Chen, Z. et al. (2017). Predictive control of indoor environment using occupant number detected by video data and CO_2 concentration. Energy and Buildings 145: 155–162.

52. Yang, Z., Li, N., Becerik-Gerber, B., and Orosz, M. (2012). A non-intrusive occupancy monitoring system for demand driven HVAC operations. Construction Research Congress: Construction Challenges in a Flat World, pp. 828–837.

53. Guo, X., Tiller, D., Henze, G.P., and Waters, C.E. (2010). The performance of occupancy-based lighting control systems: a review. Lighting Research and Technology 42 (4): 415–431.

54. Agarwal, Y., Balaji, B., Dutta, S. et al. (2011). Duty-cycling buildings aggressively: the next frontier in HVAC control. *Proceedings IPSN'11*, Chicago, IL, pp. 246–257.

55. Erickson, V., Achleitner, S., and Cerpa, A. (2013). POEM: power-efficient occupancy-based energy management system. *Proceedings IPSN'13*, Philadelphia, PA.

56. Nguyen, T. and Aiello, M. (2012). Beyond indoor presence monitoring with simple sensors. *Proceedings PECCS'12*, Rome, Italy, pp. 1–10.

57. Sruthi, M.S. (2019). IOT based real time people counting system for smart buildings. International Journal of Emerging Technology and Innovative Engineering 5 (2) Available at SSRN: https://ssrn.com/abstract=3340446.

58. Verma, N.K., Dev, R., Maurya, S. et al. (2018). People counting with overhead camera using fuzzy-based detector. In: Computational Intelligence: Theories, Applications and Future Directions - Volume 1. Advances in Intelligent Systems and Computing, vol. 798 (eds. N. Verma and A. Ghosh). Singapore: Springer https://doi.org/10.1007/978-981-13-1132-1_46.

59. Zou, J., Zhao, Q., Yang, W., and Wang, F. (2017). Occupancy detection in the office by analyzing surveillance videos and its application to building energy conservation. Energy and Buildings 152: 385–398.

60. Garca, J., Gardel, A., Bravo, I. et al. (2013). Directional people counter based on head tracking. IEEE Transactions on Industrial Electronics 60 (9): 3991–4000.

61. del-Blanco, C.R., Jaureguizar, F., and Garcia, N. (2012). An efficient multiple object detection and tracking framework for automatic counting and video surveillance applications. IEEE Transactions on Consumer Electronics 58 (3).

62. Hou, Y.L. and Pang, G.K.H. (2011). People counting and human detection in a challenging situation. IEEE Transactions on Systems, Man, and Cybernetics - Part A: Systems and Humans 41 (1): 24–33.

63. Wang, H., Jia, Q., Song, C. et al. (2010). Estimation of occupancy level in indoor environment based on heterogeneous information fusion. *Proceedings of 49th IEEE Conference on Decision and Control*, CDC 2010 (15–17 December 2010), IEEE, pp. 5086–5091.

64. Benezeth, Y., Laurent, H., Emile, B., and Rosenberger, C. (2011). Towards a sensor for detecting human presence and characterizing activity. Energy and Buildings 43 (2–3): 305–314.

65. Erickson, V.L., Carreira-Perpinan, M.A., and Cerpa, A.E. (2011). Observe: occupancy-based system for efficient reduction of HVAC energy. IEEE International Conference on Information Processing in Sensor Networks (IPSN), pp. 258–269.

66. Fleuret, F., Berclaz, J., Lengagne, R., and Fua, P. (2008). Multicamera people tracking with a probabilistic occupancy map. IEEE Transactions on Pattern Analysis and Machine Intelligence 30 (2): 267–282.

67. Teixeira, T., Jung, D., and Savvides, A. (2010). Tasking networked CCTV cameras and mobile phones to identify and localize multiple people. Ubicomp '10 Proceedings of the 12th ACM International Conference on Ubiquitous Computing (26–29 September 2010), pp. 213–222.

68. Manku, T. Method and system for locating wireless devices within a local region. US Patent 9,958,533, 1 May 2018, filed 21 February 2014. Uncopyrighted.

69. Minoli, D. and Occhiogrosso, B. (2020). Constrained average design method for QoS-based traffic engineering at the Edge/Gateway boundary in VANETs and cyber-physical environments. In: Managing Resources for Futuristic Wireless Networks (ed. M. Rath). IGI Global.

70. Pahlavan, K. and Eskafi, F. Local indoor positioning and navigation by detection of arbitrary signals. US Patent 10,151,844, 11 December 2018, filed 30 April 2015. Uncopyrighted.

71. Staff. Ultimate 2019 real time location system (RTLS) tech guide. Real Time Networks, 29 May 2019. https://www.realtimenetworks.com/blog/ultimate-2019-real-time-location-system-rtls-tech-guide.

72. Cisco Staff. Cisco unified wireless location-based services. https://www.cisco.com/en/US/docs/solutions/Enterprise/Mobility/emob30dg/Locatn.html.

73. Staff, Clarinox Technologies. Real time location systems. November 2009. https://www.clarinox.com/docs/whitepapers/RealTime_main.pdf.

74. Holler, R.E., Peters, S.E., Ptasienski, L.J. et al. Tracking compliance of personal protection articles. US Patent 9,536,209, 3 January 2017, filed 3 February 2010. Uncopyrighted.

75. EPCglobal®. EPC™ radio-frequency identity protocols, class-1 generation-2 UHF RFID, protocol for communications at 860 MHz–960 MHz, Version 1.0.9 January 2005.

76. Skyrfid Staff. Mid-range technologies. https://skyrfid.com/Mid-Range_RFID.php

77. SEGD staff. What is wayfinding?. https://segd.org.

78. Pahlaven, K. and Eskafi, F.H. Radio frequency tag and reader with asymmetric communication bandwidth. US Patent 7,180,421, 20 February 2007. Uncopyrighted.

79. Sadr, R. and Carano, B.L. Location based services for RFID and sensor networks. US Patent 10,587,993, 10 March 2020, filed 21 December 2017. Uncopyrighted.

80. Boyd, R.W. Location system for wireless local area network (WLAN) using RSSI and time difference of arrival (TDOA) processing. US Patent 7,899,006, 1 March 2011. Uncopyrighted.

81. Reed, C. What is WiFiSLAM?. Landon Technologies. https://www.landontechnologies.com/blog/what-is-wifislam/

82. ISO/IEC 24730:2014 information technology — real-time locating systems (RTLS), Second edition 2014-02-15, specifically ISO/IEC 24730–1:2014(E).

83. Amir, I., Annamalai, K., and Naim, A. System and method for multimode Wi-Fi based RTLS. US Patent 9,341,700, 17 May 2016, filed 1 February 2013. Uncopyrighted.

84. Bares, J., Daute, C., Tallon, J. et al. System and method to track time and attendance of an individual at a workplace. US Patent 10,679,158, 9 June 2020, filed 10 September 2018. Uncopyrighted.

85. CenTrak Staff. Real-time location system for hospitals: improving facilities for patients and staff. https://www.centrak.com/intro-to-rtls/ (accessed 8 August 2018).

86. Nishimura, T., Koji, K., Nishida, Y., and Mizoguchi, H. (2015). Development of a nursing care support system that seamlessly monitors both bedside and indoor locations. Procedia Manufacturing 3: 4906–4913.

87. Casilari, E., Luque, R., and Morón, M.-J. (2015). Analysis of android device-based solutions for fall detection. Sensors 15 (8): 17827–17894.

88. Yu, X., Weller, P., and Grattan, K.T.V. (2015). A WSN healthcare monitoring system for elderly people in geriatric facilities. Studies in Health Technology and Informatics 2010: 567–571.

89. Santoso, F. and Redmond, S.J. (2015). Indoor location-aware medical systems for smart homecare and telehealth monitoring: state-of-the-art. Physiological Measurement 36 (10): R53–R87.

90. Minoli, D. and Occhiogrosso, B. (2019). Practical aspects for the integration of 5G networks and iot applications in smart cities environments. Special Issue titled "Integration of 5G Networks and Internet of Things for Future Smart City," Wireless Communications and Mobile Computing". Vol. 2019, Article ID 5710834, p. 30. Hindawi and John Wiley & Sons. https://doi.org/10.1155/2019/5710834.

91. Minoli, D. and Occhiogrosso, B. (2018). Internet of things applications for smart cities. In: Internet of Things A to Z: Technologies and Applications (ed. Q. Hassan) Chapter 12. IEEE Press/Wiley ISBN-13: 978-1119456742.

TABLE A.1   Basic Terms and Concepts Relative to Positioning Technologies

Term	Definition/Concept
Access Control	Mechanisms to control access to secure, hazardous, and critical industrial or business spaces.
Active tag	Devices that primarily rely on batteries or line-power. Some examples include, but are not limited to, Active Radio Frequency Identification (Active RFID) tags; Active RFID - Infrared hybrid (Active RFID-IR) tags, Ultra-Wideband (UWB) tags, Wi-Fi tags (and/or devices), Bluetooth beacons.

TABLE A.1   (Continued)

Term	Definition/Concept
Angle of Arrival (AoA)	The angle from which a signal arrives at a receiver. A method based on this principle requires directional antennas or an array of antennas; it is a distance/angle computation-based method. The method can be used in combination with Received Signal Strength Indicator (RSSI) and Time of Arrival (ToA) distance estimation techniques to reduce the error in position estimation.
Asset Tracking	Using tags attached to assets to track movement or status.
Beacon	A device that propagates signals to make others in its environment aware of its existence, identity, location, data, etc. For example, a beacon transmits Radio Frequency (RF) data packets.
Choke point	A physical location in a space where short-range ID signals from a moving tag is received by a single fixed reader, thus indicating the location the tag as being in the proximity of the reader and tag. A funneling point such as an access turnstile to an elevator bank or an office reception area.
Cisco Compatible Extensions (CCX)	A Cisco-developed mechanism that provides a middleware allowing users of Cisco-compatible network equipment to utilize extensions developed by third-party developers. It is used in Cisco Wi-Fi-based RTLS. Other wireless networking vendors may also offer one-directional networking protocols that are similar to CCX and use the Wi-Fi physical layer.
Condition of a signal	One or more qualities of a received signal, including the Time of Arrival of the signal at a receiving antenna, the signal strength, the phase of arrival, the polarization of the signal at the Time of Arrival, the angle of arrival, the frequency of the signal, the encoding of the signal, and any condition in general that shapes the physical, temporal, and electrical characteristics of the signal including the data it represents at the point of reception of the signal. Same signal can have different conditions at two different receiving antennas, e.g. the signal from the same transmitter may reach two differently located antennas at two different instances of time at two different signal strengths, phases, frequencies, and polarizations, or signals from two different transmitters can have different conditions when arriving at the same receiver in a similar manner.
Coordinate-based location	Various methods by which ID signals from a tag are received by several readers and a position is estimated using one or a number of algorithms, such as triangulation, trilateration, or multilateration. Contrast with choke point location.
Indoor Positioning System (IPS)	Another term for a Real-time Location System (RTLS), but one with localized scope. A set of networked devices utilized to locate people or objects where Global Positioning System (GPS) lacks precision or is unavailable, such as deep inside high-rise buildings, airports, parking garages, and underground locations such as transportation sites or below-ground space/office locations (for example, in places such as Washington, DC). Can use wireless technologies with active or passive tags; or it can use other technologies such as, but not limited to, Infrared (IR), optical/Li-Fi, Ultrasound Identification (US-ID), and Ultrasonic ranging (US-RTLS); combinations are also possible (called "sensor fusion").
Item Finding	Using tags attached to keys, wallets, purses, and other personal property to help people them locate lost items.
Mobile/handheld device	An electronic device (such as a cellphone, smartphone, tablet computer, etc.) that can be carried or dislocated by a user, or attached/strapped or otherwise transported by a mobile entity.

(Continued)

TABLE A.1 (Continued)

Term	Definition/Concept
Multilateration	The process of locating a signal source by solving an error minimization function of a location estimate determined by the Difference in Time of Arrival (DToA) between Time of Arrival (ToA) signals received at multiple receivers. A navigation methodology based on measurement of the Times of Arrival of electromagnetic waves having a known propagation speed.
Passive tag (aka Labels)	Devices that acquire their power mainly by scavenging incident or ambient sources of energy that may, or may not, be proactively generated. Examples of such sources of energy include incident or ambient electromagnetic radiation (e.g. Radio Frequency [RF] radiation, infrared [IR] light, visible light, etc.), mechanical energy (e.g. vibration, acoustic energy, pressure, etc.) and thermal (e.g. by conduction, convection, etc.). When the tag passes within a defined range, a reader generates electromagnetic waves; the tag's integrated antenna receives the signal and activates the chip in the tag and a wireless communications channel is set up between the reader and the tag enabling the transfer of pertinent data.
Ranging methods	Algorithms used to determine location in an RTLS system. Examples include Angle of Arrival (AoA), Angle of Departure (AoD), Time of Arrival (ToA), Time Difference of Arrival (TDoA), Multilateration, Time-of-Flight (ToF), Two-way Ranging (TWR), Symmetrical Double Sided - Two Way Ranging (SDS-TWR), and Near-field Electromagnetic Ranging (NFER).
Real-Time Location System (RTLS)	Systems used to automatically identify and track the location of objects or people in real time, particularly in a building or other contained area. The position may be with respect to landmarks of the indoor environment, e.g. particular rooms, floors, building wings, and so forth. An RTLS system provides a function similar to what a Global Positioning System (GPS) does for an outdoor environment.
Received Signal Strength (RSS)	An absolute measurement of power in dBm, as contrasted Received Signal Strength Indicator (RSSI), which is a relative indicator on a finite-scope quantized scale.
Received Signal Strength Indicator (RSSI)	A measurement of the power level received by sensor, as expressed in terms of a discrete point scale (not absolute power *per se*). A method based on this principle makes use of the fact that the received signal strength is inversely proportional to the square of the distance. It is a distance/angle computation-based method. RSSI is a commonly implemented technique due to its low cost, practicality, and availability. This method requires tags or fixed transceivers to measure the received power (signal strength) of the incoming signals; then, using either known variations of signal strength versus distance from transmitters, or by measuring the signal strengths at various locations and matching these measured strengths to the measured strengths, position can be determined.
RF Fingerprinting	A method based on the principle that every location has a (relatively) unique radio frequency (RF) signature, so that a location can be identified by a unique set of values including measurements of neighbor emitters (e.g. readers, wireless access points, cell towers). However, because of RF propagation principles, the RF signature may not be completely unique and may repeat at a number of different locations within the zone of interest. The method requires a relatively large database (size depending on the size of the space under consideration) and a long training phase. Furthermore, the database may become stale (especially in outdoor application, such as E911), namely, the signature ages quickly as the environment changes; this makes the task of maintaining the database somewhat complex.

Term	Definition/Concept
RTLS infrastructure node	A device that emits an identification ("ID") signal, usually by use of a secondary technology such as IR, LF, or ultrasonic signals, which can be received by corresponding receivers in tags, in order to improve localization capabilities. An infrastructure node transmitter generally operates by periodically transmitting its identity by use of a signal. When a tag on a patient receives the signal from the infrastructure node, the tag will send a transmission to a central monitoring system through access points, in order to indicate that the tag is near the infrastructure node. For example, an infrastructure node transmitter may be coupled to a wheelchair and be used to help associate a tagged patient to the wheelchair that the tagged patient is sitting in; or an infrastructure node transmitter may be coupled to an office chair, and office desk, or office amenity (e.g. a coffee machine in a breakout area).
RTLS reference points	Transmitters or receivers that are deployed throughout a building (or zone of interest) to provide tag tracking. To achieve better location accuracy, a dense set RTLS reference points must be installed in the space under consideration.
Tags/RTLS Tags	(Typically) wireless, battery-operated small devices that are designed to work with an RTLS system. RTLS tags may be attached to objects (e.g. people, moveable physical assets, etc.) that need tracking. RTLS tags can also be equipped with a variety of supplementary sensors, such as: sensors to monitor vital signs of a patient (e.g. skin temperature, pulse rate, respiration rate, etc.); sensors to control movement of people (components of a security access systems, such as a smart badge); sensors to monitor environmental factors (e.g. room temperature); sensors to improve power efficiency (e.g. a motion sensor such that processing speeds, sampling rates or the like may be increased during periods of relatively greater motion); sensors to determine localization (e.g. IR and Low Frequency [LF] sensors – operating at 125 KHz); and so forth.
Time of Arrival (ToA)/Time of Flight (ToF)/Time (Difference) of Arrival (TDoA)	The propagation time a signal incurs from the transmitter to the receiver; a method based on this principle makes use of the fact that the distance between the tag and the reader is directly proportional to the time taken by the signal to travel between the two devices; it is a distance/angle computation-based method. In one implementation of this method, the smart tag may broadcast a signal to multiple wireless receivers at known locations; the time at which the signal is received by each receiver is measured, and a set of equations can be used to determine the position of the smart tag.
Triangulation	The tracing and measurement of a series or network of triangles to determine the distances and relative positions of points spread over an area or a region, typically by measuring the length of one side of each triangle and deducing its angles and the length of the other two sides by observation from this particular baseline. In RF, especially in IPS applications, triangulation utilizes the angles at which the signals arrive at multiple receivers to estimate the location of a source or tag.
Trilateration	The measurement of the lengths of the three sides of a series of overlapping or touching triangles to determine the relative position of points utilizing geometrical techniques; a method of determining the relative positions of three or more points by treating these points as vertices of a triangle or triangles of which the angles and sides can be measured. In IPS applications, it utilizes estimated ranges from multiple receivers to endeavor to establish the (approximate) location of a tag.
Wayfinding	Mechanisms and/or methods to help visitors navigate their way through complex facilities. In complex environments people need (or benefit from) assistance to guide them to their local destinations. Wayfinding are automated methods, implemented with information systems and portable devices (such as smartphones), that guide people through a physical environment; additional features can further enhance the person's understanding and experience of the space. Wayfinding is important in complex environments such as campuses (for example, universities, museums, hospitals), urban centers/public spaces, and transportation facilities (for example, airports). They can also be utilized in conjunction with Location-Based Services. However, in addition to automated systems, visual cues such as displays, maps, signage, directions, and symbols are also important.

TABLE A.2   Basic RFID Terms and Concepts Relative to Positioning Technologies

Term	Definition/Concept
Air interface	(RFID context) The complete communication link between an interrogator and a tag including the physical layer, collision arbitration algorithm, command and response structure, and data-coding methodology.
Continuous wave (CW)	(RFID context) Typically, a sinusoid at a given frequency, but more generally any interrogator waveform suitable for powering a passive tag without amplitude and/or phase modulation of sufficient magnitude to be interpreted by a tag as transmitted data.
Cover-coding	(RFID context) A method by which an interrogator obscures information that it is transmitting to a tag.
Electronic Product Code (EPC)	(RFID context) A unique identifier for a physical object, unit load, location, or other identifiable entity playing a role in business operations. Electronic Product Codes are assigned following rules designed to ensure uniqueness despite decentralized administration of code space, and to accommodate legacy coding schemes in common use. EPCs have multiple representations, including binary forms suitable for use on RFID tags, and text forms suitable for data exchange among enterprise information systems. More broadly, EPC is the trade name associated with the ISO/IEC 18000-6C RFID standards; it is also the best-known UHF (Ultra High Frequency) RFID technology.
EPC Class 0 Tags	(RFID context) Generation 1 tags are tags that one can write once and read many times (WORM): tag is factory programmable and not field programmable.
EPC Class 1 Generation 2 Standard (aka EPC Gen 2 or just Gen 2)	(RFID context) An EPC Global Consortium standard for Class-1 Generation-2 UHF RFIDs. Gen 2 represents a major step in standardization, performance, and quality. The Gen 2 standard defines the physical and logical requirements for a passive-backscatter, interrogator-talks-first (ITF), RFID system operating in the 860–960 MHz frequency range; the system comprises interrogators (also known as Readers), and tags (also known as Labels). Key features include: 96-bit EPC number support; 32-bit access password to lock the read–write characteristics of the tag as well as set the tag for disabling; some tags include user memory of up to 2048 bits or more depending on the tag. EPC Gen 2 Version 2 will support, among other capabilities: anti-counterfeiting through encrypted verification of the tags 128-bit key; data encryption through inclusion of AES-128; and Tag Hiding which allows "privileged" readers to render tags "untraceable" and hide the tags memory.
EPC Class 1 Tags	(RFID context) EPC Class Generation 1 tags are WORM tags but can be read by readers from other companies; EPC Class 1 Generation 2 (EPC Gen 2 or just Gen 2) are WMRM (Write Many Read Many) tags that have a minimum memory of 256 bits (96 bits is for the EPC number). Gen 2 tags read up to 10 times faster than Gen 1 and provide extremely high read rates. The tag can be read at any frequency between 860 and 960 MHz.
EPC Class 3 Tags	(RFID context) Battery-assisted passive tags sometimes called semi-passive tags in UHF Gen 2. These tags have not yet been fully defined but expected features are a power source to supply power to the tag and/or its sensors and or sensors with optional data logging capabilities.
EPC Class 4 Tags	(RFID context) Active tag systems. This Class 4 tag is still in the early definition stage. The UHF tag will contain a battery and can initiate communications with a reader or with another tag. Class 4 tags with contain an EPC identifier, and extended Tag ID, authenticated access control, a power source, communications via autonomous transmitter, have optional user memory and optional sensors with or without data logging capabilities. Class 4 active tags will not interfere with the communications protocols of other classes.

TABLE A.2   (Continued)

Term	Definition/Concept
EPC Tag Classes	(RFID context) EPC tag capabilities are broken down into classes and each class has specific capabilities and is backward compatible to the preceding class. Each higher class maintains the previous capabilities and characteristics and adds new capabilities. The classes are as follows: EPC Class 0 Tags, EPC Class 1 Tags, EPC Class 3 Tags, EPC Class 4 Tags.
EPCglobal	(RFID context) A neutral, not-for-profit RFID standards organization consisting of manufacturers, technology solution providers, and retailers.
EPCglobal Architecture Framework	(RFID context) A collection of interrelated standards ("EPCglobal Standards"), together with services operated by EPCglobal, its delegates, and others ("EPC Network Services"), all in service of a common goal of enhancing business flows and computer applications through the use of Electronic Product Codes (EPCs).
Interrogator (aka Reader)	(RFID context) A device that modulates/transmits and receives/demodulates a sufficient set of the electrical signals defined in the signaling layer to communicate with conformant tags, while conforming to all local radio regulations. A typical interrogator is a passive-backscatter, Interrogator-Talks-First (ITF), system operating in the 860–960 MHz frequency range. An interrogator transmits information to a tag by modulating an RF signal in the 860–960 MHz frequency range. The tag receives both information and operating energy from this RF signal. Passive tags receive their operating energy from the interrogator's RF waveform. An interrogator receives information from a tag by transmitting a continuous-wave (CW) RF signal to the tag; the tag responds by modulating the reflection coefficient of its antenna, thereby backscattering an information signal to the interrogator. The system is ITF, meaning that a tag modulates its antenna reflection coefficient with an information signal only after being directed to do so by an interrogator. Interrogators and tags are not required to talk simultaneously; rather, communications are half-duplex, meaning that interrogators talk, and tags listen, or vice versa.
Operating procedure	(RFID context) Collectively, the set of functions and commands used by an interrogator to identify and modify tags. (Also known as the Tag-identification layer.)
Physical layer	(RFID context) The data coding and modulation waveforms used in interrogator-to-tag and tag-to-interrogator signaling.
RFID system layers	(RFID context) A RTLS is logically comprised of several functional layers: the tag layer, the air interface (also called media interface) layer, and the reader layer. The Tag (device) Layer deals with architecture and EPCglobal Gen2 Tag finite state machines. The Media Interface Layer deals with frequency bands, antennas, read range, modulation, encoding, and Data Rates. The Reader Layer also deals with architecture, antenna configurations, (Gen2) sessions. Additionally, there are cloud/network, middleware, and applications aspects.
Singulation	(RFID context) Identifying an individual tag in a multiple-tag environment.
Slotted random anticollision	(RFID context) An anticollision algorithm where tags load a random (or pseudo-random) number into a slot counter, decrement this slot counter based on interrogator commands, and reply to the interrogator when their slot counter reaches zero.
UHF RFID, EPC Gen2 and ISO 18000-6C	(RFID context) All three terms usually mean the same thing. EPC, or Electronic Product Code, is the trade name associated with the ISO/IEC 18000-6C RFID standards; it is also the best-known UHF (Ultra High Frequency) technology which is why it is also referred to just as 'UHF'.

7 UWB-Based De-densification of Spaces and Work Environments

This chapter focuses on one of the various technical approaches that can be used for Office Social Distancing (OSD) and Office Dynamic Cluster Monitoring and Analysis (ODCMA): Ultra-Wideband (UWB) Smart Tags. In general, UWB technology provides relatively high throughput at short distances by utilizing low-power pulse transmission as a coexisting radio service in the C/X-band portion of the electromagnetic spectrum. A key feature is that UWB allows for precise timing resolution in defined topological zones (in the range of 10000 square feet), thus making this technology a good candidate for Indoor Positioning Systems (IPSs); with UWB, one is capable of localizing a node with a high certainty of 3–10 cm. UWB has several unique characteristics that make it suitable for many emerging and evolving Internet of Things (IoT) applications in the Wireless Personal Area Network (WPAN) arena, in addition to possible utilization as an IPS in the OSD/ODCMA context for presence monitoring and tracking of employees.

UWB-based communication provides near-gigabit per second data rate connectivity in conjunction with immunity from noise and multipath interference while also enjoying low power consumption. The ongoing technical efforts toward the development of simple and inexpensive transceivers make this technology a good choice for many short-range IoT applications, including OSD/ODCMA applications, Body Area Networks applications, smart city applications dealing with autonomous vehicles and precise location, Positive Train Control in the context of rail transportation, and other near field communications. UWB radios tags are now very popular for accurate indoor localization [1]; UWB's use of a large frequency band to transmit trains of pulses with very short duration (less than 1 ns) enables Real-Time Location System (RTLS) tag readers to precisely differentiate between pulses that are reflected from different objects and achieve high timing resolution, which, in turn, supports very accurate positioning measurements, down at the centimeter level. However, there also are a number of competing technologies that should also be assessed when deploying near field communications for IoT and OSD/ODCMA, such as Bluetooth Low Energy, ZigBee, Wi-Fi, and IEEE 802.15.3c Millimeter-Wave-Based (60 GHz) WPANs (some of these are discussed in Chapter 8).

UWB technology has been around for a few decades, but it received an impetus in the mid-2000s when the Federal Communications Commission (FCC) allowed its regulated use in the United States in a large portion of the electromagnetic spectrum. There is extensive research underway in the UWB space, particularly in the context of antenna design; the challenge is to translate this large volume of theoretical research into cost-effective, easy to deploy systems, especially for IoT and OSD/ODCMA applications. Some products, especially in the localization arena, have emerged of late; the expectation is that a lot more technology will soon reach the marketplace.

The chapter starts with a general review of UWB technology (Sections 7.1–7.4); the final subsections (Sections 7.5 and 7.6) address the smart tag application of UWB technology for work environment (office, factory floor) applications. The addition of sensors, for example, tracking

High-Density and De-Densified Smart Campus Communications: Technologies, Integration, Implementation, and Applications, First Edition. Daniel Minoli and Jo-Anne Dressendofer.
© 2022 John Wiley & Sons, Inc. Published 2022 by John Wiley & Sons, Inc.

chips to an object is clearly seen as an IoT implementation; the addition of a tracking tag or chip to an ID badge can also be considered an IoT application, although the sensor is not permanently attached to the wearer.

7.1   REVIEW OF UWB TECHNOLOGY

UWB[1] technology is formally known as Impulse Radio Ultra-Wideband (IR-UWB) technology [2]. UWB technology supports (relatively) high-speed transmission, low power consumption, simple (on paper) implementation (simple transceiver architecture), and, thanks to its very low power spectrum density, coexistence with other narrowband or wideband communication systems. UWB is also described as being carrier-less. Among other features, the high bandwidth achievable with UWB enables high-definition video streaming; the low transmit power results in low probability of interception or detection, while at the same time making it resistant to jamming, multipath interference, and other challenging field conditions. UWB can be used for a number of military, business, and consumer applications, including a number of IoT applications. Some traditional applications of UWB include through-wall imaging, ground-penetrating radar, medical imaging, surveillance, vehicular radar applications for Intelligent Transportation Systems (ITSs), localization applications (decimeter-level precision), measurement applications, and communications applications. Common IoT-type applications include tracking, high-precision locating, sensor-generated data collection, and WPANs.

UWB enjoys a number of advantages over other technologies which make it suitable, even attractive, for commercial communications applications. These advantages include but are not limited to the following: low device complexity and cost; transmissions being impervious to severe multipath and jamming; signals providing good time-domain resolution; and the operation able to coexist with other wireless services such as Wireless LANs (WLANs), cellular/UMTS/LTE and commercial/military (C-band/X-band) satellite transmission. UWB can help address spectrum exhaustion by allowing shared use of the spectrum (without interfering with other WPAN, WLAN, or cellular systems.)

UWB system transmit streams of extremely short pulses (10–10000ps) that are spread out over a wideband portion of the Radio Frequency (RF) spectrum. Because of the low emission levels permitted by the various regulatory agencies, UWB communications systems tend to be best suited for short-range indoor applications. Some of these indoors applications include peripheral connectivity (e.g. wireless printers, although these can also use Wi-Fi), and transfer of images from camcorders and media players. However, signals can be blocked by large metallic objects. There is a large body of literature on the UWB topic, [3–20] being just a short list describing the technology and some applications (reference [3] provides a detailed history of the technology.)

In the United States, the FCC instituted license-free UWB bands offering bandwidth of 7.5 GHz (from 3.1 to 10.6 GHz) and UWB at millimeter wave frequency centered at 60 GHz (57–64 GHz). UWB is now rapidly advancing as a broadband wireless communication technology. Basic UWB concepts go back to over a century. More applicable work took place in the 1960s through the 1990s, when the technology was being used for military radar and military communications applications (where the probability of intercept and the probability of detection must be low). In 2002, the FCC approved UWB for commercial use; its use became more practical (and economical) with the development of faster digital signal processing hardware (chipsets).

---

[1]Some of this material is broadly based on the reference Minoli, D. and Occhiogrosso, B. Ultrawideband (UWB) technology for smart cities IoT applications. *2018 IEEE International Smart Cities Conference (ISC2) - IEEE ISC2 2018-Buildings, Infrastructure, Environment Track, Kansas City* (16–19 September 2018).

UWB signals are defined as encompassing 0.5 GHz of spectrum or having a fractional bandwidth

$$B_f = 2\{(f_h - f_l) / (f_h + f_l)\}$$

f_h = the highest –10 dB frequency of the signal (Hz)
f_l = the lowest –10 dB frequency of the signal (Hz)

exceeding the center frequency by 20%. Depending on the application, and as specified by the FCC (in the United States – other agencies internationally), the spectrum of operation can be below 960 MHz, or in the 3.1–10.6 GHz band (7.5 GHz). The usable spectrum is regulated by various agencies across the world, such as the FCC in the United States, CEPT in Europe, Ofcom in the United Kingdom, APT in Asia. For ground penetrating radar applications, UWB is restricted to operation below 960 MHz; for vehicular applications, the power must be –10 dB in the 22–29 GHz spectrum. To preclude interference with GPSs, UWB operations must be above 3.1 GHz; the upper limit was initially set at 10.6 GHz. See Figure 7.1. Notice that in this band, the spectrum allocation overlaps other services in the same frequency of operation; therefore, the low power emissions are critical. UWB's transmit power is below the noise floor, implying that UWB does not interfere with other wireless communication systems.

Shannon's well-known relationship for the channel capacity (in bps) asserts that the capacity is linearly related to the (analog) bandwidth of the channel, but logarithmically related to the received power. Therefore, the use of a large channel (spectrum) translates in high capacity, as is the case in UWB. Because of the of the large channel bandwidths embodied in UWB systems, large digital channel capacities can be achieved in principle, per Shannon's equation (given sufficient signal-to-noise ratio [SNR]). Additionally, very low power is used in UWB. However, due to these low power levels (dictated by the FCC), the transmission distance with UWB is limited to meters.

Figure 7.2 depicts positioning of the UWB technology in the bandwidth-distance continuum, when used to transfer information (as contrasted to positioning usage). The power mask allowed by the FCC allows a maximum of 75 nW (–41.25 dBm/MHz); this level is selected to make sure that UWB transmissions do not interfere with other communication even at the Bluetooth® distances (10 m) and WLAN/Wi-Fi distances (100 m). A traditional receiver will perceive only

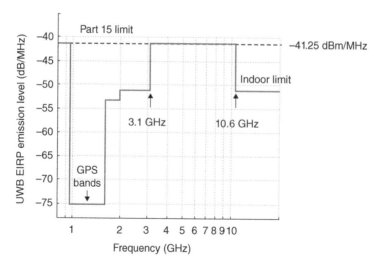

FIGURE 7.1   FCC mask for UWB.

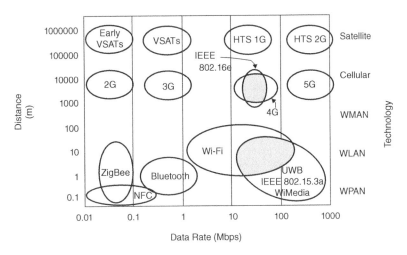

FIGURE 7.2   Approximate positioning of UWB technology.

the noise power within its own assigned system bandwidth, which is only a subset of the total UWB bandwidth; the received power is at or below the specified low-power spectral density allowed for the UWB transmitters. As a side note, emerging ka band-based High Throughput Satellite (HTS) that makes use of DVB-S2X modulation standards may modulate a large, "wideband" bandwidth, but at this time 500 MHz (0.5 GHz) is the practical limit. The so-called "wireless fiber" E/V-band microwave systems typically operate with a 500, 1250, or 2000 MHz band; thus, they can utilize UWB methods, if desired. Multipath propagation is an issue in indoor communications. UWB bursts are of very short temporal duration, mitigating the impact of multipath. UWB can coexist with other uses of the spectrum, due to the low emitted power (e.g. 20 dB below the power emission of WLANs and WPANs).

As noted, UWB enables the transmission of data at relatively high data rates over short distances by utilizing a wide spectrum. In pulse-based systems, each transmitted pulse occupies the entire UWB spectrum (or assigned bandwidth) for a very small amount of time. Pulse-based UWB radars and imaging systems typically use low repetition rates (e.g. in the range of 1–100 megapulses per second). Pulse-based UWB communications systems typically use high repetition rates (e.g. 1–2 gigapulses per second); these systems, therefore, enable, in principle, short-distance Gbps-range communications (speeds in the 600 Mbps range have been demonstrated). Additionally, UWB can carry signals through lightweight walls, doors, furniture, and other obstacles, making it a good technology for any number of confined spaces such as indoors, train cars, and vehicle-to-infrastructure. UWB systems are ideal for applications that require exact distance or positioning measurements and high-speed wireless connectivity. Forward Error Correction (FEC) techniques can also be employed in UWB systems. UWB's can achieve high throughput in short-range networks with multi-gigabit-per-second data rates; UWB can be seen as a Near Field Communication (NFC) system, but at much higher data rates.

The large spectrum bandwidth allows a transmission of high data (>100 Mbps); however, the transmission can be achieved only over short WPAN-like distances (~10 m) being that only very low power is assignable to each bit. Alternatively, low-data rate communication (e.g. <1 Mbps) is possible over longer distances by utilizing the large spreading factor. A point of observation is that the large spectrum bandwidth makes it somewhat difficult to design and engineer inexpensive transceivers. Accurate timing along with high sampling rate are required; furthermore, there is complex computational effort in the receiver in order to process the samples. Traditional radio systems (including Bluetooth or Wi-Fi) utilize Received Signal Strength Indicator (RSSI) mechanisms for localization, while UWB utilizes the signal's Time of Flight (ToF) mechanisms; this enables the technology to be used to make very exact ranging measurements.

7.2   CARRIAGE OF INFORMATION IN UWB

UWB transmitters can transmit digital pulses at precise time intervals, utilizing a wide spectrum and low power. The receiver and the transmitter must be time-synchronized to the picosecond to receive sent pulses. The basic idea is to create, transmit, and receive a short burst (in picoseconds-to-nanoseconds duration), while spreading the signal energy over a wide spectrum.

7.2.1   Pulse Communication

A distinguishing difference between UWB and traditional radio transmissions is that traditional systems transmit information by altering the amplitude, frequency, and/or phase of a sinusoidal wave, while UWB systems transmit data by emitting radio energy at specific time intervals while at the same time having the signal occupying a large bandwidth. This UWB methodology give rises to time modulation or pulse-position (low pulse rates are utilized to support time or position modulation, but gigapulse rates can also be employed). Additionally, the data can also be modulated on UWB pulses by encoding the polarity, specifically the phase, of the pulse, the amplitude, and/or by utilizing orthogonal pulses, as further discussed in Section 7.2.2.

UWB has a simpler architecture because high-quality stable oscillators and tuning circuits typically used in traditional (call it "narrowband") transmission to modulate and demodulate signals are not needed. On the transmit side, UWB uses direct modulation: it directly modulates the baseband signal onto the carrier, thus eliminating a number of transceiver components. However, the receivers are more complex and also may require digital signal processing capabilities; the compensating factor is that advanced digital signal processing can now be easily achieved with Field Programmable Gate Arrays (FPGAs), Application-Specific Integrated Circuits (ASICs), or Systems On a Chip (SOCs).

Make note that, the classical superheterodyne method of transmission and reception is still in wide use in many wireless systems. Here, at the receiver a local oscillator generates a signal that is mixed with (subtracted from) the incoming signal. However, there are some disadvantages with this approach being that it entails multiple downconversion, Local Oscillator (LO) costs, and the injection of "image" signals that can cause interference to the received signal (thus impacting bit error rates). Another method is direct modulation/conversion. Here the LO frequency is set to be the same as the incoming signal frequency. This process allows the recovered signal to be directly at baseband (no downconversion needed, eliminating circuit costs, and making the analog-to-digital sampling simpler and more robust – no signal images are generated, but there still are some challenges with dc offsets and LO leakage). An architecture that is becoming popular entails direct RF sampling, namely the direct digitization of the received signal without downconversion). Here, some filtering and RF amplification are needed, but the design is simple. This technique requires very fast analog-to-digital sampling (a rate of 2.5x the highest frequency is required).

UWB utilizes a train of temporally narrow (e.g. 2 ns) impulses rather than a modulated carrier to transmit information. Since the pulse occupies a wide frequency band, the signal's rising edge is very steep which allows the receiver to accurately establish the arrival time of the signal (i.e. recover clock). UWB pulses can be identified even in noisy environments, making them resistant to multipath effects. These characteristics make UWB ideal for precise ranging applications [21]. See Figure 7.3.

The basic difference between the Spread Spectrum (SS) and UWB systems is that the wide bandwidth in a UWB waveform is produced by pulse duration and pulse shaping, not by spreading with a chipping or hopping sequence as is the case in Direct Sequence Spread Spectrum (DS-SS) or in Frequency Hopping Spread Spectrum (FH-SS). A SS system operates on a narrowband baseband signal and modulates it with a wideband encoding signal that distributes the underlying baseband signal over a much larger bandwidth. UWB is a time-domain process that

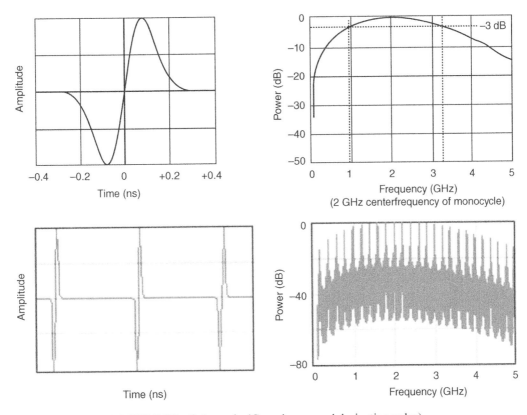

FIGURE 7.3   Pulse train (Gaussian second derivative pulse).

utilizes time frequency distribution. A UWB system can directly generate wide-bandwidth signals using Fourier transforms from very short pulses by using its pulse duration and pulse shaping. Unlike SS unity-constant duty cycles, UWB pulses have very small duty cycle [20].

Because UWB is a baseband-oriented technology, pulse shape is understandingly important; the spectrum of the transmitted/received signal is determined by the pulse width and pulse shape. The pulse shape is selected in order to spread the over-the-air energy in frequencies that minimize the power spectral density and the interference; additionally, pulse shape is selected to avoid having a dc component, in order to achieve antenna radiation efficiency. Among several others, three common monopulse pulse shapes are:

- Gaussian monopulse.
- Gaussian second derivative (also known as "Rayleigh monocycle" or "Scholtz monocycle").
- RZ Manchester.

Notice that these pulses do not have dc component; this is achieved by having balanced positive and negative excursions. The Gaussian pulse is the most widely used pulse in UWB communication. The Gaussian pulses are frequently used because they are easily generated (compared with having to generate a rectangular pulse). See Figure 7.4. There are also a number of other UWB waveforms (not discussed here), such as the Laplacian, the Hermitian, and so on.

The received pulse is typically different from the pulse that was transmitted due to various antenna and propagation properties. The pulse shape determines the center frequency of operation; the pulse length (and other matters such as the rise time, the filtering, and so on)

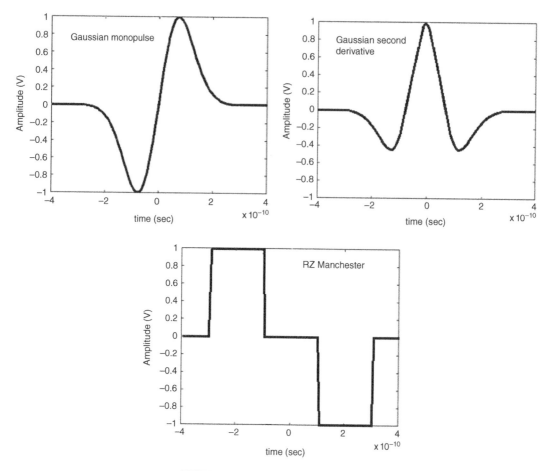

FIGURE 7.4   Various pulse shapes.

determines the spectral shape and the system bandwidth. Very short pulses can counteract channel effects (such as multipath reflections) in dense environments. The data rate (which is based on the number of pulses transmitted per bit) can be traded-off with power usage (up to the allowed regulatory maximum) and with distance of transmission. Because of the wide spectrum of use (say 6 GHz), impulse responses become "sparse," i.e. resolvable Multipath Components (MPCs) are temporally separated by delay regions (in the delay domain) that do not contain any significant energy content. See Figure 7.5. Transmission over large bandwidth provides fine-grain time–space resolution; these features can be used for localization of devices (with UWB transceivers). Another application is the distribution of time stamps to network elements.

7.2.2   UWB Modulation

As noted, there are a number of modulation methods that can be used in UWB. The impulse UWB radio is carrierless (operates at baseband); no mixer or upconversion is needed. Impulse Radio modulation schemes utilize monopulses that spread the energy of the signal over several gigahertzes of spectrum. Monopulse pulses are spaced apart in time uniformly, say about a 100–1000 times the pulse duration, resulting in pulse train that has a low duty cycle. The lower the duty cycle, the lower transmission power required, making UWB useful for some IoT applications where power conservation and unattended sensor longevity are important.

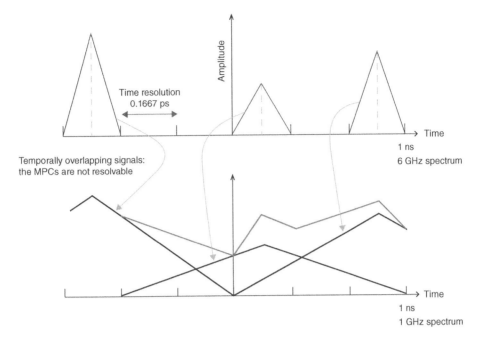

Resolvable Multipath Components (MPCs) are separated by delay regions (in the delay domain) that do not contain any significant energy contribution

FIGURE 7.5   Resolvable and unresolvable MPCs.

The monopulse pulse train carries no information per se: additional processing is necessary to modulate the monopulse pulse train in order to transmit actual information [20]. Three basic methods are summarized below (also see Figure 7.6), but other methods exist.

- Pulse Position Method (PPM): A time hopping sequence with a specified temporal time shift is used. A "1" is encoded by a time shift and a "0" is encoded with no time shift. The sender and receiver have to be precisely synchronized. In Figure 7.6, the left-most diagram depicts the PPM method when the waveform is a Gaussian second derivative.
- On–Off Keying (OOK): Here a "1" is indicated with the presence of a pulse and "0" by the absence of a pulse at the specified reference time slot. The sender and receiver have to be precisely synchronized. In Figure 7.6, the middle diagram depicts the OOK method when the waveform is a Gaussian second derivative. Several position systems on the market use OOK.
- Binary Phase Shifting Keying (BPSK): Here there is a phase shift to encode bits: the information (bit) is carried by the polarity of the pulse. A phase shift of 0° codes the bit "1"; a phase shift of 180° codes the bit "0." The sender and receiver have to be precisely synchronized. In Figure 7.6, the right-most diagram depicts the BPSK method when the waveform is a Gaussian second derivative.

Modulation in the UWB environment entails the use of oscillators that generate a Pulse Repetition Frequency (PRF). Refer to Figure 7.7, which compares traditional systems (Part a) with UWB systems (Part b). As noted above, in Impulse UWB (aka "Time Modulated UWB" [TM-UWB]), pulses shorter than a nanosecond are utilized with a variable pulse-to-pulse interval (aka pulse position modulation); the interval variation is measured and used to recover the information transmitted across the link – this includes the information and a channel code. An information bit may be spread over multiple pulse pairs and coherently added in the

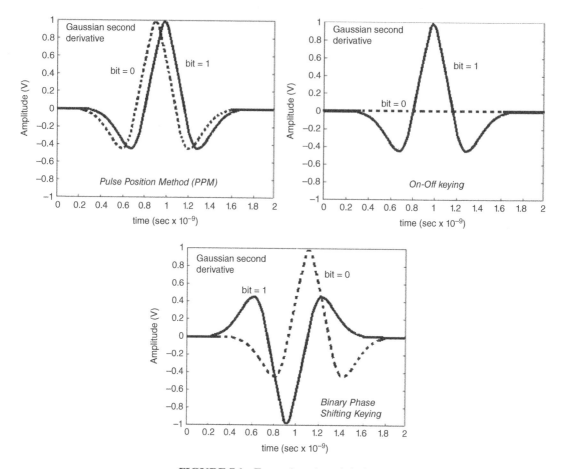

FIGURE 7.6   Examples of modulation.

receiver [22, 23]. Note that the precise timing of the system enables distance determination (which is a key application of UWB).

UWB also operates in Direct Sequence (DS) mode known as (DS-UWB). Here, the data are modulated using a "signature waveform." The Gaussian pulse is modified by the signature waveform. This is followed by modulation, typically Phase Shift Keying (PSK). The transmitter is straightforward, but the receiver is somewhat complex because of the need for correlation recovery circuitry (refer back to Figure 7.7.).

A Multiband Orthogonal Frequency-Division Multiplexing (MB-OFDM) method can also be used. In one implementation, MB-OFDM segments the UWB spectrum into multiple 528-MHz wide bands; each of these bands includes 128 carriers modulated utilizing OFDM or Quadrature Phase Shift Keying (QPSK) [24]. The composite signal occupies the 528 MHz band for approximately 300 ns before switching to another band. The MB-OFDM radio utilizes well-known coding, scrambling, and inverse fast Fourier transform to generate the signal to be transmitted. Contemporary UWB systems for wireless communications utilize OFDM techniques. The UWB communications system, promoted by WiMedia® Alliance, utilizes a MB-OFDM structure defined in the ECMA-368 standard (see Section 7.3 below). WiMedia-defined UWB technology has been adopted by the USB Implementers Forum for wireless USB and by the Bluetooth SIG, Inc. for high-speed Bluetooth communication.

The energy on each path is low, requiring some new techniques for proper detection, reception, and demodulation. Antennas play a big role in UWB. Antenna designs that result in the

(a) Traditional Transmission System

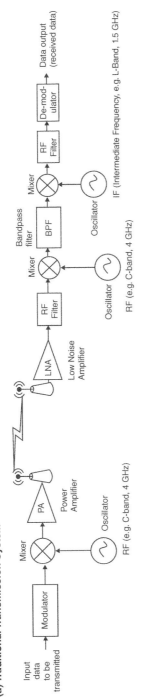

(b) UWB Transmission system

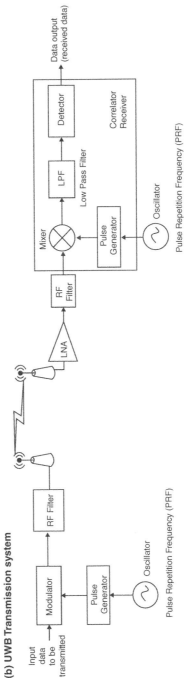

FIGURE 7.7 Transmission systems. Part a: Traditional system. Part b: UWB system.

lowest practical distortion (especially in the phase component) are sought. The behavior of the antenna must be coherent over the entire spectrum of operation; signal variations and distortions in the received radiation patterns must be avoided or completely minimized.

7.3 UWB STANDARDS

Lack of complete progress in UWB standards and disappointing performance of the early implementations have resulted in limited deployment of UWB in commercial and consumer products to date. The handful of UWB-related standards that have been developed in the past few years include the IEEE 802.15.4.f, the ISO-24730-61 Draft International standard (mentioned in Chapter 6), ETSI (European Telecommunications Standards Institute) EN 302065-1 Part 2 as well as ECMA-368 (this organization now known as "Ecma," originally as the European Computer Manufacturers Association [ECMA]).

UWB has been positioned as a WPAN technology for example, in the IEEE 802.15.3a (low rate) draft standard. This effort endeavored to provide a higher speed UWB Physical Layer (PHY) enhancement to IEEE 802.15.3 for applications such as imaging and multimedia. The IEEE 802.15.3a task group dissolved in 2006 before completing the standardization effort: the task group was unable to reach an agreement regarding two technology proposals MB-OFDM and DS-UWB. The work was picked up and completed by the WiMedia Alliance and the USB Implementer Forum.

There has been some related IEEE work. IEEE 802.15.4a standard (2007) is an amendment to the original 2006 IEEE 802.15.4 standard (2006) that defines the Medium Access Control (MAC) (link level) and PHY (physical level) layers for small WPAN with short range and low data rates. The IEEE 802.15.4a amendment added two new PHYs into the standard: Chirp Spread Spectrum (CSS) and UWB. These amendments added higher data throughput and precise ranging methods in the UWB PHY.

Further along, in 2011 the IEEE 802.15.4–2011 revision endeavored to integrate these standards into a single document. The IEEE 802.15.4–2011 specifies three frequency bands to define 16 radio channels (the bands being the sub-gigahertz, the low, and the high band). The standard specifies four data rates for the UWB PHY: 27 Mbps, 6.8 Mbps, 850 kbps, and 110 kbps. See discussion in Section 7.2.2.

IEEE 802.15.3c WPAN is a millimeter-wave-based (60 GHz WPAN) alternative PHY for the existing 802.15.3-2003 WPAN Standard. The goal is to operate up to 5 Gbps at 10 m. This system makes use of unlicensed millimeter-wave band providing small, line-of-sight, coverage area but high data rate while at the same time providing reliable coexistence with the existing microwave band wireless systems. IEEE 802.15.3c-2009 was published in 2009. This mmWave WPAN operates in the 57–66 GHz range. Target applications include wireless data bus for cable replacement, for high-speed Internet access, and for streaming content download for video on demand, Ultra HD (UHDTV), and Over The Top (OTT). Three PHY modes were defined (these are not UWB methods per se):

- Single carrier mode (up to 5.3 Gbps).
- High-speed interface mode (single carrier, up to 5 Gbps).
- Audio/visual mode (OFDM, up to 3.8 Gbps).

IEEE Std 802.15.4™ (IEEE 802.15.4-2020 – Standard for Low-Rate Wireless Networks)
IEEE Std 802.15.4 is a standard for Low-Rate Wireless Networks briefly described in Chapter 4. The protocol and compatible interconnection for data communication devices using low data-rate, low-power, and low-complexity short-range RF transmissions in a WPAN are defined in this standard. A variety of PHYs have been defined that cover a wide variety of

frequency bands [25]. In addition, the standard provides modes that allow for precision ranging, where some of the alternate PHYs provide precision ranging capability that is accurate to one meter. The short discussion that follows is based on the standard; interested readers should consult the full standard for the relative details.

The massive specification is 707 pages long. The standard has gone through three revisions. It aims at very low-cost, low-power communications. The initial standard, IEEE Std 802.15.4-2003, defined two optional PHYs, operating in different frequency bands with a simple and effective MAC.

In 2006, the standard was revised and added two more PHY options. The MAC remained backward compatible, but the revision added MAC frames with an increased version number and a variety of MAC enhancements, including the following: (i) support for a shared time base with a data time stamping mechanism; (ii) support for beacon scheduling; (iii) synchronization of broadcast messages in beacon-enabled Personal Area Networks (PANs); and (iv) improved MAC layer security.

In 2011, the standard added four more PHY options along with the MAC capability to support ranging. Additionally, the organization of the standard was changed so that each PHY would have a separate clause, and the MAC clause was split into functional description, interface specification, and security specification.

The 2015 revision added the following features, among others: (i) low-energy mechanisms; (ii) a variety of new PHY modulation, coding, and band options to support a wide variety of application needs including Radio Frequency Identification (RFID), Smart Utility Networks (SUNs), Television White Space (TVWS) operation, Low-Energy Critical Infrastructure Monitoring (LECIM), and Rail Communications and Control (RCC). It describes Active RFID as "devices are used to identify and often locate people or objects in industrial or commercial environments. Typical applications include asset management, inventory management, process control and automation, safety and accountability. In its simplest form an active RFID system comprises a number of transmit-only tags that periodically transmit a packet containing a unique ID and a small amount of data. The packet is received by one or more readers that may simply register the tag as present, may employ further processing to determine the location of the tag, or forward data to an application server. More complex active RFID systems might employ two-way communications with the tag for control, communication, and coordination." [25].

The standard addresses the following UWB PHY issues in section 16 of the standard (the PHY uses the common IEEE Std 802.15.4-MAC mechanism)

- High-Rate Pulse Repetition Frequency (HRP) UWB PPDU format
  - PPDU encoding process
  - Symbol structure
  - PSDU timing parameters
  - Preamble timing parameters
  - SHR field
  - PHR field
  - PHY Payload field
- Modulation
  - Spreading
  - FEC
- RF requirements
  - Operating frequency bands
  - Channel assignments

- ○ Regulatory compliance
- ○ Operating temperature range
- ○ Baseband impulse response
- ○ Transmit PSD mask
- ○ Chip rate clock and chip carrier alignment
- ○ TX-to-RX turnaround time
- ○ RX-to-TX turnaround time
- ○ Transmit center frequency tolerance
- ○ Receiver maximum input level of desired signal
- HRP UWB PHY optional pulse shapes
  - ○ HRP UWB PHY optional Chirp on UWB (CoU) pulses
  - ○ HRP UWB PHY optional Continuous Spectrum (CS) pulses
  - ○ HRP UWB PHY Linear Combination of Pulses (LCP)
- Extended preamble for optional CCA mode 6
- Ranging

The HRP UWB PHY waveform covered by the standard is based upon an impulse radio signaling scheme using band-limited pulses. The HRP UWB PHY supports three independent bands of operation:

- The sub-gigahertz band, which consists of a single channel and occupies the spectrum from 249.6 to 749.6 MHz.
- The low band, which consists of four channels and occupies the spectrum from 3.1 to 4.8 GHz.
- The high band, which consists of 11 channels and occupies the spectrum from 6.0 to 10.6 GHz.

A combination of Burst Position Modulation (BPM) and Binary Phase-Shift Keying (BPSK) is used to support both coherent and noncoherent receivers using a common signaling scheme. The combined BPM-BPSK is used to modulate the symbols, with each symbol being composed of an active burst of UWB pulses. Figure 7.8 shows the sequence of processing steps used to create and modulate an HRP UWB PHY Protocol Data Unit (PPDU), as described in the standard [25]. The various data rates are supported through the use of variable-length bursts. Figure 7.9 shows the protocol layers covered by the standard. Figure 7.10 shows the format for the HRP UWB PPDU, where the PHY Service Data Unit (PSDU) is received from the MAC layer via the PHY Service Access Point (SAP).

The structure and timing of a symbol is illustrated in Figure 7.11. Each symbol consists of an integer number of possible chip positions, N_c, each with duration T_c. The overall symbol period denoted by T_{dsym} is given by $T_{dsym} = N_c T_c$. Furthermore, each symbol is divided into two Burst Position Modulation (BPM) intervals, each with duration $T_{BPM} = T_{dsym}/2$, which enables binary position modulation. A burst is formed by grouping N_{cpb} consecutive chips and has duration $T_{burst} = N_{cpb} T_c$. The fact that burst duration is typically much shorter than the BPM duration, i.e. $T_{burst} << T_{BPM}$, provides for some multiuser access interference rejection in the form of time hopping. The total number of burst durations per symbol, N_{burst}, is given by $N_{burst} = T_{dsym}/T_{burst}$. In order to limit the amount of inter-symbol interference caused by multipath, only the first half of each T_{BPM} period shall contain a burst. Therefore, only the first $N_{hop} = N_{burst}/4$ possible burst positions are candidate hopping burst positions within each BPM interval. Each burst position can be varied on a symbol-to-symbol basis.

Interested readers should consult the full standard for the relative details.

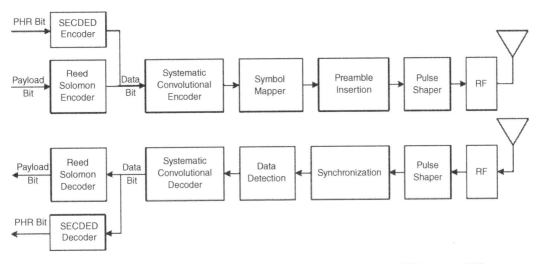

FIGURE 7.8   IEEE Std 802.15.4 HRP UWB PHY signal flow (top: TX; bottom: RX).

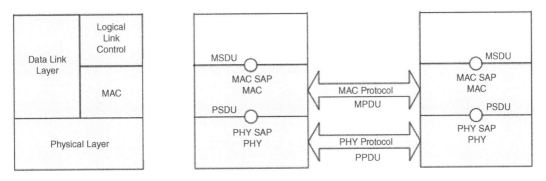

MSDU: MAC Service Data Unit

PSDU: PHY Service Data Unit

MPDU: MAC Protocol Data Unit

PPDU: PHY Protocol Data Unit

FIGURE 7.9   Protocol layers covered by various UWB standards.

ETSI EN 302065-1 Part 2

ETSI EN 302065-1 Part 2, *"Short Range Devices (SRD) using Ultra-Wideband technology (UWB); Harmonised Standard covering the essential requirements of article 3.2 of the Directive 2014/53/EU; Part 3: Requirements for UWB devices for ground based vehicular applications"* [26], defines requirements for UWB location tracking devices. Specifically, it defines operating principles for transceivers, transmitters, and receivers UWB technologies and is used for short-range applications in road and rail vehicles, which includes devices mounted inside or at the surface; the specification applies to impulse, modified impulse, and RF carrier-based UWB technologies in the main operating frequency ranges from 3.1 to 4.8 GHz or from 6 to 9 GHz. In ETSI EN 302065-1, UWB equipment is categorized based on the operational frequency range and the type of tracking it performs:

- LT1 systems: these systems, operating in the 6–9 GHz region are intended for general location tracking of people and objects. They operate on an unlicensed basis. The transmitting terminals in these systems are mobile (indoors or outdoors) or fixed (indoors only). Fixed

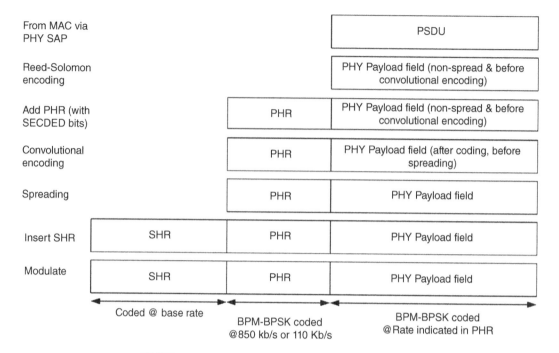

FIGURE 7.10   IEEE Std 802.15.4 HRP UWB PPDU.

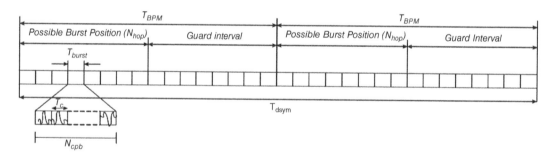

FIGURE 7.11   IEEE Std 802.15.4 HRP UWB PHY symbol structure.

outdoor LT1 transmitters are not permitted. Typically, LT1 transmitters are mobile location tracking tags which are attached to people or objects, and tags are tracked using a fixed receiver infrastructure to only receive the UWB emission emitted by the tags.

- LT2 systems: these systems, operating in the 3.1–4.8 GHz region are intended for person and object tracking and industrial applications at well-defined locations. The transmitting terminals in these systems may be located indoors or outdoors and may be fixed or mobile. They operate at fixed sites and may be subject to registration and authorization, provided local coordination with possible interference victims has been performed.

- Location Application for Emergency Services (LAES) systems: these systems, operating in the 3.1–4.8 GHz region, are intended for tracking staff belonging to the fire and other emergency services, who need to work in dangerous situations. Being able to track such people, even when deep inside a building, provides an important enhancement to command and control and to their personal safety. Typically, an LAES system is deployed temporarily at the scene of a fire or other emergency in a building. Licenses may be required for user organization.

ECMA-368 Standard

This International Standard specifies the UWB PHY and MAC sublayer for a high-speed, short-range wireless network, utilizing all or part of the spectrum between 3100 and 10 600 MHz supporting data rates of up to 480 Mbps [27]. It is a massive 344-page specification. This standard divides the spectrum into 14 bands, each with a bandwidth of 528 MHz. The first 12 bands are then grouped into four band groups consisting of three bands. The last two bands are grouped into a fifth band group. A sixth band group is also defined within the spectrum of the first four, consistent with usage within worldwide regulatory regulations. A MB-OFDM scheme is used to transmit information. A total of 110 subcarriers (100 data carriers and 10 guard carriers) are used per band. In addition, 12 pilot subcarriers allow for coherent detection. Frequency-domain spreading, time-domain spreading, and FEC coding are provided for optimum performance under a variety of channel conditions. The MAC sublayer is designed to enable mobility, such that a group of devices may continue communicating while merging or splitting from other groups of devices. To maximize flexibility, the functionality of this MAC is distributed among devices. These functions include distributed coordination to avoid interference between different groups of devices by appropriate use of channels and distributed medium reservations to ensure Quality of Service. The MAC sublayer provides prioritized schemes for isochronous and asynchronous data transfer. To do this, a combination of Carrier Sense Multiple Access (CSMA) and Time Division Multiple Access (TDMA) is used. A Distributed Reservation Protocol (DRP) is used to reserve the medium for TDMA access for isochronous and other traffic. For network scalability, Prioritized Contention Access (PCA) is provided using a CSMA scheme. The MAC has policies that ensure equitable sharing of the bandwidth [27]. Taken together, the PHY and MAC specified in this Ecma standard are suited to high rate, zero infrastructure communications between a mixed population of portable and fixed electronic devices.

7.4 IoT APPLICATIONS FOR UWB

UWB's ability is to provide relatively high data rate links with immunity from noise and multipath interference, low power consumption, along with the expectation of simple and eventually inexpensive transceivers, make it a candidate for short-range IoT applications. While the current data rate however is not "ultra" or "huge," being around half-to-a-gigabit per second, still many non-video-based IoT applications can operate at these data rates. The technology can be used indoors in home, offices, vehicles of all types, and institutional setting (for example, hospitals [28]). While the focus of this discussion is on tracking moving entities, specifically people, UWB-powered IoT applications include many other use cases [1, 11, 13, 14, 17–19, 21, 29–38]:

- Announcement services.
- Assistance services.
- Boat docking.
- Body implants; Wireless Capsule Endoscopy Device.
- Content download (home applications).
- Discovering other users in proximity.
- Emergency support, finding victims or objects.
- Fleet management.
- Healthcare biomedical instrumentation.
- Healthcare Body Area Networks (BANs).
- Home connectivity/home theater.
- Hospital telemetry.
- Hotspots (indoors and outdoors).

- Indoor tacking of people and/or assets.
- Intelligent ambient sensing.
- Local advertisements.
- Location-based information services.
- Medical applications.
- Membership management and positioning of multiple objects.
- Museums walks.
- Pedestrian route guidance.
- Physical layer signaling scheme for Wireless Sensor Networks.
- Public safety.
- Radar for imaging/sensing.
- Ranging.
- Real-time location sensing.
- RFID tags for low cost identification of items.
- Robot guidance (in-building).
- Route guidance for the blind.
- Smart city applications, including location-based services.
- Smart warehouse asset tracking.
- Sports (e.g. embedding Inertial Measurement Unit sensors and UWB radios inside balls and players' shoes).
- Tag tracking (building whereabouts).
- Tracking elderly folks; tracking kids; tracking pets.
- Tracking services, RTLSs.
- Traffic warning.
- Vehicle-to-Vehicle (V2V) and Vehicle-to-Infrastructure communication (V2I).

As noted, UWB allows a more efficient use of the available spectrum while causing no interference to pre-existing narrowband applications. The basic range is 10–20 m (in some implementations, slightly more), covering WPAN applications or localization. UWB signals have good material penetration capabilities; however, metal obstacles may degrade the performance of this technology.

Even beyond OSD/ODCMA, there may be IoT applications that require distance estimations. Some of these location tracking requires centimeter accuracy (e.g. route guidance for the blind), some need meter-level accuracy (e.g. pedestrian route guidance), others can deal with 10- to 50-m accuracy (e.g. location-based services). In broad terms, there are two approaches to determine the distances between two devices. One method entails using signal strength as a tool, specifically using the RSSI. One knows from Coulomb's law that the signal strength decreases with increasing distance from the transmitter in a mathematically defined manner. Using the electromagnetic equations, one can estimate the distance between a receiver and transmitter. This approach has a number of limitations, including the fact that the radio channel changes instantaneously, impacting the RSSI parameter; the RSSI parameter can be also degraded by multipath propagation. Systems such as Wi-Fi, Bluetooth, Bluetooth Low Energy (BLE), and Active RFID utilize this distance estimation process. The other method (used in UWB) utilizes the signal's ToF. This method achieves much more accuracy – up to a centimeter level, depending on the frequency and nature of the signal. By aggregating the ToF measurements from several devices, one can obtain an accurate position of the device (transmitter) in question. There are a number of other methods, each with pros and cons.

Although UWB has some limitations of its own, it enjoys a positioning accuracy that can be useful in a number of IoT applications.

For indoor navigation and tracking using UWB signals, multipath propagation processing can be successfully utilized. The MPCs need to be extracted from the measured Channel Impulse Response (CIR). In indoor localization applications, objects such as furniture, walls, windows typically give rise to dense multipath channels; that is, the CIR at a given position consists of many reflections and diffuse scattered components. The analysis of any wireless communication system requires a reliable model of CIR. The work by Saleh and Valenzuela shows that MPC arrivals in CIRs appear at the receiver in clusters [39]. The UWB channel can thus be characterized based on its MPCs obtained from its CIR:

$$H(n) = \sum\sum a_{i,l} \exp\left(j\phi_{i,l}\right)\delta\left(n - \Gamma_l - \tau_{i,l}\right)$$

(with the first summation going from $i = 1$ to $i = N$ and the second summation going from $l = 1$ to $l = L$), where $H(n)$ is the discrete UWB channel response in time domain; L is the total number of MPCs in the i-th cluster; N represents the total number of clusters during the scan interval; $a_{i,l}$, $\phi_{i,l}$, $\tau_{i,l}$ represent the amplitude, phase, and delay of the l-th MPC in the i-th cluster, respectively; and Γ_y is the delay of the y-th cluster [40].

Pedestrians may use navigation tools implemented in smartphones; navigation data can also be utilized in crowdsensing applications. In general, multipath reception degrades the accuracy of the positioning calculations. Strategies to mitigate multipath effects on location determinations are, in general, based on the estimation of the CIR in order to remove the influence on the estimate of the line-of-sight path delay. Some new methods treat MPCs as signals from virtual transmitters that are time-synchronized to the physical transmitter and fixed in their position. To use the information of the MPCs, these new methods estimate the user position and the position of the virtual transmitters simultaneously and do not require any prior information such as room-layout or a database for fingerprinting [41]. While several variants exist, multicarrier UWB (MC-UWB), Orthogonal Frequency Division Multiplexing UWB (OFDM-WBD), and Frequency Modulation UWB (FM-UWD) are the schemes likely to see active implementation and deployment.

7.5   UWB APPLICATIONS FOR SMART CITIES AND FOR REAL-TIME LOCATING SYSTEMS

7.5.1   Applications for Smart Cities

In the specific context of smart cities, documented applications of UWB have included the following, among several others: transportation (e.g. rail, automotive); car and driver safety; public safety (e.g. highway/tunnel safety, tags for blind people); tracking of people; hospital operations and safety; tracking of goods and assets (within a centimeter or millimeter – depending on the application); Automatic Vehicle Location (AVL) for fleet management; vehicle security; industrial manufacturing (including high-precision robotics).

For example, there have been efforts to accelerate the deployment of modern train signaling technologies in the subway system, to increase the number of trains at peak periods and promote faster and more reliable service. The UWB technology aims at eliminating the need to acquire and install expensive, cumbersome equipment required by Computer Based Train Control (CBTC) signal technology, since the UWB-based network has the potential to provide precise and accurate locations for subway cars within centimeters [42].

Exact spatial/location data are needed in robotic manufacturing, where UWB is an ideal solution, also in conjunction with less spatially precise tracking of the movements or confluence of assets, inventory, and supply chain-based components used in the manufacturing process. With UWB systems, a physical component in the manufacturing process can be tracked to the millimeter level, allowing complex assembly to take place without human involvement. Object tracking is also used for monitoring safety conditions (especially around possibly dangerous machinery, construction sites, forklifts, furnaces [43], cleaning equipment [44] or where enable applications in which people and robots work together).

7.5.2   UWB Applications to Real-Time Location Systems

As discussed in Chapter 6, locating systems are used in variety of environments and applications to locate and/or track people, machines, or objects; these systems are applicable to OSD/ODCMA management to address return-to-work considerations in the context of the COVID-19 pandemic.

UWB has increasingly gained interest in RTLS applications due to its capability to obtain typical location precision of 30 cm or better, making it more accurate than other wireless technologies. A number of UWB applications to RTLS environments have been documented in the literature, including, but not limited to [1, 45–75]; as an example, RTLSs based on UWB are becoming increasingly popular in sporting venues (e.g. football stadiums) for tracking players; in this application, football players wear tags that periodically send out over-the-air packets as consecutive UWB RF pulses [50]. Documented UWB localization solutions of the 2010s have been aimed, in practical terms, at localizing at most tens of mobile tags (say 30–200) and have been evaluated mostly in smaller office environments or open zones [1] (for example, [61, 62] describe high accuracies – 11 and 3 cm respectively – but they are limited to localizing one or two tags in a small area). More recent commercial solutions becoming available at press time advertise high update rates and tag counts, but the deployment venue is not always clearly identified.

Some UWB location systems provide real-time collection of data and, in some instances, real-time analysis of the data; they may generate data related to location, change in location, velocity, change in acceleration, and/or orientation for one or more people, machines, or objects at a site at which the location system is deployed. Some systems provide simultaneous tracking of a multiplicity of people, machines, or objects and provide real-time (or near real-time) data to upstream analytics systems. Such indications may be uploaded to a variety of management systems in the Intranet or in the cloud, including a visualization dashboard system, an operations support system, a control system, an analytics system, and/or a statistics system. Some systems can calculate statistical data and analytics data based on location data associated with people or objects using various in situ sensors. In some instances, the location of people, machines, or objects over time and analytical/statistical data may be uploaded to a combined visualization dashboard depicting "heat maps." Organizations may utilize the data and/or the analyses provided by the location system to control, assess, and/or manage operations and/or processes associated with the corresponding site [76].

Some observers believe that UWB has still some technical challenges before becoming a ubiquitous indoor localization choice; specifically, as implied in the early part of this chapter, the regulatory constraints impose certain PHY guidelines on transmission power, that may impact the scalability of the system due, theoretically, to potential interference with or from other technologies, swamping out its signals in high-density environments; on the other hand, UWB technology is intrinsically very robust to multipath fading, as noted, which is a critical feature in confined situations such as indoor localization. In spite of these concerns, it turns out that while scalability has been a concern in the past due to the just-cited technical requirements that are part of the regulation regarding power, making conventional medium access strategies somewhat complex to analyze and implement, UWB has been shown of late to be usable to simultaneously localize thousands of nodes in a dense network; however, the number of supported devices varies greatly depending on the MAC and PHY configuration choices, of which there are several [1].

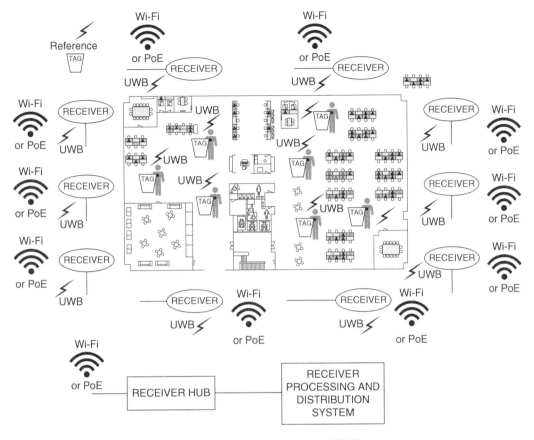

FIGURE 7.12 UWB-based RTLS.

Figure 7.12 provides a pictorial view of a possible business UWB office setup. The figure illustrates an example of a UWB-based RTLS that can be utilized for calculating a location by an accumulation of Time of Arrivals (ToAs) at a central processor/hub; the ToAs represent a relative ToF measure from the RTLS tags, as recorded at each UWB receiver/reader. Receivers are positioned at predetermined coordinates within and/or around the monitored area or office and are configured to sense signals transmitted by the location tags as well as by the reference tags; the receivers are connected to the receiver hub via Ethernet cables, for example, using Power over Ethernet (PoE), or are connected wirelessly via Wi-Fi links.

In the example of Figure 7.12, each of the RTLS receivers includes (i) a receiver for UWB transmissions and (ii) a packet decoding circuit that extracts a ToA timing pulse train, the transmitter ID, the packet number, and/or other information that may have been encoded in the tag transmission signal (e.g. personnel information, and so on). Each of the receivers includes a time measuring circuit that measures the ToA of tag bursts, with respect to its internal counter. The time measuring circuit is phase-locked with a common digital reference clock from the receiver hub. The reference clock signal establishes a common timing reference for the receivers. Thus, multiple time measuring circuits of the respective receivers are synchronized in frequency (but not necessarily in phase). The receivers transmit measurement to the receiver hub. In some examples, measurement data are transferred to the receiver hub at regular polling intervals. The receiver hub computes the tag's location by processing ToA measurements relative to multiple data packets detected by the receivers [49, 76]. In some examples, ToA measurements from multiple receivers are processed by the receiver hub to determine a location of the RF location tag by a Differential Times of Arrival (DToA) analysis of the multiple ToAs.

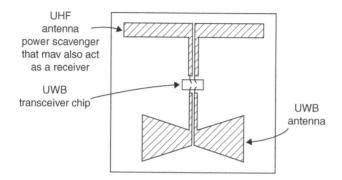

UHF
antenna
power scavenger
that mav also act
as a receiver

UWB
transceiver chip

UWB
antenna

Passive tag that comprises a transceiver chip
composed of a UWB radio transmitter and a UHF
power scavenger

FIGURE 7.13   UWB transceiver that is powered by an RF input through an antenna interface [45].

As noted in Section 7.1, UWB is a radio technology that employs high-bandwidth communications using a large portion of the radio spectrum; however, notwithstanding the high operating bandwidths, UWB communications are capped by a channel capacity that defines a theoretical maximum possible number of bits per second of information that may be conveyed through one or more links in an area. As such, channel capacity may limit the number of UWB devices that can concurrently communicate within a particular area. That may be an issue when UWB is used as a generic PAN technology; however, it is generally not an issue for RTLS/OSD applications. Generally, UWB tags require battery power; thus, another consideration related to UWB, noted in Chapter 6, and hinted earlier in this chapter, is the power consumption rate of the wearable tags. To address this issue, a number of transmission trade-offs have been proposed (more on this below); furthermore, some manufacturers have developed methods and solutions for making a passive UWB radio device that can be powered by incident electromagnetic energy and which communicates its data through UWB impulses [45, 46]; Figure 7.13 illustrates an example of a passive tag with a UWB transceiver that is powered by an RF input through an antenna interface. For battery-operated UWB tags, some trade-offs, described in Section 7.1, can be employed to extend battery life.

With UWB RTLSs, there are two types of positioning techniques that can be utilized, both of which are ToF-based [77] (also see Figures 7.14 and 7.15):

- ToA/TDoA: this technique is based on measuring the time difference between received signals arriving at a set of anchor sensors (also variously known as "receiver nodes" or "reader nodes"). To make use of this technique, the anchor sensors must be accurately synchronized to operate on the same clock (sub-nanosecond synchronization is needed since a 1 ns timing error translates in 30 cm position inaccuracy). Tags transmit at regular intervals (aka "refresh rates") using brief "Blink messages" (aka "blinks"). The blinks are processed by all the anchors within the communication range. Regardless of their synchronization status, the anchors will then send all the time stamps to the RTLS server. In order to calculate the position of the tag, the RTLS server considers only the time stamps coming from at least four anchors with the same clock base (these anchors are designated as "Primary Anchors" and are so designated automatically).

- Two-Way Ranging (TWR): this technique determines the ToF of the UWB radio frequency signal, then calculates the distance between the nodes, since the "free space" propagation speed of electromagnetic waves is known, being the speed of light. The TRW technique requires bidirectional communication between the tag and each of the involved anchors.

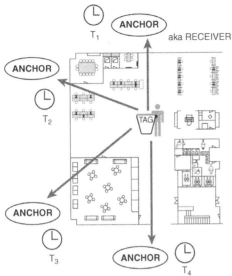 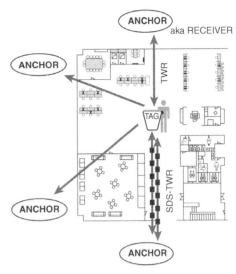

Time Difference of Arrival Scheme

The tag blinks at a certain point in time (T_0) and the message is received by four anchor (receiver) points at different times ($T_1...T_4$). To determine the position of the tag, the differences among the arrival times have to be computed.

Two Way ranging Scheme

The tag initiates the process by sending the first message at time T0. The anchor, after a certain period (Treply), replies with a message containing the packet timestamps T1 and T2, which are the reception moments of the first packet and the reply to the tag, respectively. At this juncture, the tag is able to compute the ToF, which is in turn used to estimate the distance between the anchor and the tag. A more accurate estimation can be achieved when additional packets are exchanges, as is the case in SDS-TWR.

FIGURE 7.14 UWB positioning techniques.

Common TWR schemes are (i) the single-sided TWR and, (ii) the symmetrical double-sided TWR (SDS-TWR). The TWR process is applied between the tag and the selected anchor – only one anchor may be actively involved in TWR at a given time slot. No time synchronization is needed between the UWB devices. To support TWR, three messages are used and – based on the time of reception of the third message in the sequence – the anchor determines the ToF of the UWB signal:

- Time of Sending Poll (TSP): the tag initializes the TWR process by sending a "poll" message to the known address of the anchor.
- The anchor records the Time of Reception of Poll (TRP), and it replies to the tag with a response message at the Time of Sending Response (TSR).
- The tag, upon receiving the response message, records the Time of Response Reception (TRR) and creates the response message which contains its ID, TSP, TRR, and Time Start Final (TSF).

Notice that to undertake this handshake, there needs to be prior address binding between the tags and the anchors.

TDoA positioning is favored by some vendors since systems based on TDoA tend to have longer battery lifespans for the tags, being that the tags operate on using simple blink messages:

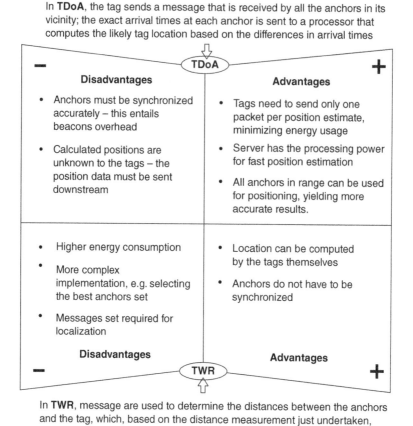

In **TDoA**, the tag sends a message that is received by all the anchors in its vicinity; the exact arrival times at each anchor is sent to a processor that computes the likely tag location based on the differences in arrival times

TDoA

−   **Disadvantages**

- Anchors must be synchronized accurately – this entails beacons overhead
- Calculated positions are unknown to the tags – the position data must be sent downstream

Advantages   **+**

- Tags need to send only one packet per position estimate, minimizing energy usage
- Server has the processing power for fast position estimation
- All anchors in range can be used for positioning, yielding more accurate results.

- Higher energy consumption
- More complex implementation, e.g. selecting the best anchors set
- Messages set required for localization

- Location can be computed by the tags themselves
- Anchors do not have to be synchronized

−   **Disadvantages**

TWR

Advantages   **+**

In **TWR**, message are used to determine the distances between the anchors and the tag, which, based on the distance measurement just undertaken, will compute its position

FIGURE 7.15   Comparison of methods.

the tags only need to send one blink message in order to be positioned; as noted, TWR requires multiple messages to be exchanged in order to locate the tag. Also, because there is no prior address binding between the tags and anchors, the number of anchors operating in a system can be large, serving a large (office) space and, thus, supporting scalability. Scalability is also supported by the fact that with TDoA tags only use a short portion of time to send a blink; therefore, a high number of tags can be transmitting signals within one refresh rate. Systems discussed in the literature have used a variety of techniques, including TDoA only, TWR only, hybrid TDoA-TWR, hybrid TDoA-TWR-Angle of Arrival (AoA), and RSSI. In terms of localization precision, the hybrid TDoA-TWR approach performs better than the standard TDoA [1].

Some UWB implementations based on TDoA that are applicable to OSD/ODCMA make use of a timing reference clock, so that at least a subset of the receivers may be synchronized in frequency; this enables the relative ToA data associated with each of the RTLS tags to be registered by a counter associated with the receivers. A reference tag, preferably a UWB transmitter, positioned at known coordinates, can be used to determine a phase offset between the counters associated with the receivers [49] (see again Figure 7.12). The RTLS tags and the reference tags reside in an active RTLS field. The tags and the receivers can be configured to provide two-dimensional and/or three-dimensional precision localization, even in the presence of multipath interference, due in part to the use of short, nanosecond duration pulses whose ToF can be accurately determined using detection circuitry in the receivers. In some implementations, this short pulse characteristic allows data to be conveyed by the system at a higher peak power, but at lower *average* power levels, than a wireless system configured for high data rate communications, yet still operating within FCC/ETSI requirements. In order to provide an acceptable performance level while complying with the regulatory restrictions, the tags may typically operate with an instantaneous −3 dB bandwidth over a spectrum of approximately 400 MHz and an average transmission rate below 187 pulses in a 1 msec interval, provided that the packet rate is sufficiently low. In such examples, the predicted maximum range of the system, operating with a center frequency of 6.55 GHz, is approximately 600 feet in instances in which a 12 dBi directional antenna is used at the receiver − in other examples, the range will depend on the receiver's antenna gain. Alternatively, the range of the system allows for one or more tags to be detected with one or more receivers positioned throughout the office space. Such a configuration satisfies constraints mandated by the FCC/ETSI related to peak and average power densities (e.g. Effective Isotropic Radiated Power density [EIRP]), while still optimizing system performance related to range and interference. Typically, tag transmissions with a − 3 dB bandwidth of approximately 400 MHz yields an instantaneous pulse width of approximately 2 ns that enables a location resolution to better than 30 cm. The tags utilize UWB transmitters that transmit blink data (e.g. multiple pulses at a 1 Mbps burst rate, such as 112 bits coded with OOK "modulation" at a rate of 1 Mbps); such data consist of an information packet that may include, but is not limited to, ID information, a sequential burst count, and/or other desired information for object or personnel identification, inventory control, and so on. In cases, the sequential burst count (e.g. a packet sequence number) from each tag can be utilized by the central processor/hub to establish correlation of ToA measurement data from various receivers [49].

As discussed, RTLS systems based on UWB waveforms are able to achieve very fine resolution because of their ultrashort pulse durations, for example, sub-nanosecond to nanosecond pulses: instantaneous pulse width of 2 ns can achieve a location resolution to better than 30 cm. The information packet transmitted over a UWB system is typically of a short length (e.g. 112 bits of OOK at a rate of 1 Mbps); in some implementations, higher packet repetition rates (e.g. 12 transmissions per second, 12 Hz) and/or higher data rates (e.g. 1 Mbps, 2 Mbps) result in larger datasets for filtering, enabling one to achieve a more accurate location estimate. There is a trade-off between higher repetition rate and the battery life, as follows [76]:

- In some systems, the shorter length of the information packets, in conjunction with different packet rates, data rates, and other system factors, results in a longer battery life (e.g. seven years battery life at a transmission rate of 1 Hz with a 300 mAh cell).
- Some applications require higher transmit tag repetition rates to track a dynamic environment (e.g. 12 transmissions per second); in such applications, the battery life may be shorter.

Additional trade-offs exist, as discussed in [76]. Signals from the RF location tags may be received more than once at receiver: a receiver may receive a pulse directly from the RF location or after being reflected along the way; reflected signals travel longer and are, thus, received later in time than the original direct signal; this delay is known as an echo delay or multipath delay. In cases where the reflected signals are strong enough to be detected by the receiver, the signals can corrupt a data transmission through inter-symbol interference. Short information packets are preferred: when the packet durations are short (e.g. 112 μs), they allow inter-pulse times sufficiently longer (e.g. 998 ns) than the expected echo delays, thus avoiding data corruption. Reflected signals can be expected to become weaker as delay increases, due to a larger number of reflections and due to the longer distances traveled. Thus, beyond some value of inter-pulse time (e.g. 998 ns), corresponding to some path length difference (e.g. 900 feet), there will be no advantage to further increases in the inter-pulse time (and, hence, lowering of burst data rate) for any given level of transmit power. In this manner, minimization of packet duration allows the battery life of a tag to be maximized, given that its digital circuitry need only be active for a brief time. Clearly, different physical environments can have different expected echo delays, so that different burst data rates – and, hence, packet durations – may be appropriate in different situations. Minimization of the packet duration also allows a tag to transmit more packets in a given time period, although in practice, regulatory average EIRP limits often provide an overriding constraint. However, short packet duration also reduces the likelihood of packets from multiple tags overlapping in time, causing a data collision. Thus, minimal packet duration allows multiple tags to transmit a higher aggregate number of packets per second, allowing for the largest number of tags to be tracked, or a given number of tags to be tracked at the highest rate. The high burst data transmission (TX) rate (e.g. 1 MHz), coupled with the short data packet length (e.g. 112 bits) and the relatively low repetition rates (e.g. 1 TX/sec), results in some advantages in specific implementations [76]:

1. A greater number of tags may transmit independently from the field of tags with a lower collision probability, and/or,
2. Each independent tag transmit power may be increased, with proper consideration given to a battery life constraint, such that a total energy for a single data packet is less than a regulated average power for a given time interval (e.g. a 1 msec time interval for an FCC regulated transmission).

In some applications, also in the OSD arena, some tangential information can be transmitted at a rate of one time per minute (for example, once every 720 times the data packet is transmitted), but the data packet – for example, the real-time location of the person – is transmitted at a higher rate (e.g. 12 transmissions per second).

The discussion that follows describing a typical tag transmission pulsing sequence and the illustrative figures is based on reference [50]. Figure 7.16 shows an example timing diagram for an RTLS tag transmission in a high-resolution ToA determination system. The timing diagram includes a TX clock, a preamble, and a data packet; the data packet includes the preamble, a sync code, a header, a transmit (TX) ID, and a Cyclic Redundancy Check (CRC). The preamble is

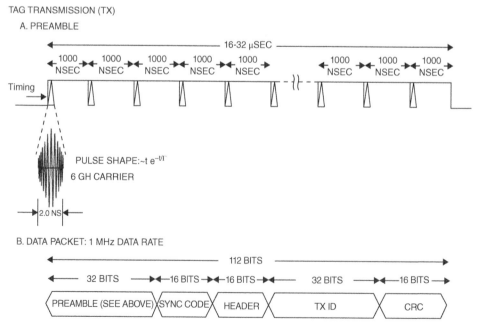

FIGURE 7.16 Example of a tag transmission pulsing sequence for a UWB-based RTLS [50].

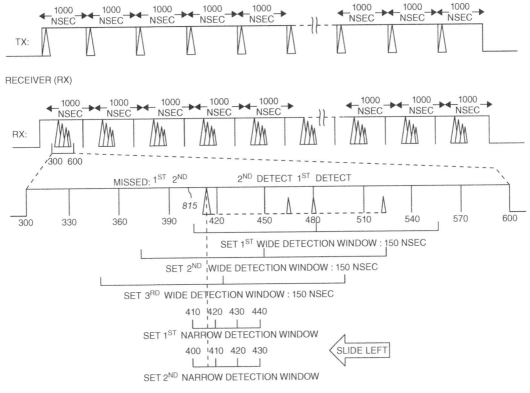

FIGURE 7.17   Example timing diagram for a receiver [50].

composed of a transmit (TX) series of pulses equally spaced in time, in accordance with a period associated with the TX clock. In some implementations, the period associated with the TX clock is approximately one microsecond (μsec) (the TX clock operates at a frequency of 1 MHz). Each individual TX pulse in the TX series of pulses is identical. In some examples, the TX pulse includes 6 GHz carrier wave modulated by a 2 ns pulse, such as a triangular or rectangular function. In some examples, the TX pulse is additionally shaped at a receiver by a transmit-and-receive antenna and any electronics associated with an amplification or pre-amplification of the TX pulse, in conjunction with the high-resolution ToA determination system. The TX series of pulses is used to provide for an iterative windowing function, such as, an adjustable coarse timing window. Figure 7.17 shows an example timing diagram for a receiver (RX) adjustable coarse timing window function. The example timing diagram of Figure 7.17 includes the TX clock and the TX series of pulses associated with the RTLS tag transmission (TX), illustrated in Figure 7.16, and an RX clock and an RX clock timing diagram. A received (RX) pulse train is composed of a series of the RX pulses, corresponds to the TX series of pulses, and is synchronized to the RX clock, which is resident at a receiver. As shown in Figure 7.17, an RX pulse signature, representing an earliest pulse and a series of echoes and possible noise pulses, is associated with the RX clock timing diagram, and is also associated with the corresponding TX pulse. In the example of Figure 7.17, the 1 MHz TX clock and the associated 1 MHz RX clock may be out of phase with respect to each other. However, the relative stability of the respective TX clock and RX clock frequencies for the short-duration TX transmit time allows for an iterative, adaptive adjustment of the RX clock phase with respect to the TX clock phase, effecting a change in the receiver RX adjustable coarse timing window function. The example RX adjustable coarse timing window function of Figure 7.17 is composed of (i) a series of detection windows, including wide detection windows and narrow detection windows, and (ii) an associated set of functions to adaptively position the series of wide and narrow detection windows to center the RX pulse in the corresponding window. In the example of Figure 7.17, there are three wide detection windows and two narrow detection windows.

7.6 OSD/ODCMA APPLICATIONS

There have been industry efforts to utilize UWB technology and supporting analytics to support COVID-19 driven OSD/ODCMA for office, factory, and warehouse applications. Smart tags can be used to ensure employee safety and can accelerate return-to-work by monitoring social distancing and person-to-person and person-to-object-to-person contact tracing. (Typical industrial applications, including warehousing, may require 3D x-y-z position determination, which is possible with UWB systems.) Many of the systems on the market combine smart algorithms with AI machine learning. The advantage of UWB, already discussed, is the precision of localization within centimeters (e.g. 4 in.), while Bluetooth/Wi-Fi has typical accuracy of 1–6 yards and GPS 7–10 yards (if even working reliably in a specific indoor setting). Vendors claim that several thousand large firms are already using these UWB systems. High-end products on the market claim unlimited area size and unlimited tracked devices.

Four specific exemplary products are highlighted in this section to give weight to the availability of the technology (no product is explicitly sanctioned herewith).

One illustrative UWB example is the system developed by Ubisense [74]. UWB tags provide high-confidence separation measurements. Tags carried by people and (optionally) attached to equipment are precisely located in three dimensions to within 15 cm; sensor base stations are deployed in the office or factory environment to enable precise location measurements. RTLS stability provides high-confidence contact traces uncontaminated by the false positives of less accurate systems. The UBsense system tracks contact events, not people, over time; tag data are completely anonymized until a potential safety risk is identified. The system uses an AI-driven contagion model that operates beyond simple separation measurement and makes use of detailed models to understand movement and interaction of people over time and within the context of their environment; specifically, the contagion model is quickly and easily adaptable as transmission mechanisms become better understood. See Figure 7.18. Analytics (cloud-based) include (i) a

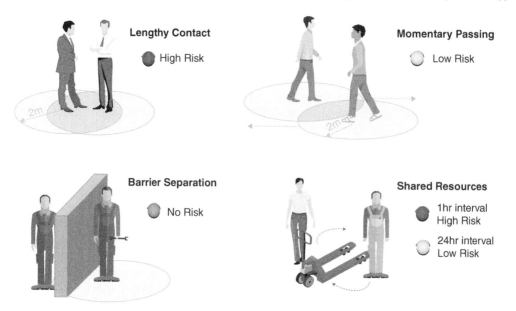

FIGURE 7.18 UWB OSD model (Courtesy, UBsense Limited).

Social Distance model; (ii) a Contagion Model based on an AI Engine to identify possible contagion events; (iii) a Reporting Engine for visualizing anonymized employee behavior and highlighting critical encounters; and (iv) an aggregation database to store encounters and produce accumulated employee risk profiles. Simulations are used to assess potential impact on workers and spaces of a contagion event. The Social Distance Model entails (i) defining pre-configured (and user-definable) contact distance and time thresholds, and (ii) defining anonymized recording of time, location, and tags breaching social distancing rules. The Contagion Model entails (i) a predefined model for transmission (person–person and person–object–person), incubation and disintegration variables that are user-definable as contagion becomes better understood and (ii) contact tracing mechanisms (from symptoms dating back through incubation time) to identify "at-risk" tag population – this is calculated using accumulated probability of infection from contagion model variables. Real-time alerts are delivered based on customer device/environment requirements (light, sound, haptic). The Reporting Engine provides a list of "at-risk" tags (individuals and objects) if infection is identified; also, it provides an anonymized contact history to identify at-risk spaces with a higher incidence of contact events for potential reconfiguration.

Reference [74] observes that

"As businesses restart in the wake of a global pandemic contact tracing from UBsense can help keep employees safe and make operations more resilient. A three-dimensional digital map of [the] facility is created; small lightweight tags carried or worn by people show precise, real-time locations using highly reliable UWB technology (Figure 7.19, top). When individuals move within a set distance of each other, a contact event is recorded and graded as either a momentary passing contact or a more significant full contact (Figure 7.19, middle two panels). UBsense contact tracing provides dashboards for authorized users to analyze these contact events. The daily overview dashboard shows the number of contact events occurring over time. The social distancing performance of teams or areas is reported with spikes or patterns easily identified. The contact events dashboard provides a more detailed view of all contact events highlighting the time location and duration of specific interactions. Areas can be easily identified that may need reconfiguration (Figure 7.19, lower two panels).

For example, a corridor with a high frequency of passing contact events requiring better traffic management or a cluster of full contact events, highlighting a process or workstation that needs a redesign. By adding the location and movement of cleaners or cleaning equipment within designated areas it is possible to be more targeted and effective in reducing transmission from surfaces and equipment. The hygiene status dashboard tracks the presence of people over time, highlighting high traffic areas that are frequently used or overdue for cleaning, to create a prioritized dynamic schedule of areas that most need hiding control. Should a person full ill or test positive for coronavirus, the contact tracer dashboard enables authorized users to identify others who are at risk of infection and require notification testing or to self-isolate. By entering the COVID-19 test date, and the date symptom started contacts are ranked as being high or medium risk of transmission, based on the history of contact events between people and the use of shared tools or equipment. Depending on the chosen privacy setting, individuals can be identified directly within the dashboard or anonymized for offline review."

A second example of a UWB-based RTLS available at press time is the Zebra Technology system [75] (see Figure 7.20). This UWB RTLS is designed for applications requiring accurate, precise, and high update rate real-time location. The system tags that feature long battery life. The system is compliant with the new International UWB Standard, IEEE 802.15.4, as well as the ISO-24730-61 Draft International standard. Features include:

- High real-time location accuracy – better than 30 cm (1 feet) line-of-sight.
- Long tag battery life – up to seven years at 1 Hz blink rate.
- Long RTLS Range – up to 200 m (650 feet) line-of-sight.
- High real-time location tag throughput – up to 3500 1 Hz tags/hub.

- Programmable, fast tag blink rates – up to 200 times/second.
- The UWB Hub runs the UWB real-time location software, capable of delivering thousands of tag blinks per second at up to 30 cm accuracy. The UWB Hub allows all UWB Sensors to be used over the entire facility, yielding maximum coverage with minimal sensor count.
- UWB Sensors are placed throughout the coverage area. They may be daisy chained to one another, providing for a simple and cost-effective installation.
- Zebra provides two types of UWB Tags: a UWB Tag and a UWB Tag Badge. UWB Tags are battery-operated devices that are affixed to assets or personnel; the UWB Tag features a small circular form factor and is generally used on assets. The UWB Tag Badge features a low profile and is generally used on personnel or on assets. UWB transmissions are extremely short in duration, providing excellent real-time location accuracy, long battery life, and high tag throughput.

FIGURE 7.19   UWB apparatus and analytics (Courtesy, UBsense Limited).

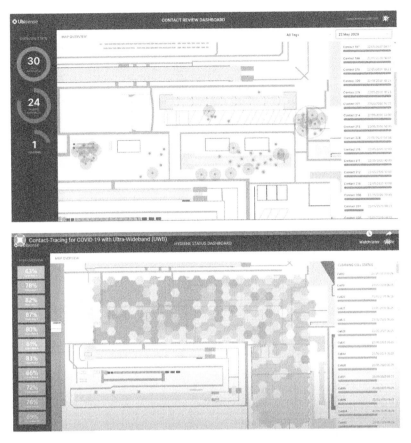

FIGURE 7.19   (Continued)

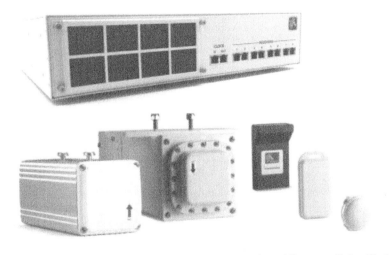

FIGURE 7.20   Example of UWB RTLS System, Zebra Technology (Courtesy Zebra Technology).

3D tracking, motion captures, infinite wireless scalability, auto setup, Artifical intelligence & behavioural learning
www.airtls.com

UWB facts and figures

Ultra-Wideband (UWB) is the best worldwide radio standard to perform localization. Locate tagged objects with sub 10 cm accuracy, at a distance of >1 km, at a cost of 1 to 6 €/m². This makes it functionally and economically viable to deploy indoor, outdoor, in volume, or in difficult-to-access locations.

19 Unique features of AIRTLS

- **Full wireless** solution, minimize cables
- **LTE proof** (5G, 2019 EU rollout) with special RF frontend design
- Distance + arrival angle (3° accuracy 2D or 3D)
- LTE proof (5G, 2019 EU rollout) with special RF frontend design
- Distance + arrival angle (**3° accuracy 2D or 3D**)
- Automatic deployment (no need for calibration)
- 2.1 cm accuracy (bias error).
- 2 to 30 cm position accuracy (depends on environment)
- Mems sensorfusion (UWB + motion analysis) for sub cm accuracy
- World smallest tag with all sensors build in
- World lowest idle current 95nA, RTC&RAM = on
- Anchors use 1 Watt (easy to solar power)
- Wired and wireless synchronisation
- Continuous signal quality and strength evaluation
- AES-256 encryption
- White labelling possible
- Opensource tag designs
- Top1 in decawave human wearable ranking
- No annual fee for RTLS software
- Highest update rate 4500 updates/sec/channel
- Longest range > 1.1 km, 6.8 Mbps (horn antenna)
- All "magic" happens inside the box (dual core 1GHz/anchor option, distributed edge computing); no need for the IT department or external servers

As a result of accurate location information, businesses can make considerable improvements in costs, service and profits. The decawave DW1000 UWB transceiver capability is unrivalled in the market and has multiple key advantages: Location precision: locating of tagged items to within 5 cm indoors, even while moving at speeds up to mach 1, 1225 km/h [340 m/s] Density: Supports the accurate location of max 7500 items per second in a > 100 m radius *Maximum amount of tracked tags: no-limit;* Communications / location range: Communicates over ranges of up to 300 m LOS (at 100 kbps, 850 kbps and 6.8 Mbps) with special antenna > 1 km; Low power: DW1000 can operate up to 10 years from 1 battery charge depending on the mode of operation; Low cost: Small package size and competitive pricing; Future proofing; solution based on IEEE802.15.4-2011 worldwide standard.

FIGURE 7.21   Airtls product features (Courtesy Airtls).

Another example of a company that produces UWB devices is Airtls [78]; it is based on a widely used UWB chip, the DW1000 (by Decawave Ltd., Dublin, Ireland). On their web page (see Figure 7.21 for a screengrab), they claim (i) no limit on the number of tags tracked, (ii) a maximum update rate of 4500 updates/sec/channel, and (iii) the ability to accurately localize max 7500 items per second in a > 100 m radius with a 2.1 cm accuracy.

A last example is a system from Pozyx NV. The firm has UWB-based TDoA and TWR positioning system that is claimed to support unlimited area size and unlimited tracked devices; over three thousand companies use this system [79]. The system can position thousands of tags over a large area with 10–30 cm accuracy. The system utilizes a local processing server (called gateway), which connects to all Pozyx anchors over Ethernet. The gateway performs the wireless synchronization between the anchors, manages the scheduling, and computes the position of all the tags. See Figure 7.22. The gateway also monitors the system and provides alerts when it is malfunctioning. The system supports TDoA-based positioning and is designed with small tags and long battery life in mind (years): the tags simply transmit a short packet that can contain custom data payloads and the complexity is therefore kept at the gateway and anchor level. The

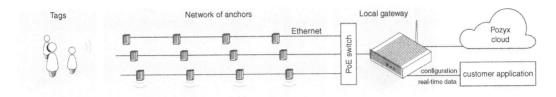

FIGURE 7.22 Pozxy system (Courtesy of Pozxy).

system can provide up to 1000 position updates per second and can handle up to 264 anchors per server; up to eight anchors can be connected in a chain with PoE+.

Additional products are available on the market. The short list above is intended only to demonstrate that the RTLSs UWB products are available off-the-shelf and to highlight some of the technology features that have seen implementation.

REFERENCES

1. Ridolfi, M., Van de Velde, S., Steendam, H., and De Poorter, E. (2018). Analysis of the scalability of UWB indoor localization solutions for high user densities. Sensors (Basel) 18 (6): 1875. Published 2018 Jun 7. https://doi.org/10.3390/s18061875.
2. Wang, J., Raja, A.K., and Pang, Z. (2015). Prototyping and experimental comparison of IR-UWB based high precision localization technologies. 2015 IEEE 12th international Conference on Ubiquitous Intelligence and Computing and 2015 IEEE 12th international Conference on Autonomic and Trusted Computing and 2015 IEEE 15th international Conference on Scalable Computing and Communications and Its Associated Workshops (UIC-ATC-ScalCom), Beijing, pp. 1187–1192. https://doi.org/10.1109/UIC-ATC-ScalCom-CBDCom-IoP.2015.216.
3. Nikookar, H. and Prasad, R. (2009). Introduction to Ultra Wideband for Wireless. Springer.
4. Lembrikov, B. (ed.) (2010). Ultra Wideband. INTECH.
5. Matin, M. (2010). Ultra wideband preliminaries. In: Ultra Wideband (ed. B. Lembrikov). Croatia: INTECH.
6. Harmuth, H.F. (1968). A generalized concept of frequency and some applications. IEEE Transactions on Information Theory 14 (3): 375–382.
7. Bennett, C.L. and Ross, G.F. (1978). Time-domain electromagnetics and its applications. Proceedings of the IEEE 66: 299ff.
8. Hussain, M.G. (1998). Ultra-wideband impulse radar-an overview of the principles. IEEE Aerospace and Electronics Systems Magazine 13: 9–14.
9. DiBenedetto, M., Kaiser, T., Molish, A.F. et al. (2006). UWB communication systems: a comprehensive overview. UERASIP Book Series on Signal Processing & Communications https://doi.org/10.1002/0470042397.
10. Niemelä, V., Haapola, J., Hämäläinen, M., and Iinatti, J. (2017). An ultra wideband survey: global regulations and impulse radio research based on standards. Communications Surveys & Tutorials IEEE 19: 874–890.
11. Zhai, C., Zou, Z., Zhou, Q. et al. (2016). A 2.4-GHz ISM RF and UWB hybrid RFID real-time locating system for industrial enterprise internet of things. Enterprise Information Systems 11 (6): 909–926, Published online: 04 Mar. https://doi.org/10.1080/17517575.2016.1152401.
12. Fang, Y., Han, G., Chen, P. et al. (2016). A survey on DCSK-based communication systems and their application to uwb scenarios. IEEE Communications Surveys & Tutorials 18 (3, Third Quarter): 1804–1837. https://doi.org/10.1109/COMST.2016.2547458.
13. Anzai, D., Katsu, K., Chavez-Santiago, R. et al. (2014). Experimental evaluation of implant UWB-IR transmission with living animal for body area networks. IEEE Transactions on Microwave Theory and Techniques 62 (1): 183–192. https://doi.org/10.1109/TMTT.2013.2291542.
14. Thotahewa, K.M.S., Redouté, J-M., and Yuce, M.R. (2014). A UWB wireless capsule endoscopy device. Engineering in Medicine and Biology Society (EMBC) 2014 36th Annual International Conference of the IEEE, pp. 6977–6980.

15. Ahmad, W., Tarczynski, A., and Budimir, D. (2017). Design of monopole antennas for UWB applications. Antennas and Propagation & USNC/URSI National Radio Science Meeting 2017 IEEE International Symposium on, pp. 2323–2324.

16. Gulam Nabi Alsath, M. and Kanagasabai, M. (2015). Compact UWB monopole antenna for automotive communications. IEEE Transactions on Antennas and Propagation 63 (9): 4204–4208. https://doi.org/10.1109/TAP.2015.2447006.

17. Sharma, S., Gupta, A., and Bhatia, V. (2017). A simple modified peak detection based UWB receiver for WSN and IoT applications. *Vehicular Technology Conference*, 2017 IEEE 85th, Sydney, Australia, 4–7 June 2017. doi:10.1109/VTCSpring.2017.8108192.

18. Silva, B. and Hancke, G.P. (2016). IR-UWB-based non-line-of-sight identification in harsh environments: principles and challenges. IEEE Transactions on Industrial Informatics 12 (3): 1188–1195. https://doi.org/10.1109/TII.2016.2554522.

19. Sahinoglu, Z. and Gezici, S. (2011). Ultra-Wideband Positioning Systems. Cambridge University Press.

20. Thakre, A.K. and Dhenge, A.I. (2012). Selection of pulse for ultra wide band communication (UWB) system. International Journal of Advanced Research in Computer and Communication Engineering 1 (9): 11.

21. Staff. UWB Technology. http://www.sewio.net/technology.

22. Molisch, A.F. (2009). Ultra-wide-band propagation channels. Proceedings of the IEEE 97 (2): 353–371.

23. Meissner, P., Gigl, T., and Witrisal, K. (2010). UWB sequential Monte Carlo positioning using virtual anchors. *Proceedings 2010 International Conference on Indoor Positioning and Indoor Navigation*, IPIN, Zurich.

24. Batra, A., Balakrishnan, J., Aiello, G.R. et al. (2004). Design of a multiband OFDM system for realistic UWB channel environments. IEEE Transactions on Microwave theory and Technology 52 (9): 2123–2138.

25. IEEE Std 802.15.4™. (2015). IEEE Standard for Low-Rate Wireless Networks. https://standards.ieee.org/standard/802_15_4-2020.html.

26. ETSI EN 302 065-1 Part 2. Short range devices (SRD) using ultra wide band technology (UWB); harmonised standard covering the essential requirements of Article 3.2 of the Directive 2014/53/EU; Part 3: requirements for UWB devices for ground based vehicular applications. ETSI, 650 Route des Lucioles, F-06921 Sophia Antipolis Cedex, France.

27. ECMA-368 Standard. (2008). High rate ultra wideband PHY and MAC standard, 3e, Ecma International, Rue du Rhône 114, CH-1204 Geneva. http://www.ecma-international.org.

28. CenTrak Staff. Real-time location system for hospitals: improving facilities for patients and staff. https://www.centrak.com/intro-to-rtls (accessed 8 August 2018).

29. Lee, Y., Kim, J., Lee, H. et al. (2017). IoT-based data transmitting system using a UWB and RFID system in smart warehouse. Ubiquitous and Future Networks (ICUFN), 2017 Ninth International Conference on, Milan, Italy, 4–7 July 2017. https://doi.org/10.1109/ICUFN.2017.7993846.

30. Rhee, W., Liu, D., Zhang, Y., and Wang, Z. (2017). Energy-efficient proprietary transceivers for IoT and smartphone-based WPAN. Radio-Frequency Integration Technology (RFIT), IEEE International Symposium on, Seoul, South Korea, 30 August–1 September 2017. https://doi.org/10.1109/RFIT.2017.8048283.

31. Crepaldi, M., Sanginario, A., and Ros, P.M. (2016). Low-latency asynchronous networking for the IoT: routing analog pulse delays using IR-UWB. New Circuits and Systems Conference (NEWCAS), 14th IEEE International, Vancouver, BC, Canada, 26–29 June 2016. https://doi.org/10.1109/NEWCAS.2016.7604842.

32. Kang, J., Rao, S., Chiang, P. et al. (2016). Area-constrained wirelessly-powered UWB SoC design for small insect localization. Wireless Sensors and Sensor Networks (WiSNet), IEEE Topical Conference on, Austin, TX, USA, 24–27 January 2016. https://doi.org/10.1109/WISNET.2016.7444310.

33. Espinoza, J.R., Padilla, V.S., and Velasquez, W. (2017). IoT generic architecture proposal applied to emergency cases for implanted wireless medical devices. *Proceedings of the International MultiConference of Engineers and Computer Scientists*, 2017 Vol II, IMECS 2017, Hong Kong (15–17 March 2017).

34. Dey, S. and Karmakar, N.C. (2017). An IoT empowered flexible chipless RFID tag for low cost item identification. Humanitarian Technology Conference (R10-HTC), 2017 IEEE Region 10, Dhaka, Bangladesh, 21–23 December 2017. https://doi.org/10.1109/R10-HTC.2017.8288933.

35. Kianpour, I. and Hussain, B. (2017). A complementary LC-tank based IR-UWB pulse generator for BPSK modulation. East-West Design & Test Symposium (EWDTS), 2017 IEEE, Novi Sad, Serbia, 16 November 2017, 29 September–2 October, 2017. https://doi.org/10.1109/EWDTS.2017.8110138.

36. Jeong, S., Park, J.B., and Cho, S.H. (2016). Membership management and positioning method of multiple objects using UWB communication. Procedia Computer Science, Elsevier, pp. 640–648. https://doi.org/10.1016/j.procs.2016.03.081.

37. Shahrestani, S. (2017). Assistive IoT: deployment scenarios and challenges. Internet of Things and Smart Environments, Springer.

38. Gowda, M., Dhekne, A., Shen, S. et al. (2017). Bringing IoT to sports analytics. 14th USENIX Symposium on Networked Systems Design and Implementation (NSDI '17), Boston, MA, USA, March 27–29, 2017.

39. Ruisi, H., Chen, W., Ai, B. et al. (2016). On the clustering of radio channel impulse responses using sparsity-based methods. IEEE Transactions on Antennas and Propagation 64 (6) https://doi.org/10.1109/TAP.2016.2546953.

40. Khawaja, W., Guvenc, I., and Chowdhury, A. (2017). Ultra-wideband channel modeling for hurricanes. IEEE 86th Vehicular Technology Conference (VTC-Fall), 24–27 September 2017, Toronto, ON, Canada. https://doi.org/10.1109/VTCFall.2017.8287903.

41. Gentner, C., Pöhlmann, R., Ulmschneider, M., et al. (2015). Multipath assisted positioning for pedestrians. *Proceedings of the 28th international Technical Meeting of Satellite Division of the Institute of Navigation (ION GNSS+ 2015)*, Tampa, Florida (September 2015), pp. 2079–2086.

42. MTA Staff. (2018). MTA selects 8 winners, 2 honorable mentions of MTA genius transit challenge, 9 March 2018. http://www.mta.info/news/2018/03/09/mta-selects-8-winners-2-honorable-mentions-mta-genius-transit-challenge (accessed 8 August 2018).

43. Tabarovskiy, O. (2018). Large scale UWB based location at blast furnace site: lessons learned. Geo IoT World EMEA 2018, June 11–13, 2018, Brussels, Belgium.

44. Celan, V., Stancic, I., and Music, J. (2017). Ultra wideband assisted localization of semi-autonomous floor scrubber. IEEE Journal Of Communications Software And Systems 2: 13.

45. Pahlavan, K. and Eskafi, F. (2015). Local indoor positioning and navigation by detection of arbitrary signals. US Patent 10,151,844, 11 December 2018, filed 30 April 2015. Uncopyrighted.

46. Pahlaven, K. and Eskafi, F.H. (2005). Radio frequency tag and reader with asymmetric communication bandwidth. US Patent 7,180,421, 20 February 2007, filed 12 November 2005. Uncopyrighted.

47. Quan, X., Choi, J.W., and Cho, S.H. (2014). In-bound/out-bound detection of people's movements using an IR-UWB radar system. *2014 International Conference on Electronics, Information and Communications (ICEIC)*, Kota Kinabalu, Malaysia, 1–2. doi:10.1109/ELINFOCOM.2014.6914407.

48. Bartoletti, S., Conti, A., and Win, M.Z. (2017). Device-free counting via wideband signals. IEEE Journal on Selected Areas in Communications 35 (5): 1163–1174.

49. Turner, B., Ameti, A., Richley, E.A., and Mueggenborg, A. (2018). System, apparatus and methods for variable rate ultra-wideband communications. US Patent 9,953,196, 24 April 2018, filed 3 January 2017. Uncopyrighted.

50. Richley, E.A. and Ameti, A. (2018). Methods and apparatus to mitigate interference and to extend field of view in ultra-wideband systems. US Patent 10,673,746, 2 June 2020, filed 14 September 2018. Uncopyrighted.

51. Richley, E.A., Turner, B., Wang, C., and Ameti, A. (2017). Receiver processor for adaptive windowing and high-resolution TOA determination in a multiple receiver target location system. US Patent 10,285,157, 7 May 2019, filed 3 November 2017. Uncopyrighted.

52. Zhang, J., Orlik, P.V., Sahinoglu, Z. et al. (2009). UWB systems for wireless sensor networks. Proceedings of the IEEE 97 (2): 313–331. https://doi.org/10.1109/JPROC.2008.2008786.

53. Fontana, R.J., Richley, E., and Barney, J. (2003). Commercialization of an ultra wideband precision asset location system. IEEE Conference on UltraWideband Systems and Technologies, Reston, VA, USA, pp. 369–373. https://doi.org/10.1109/UWBST.2003.1267866.

54. Cheong, P., Rabbachin, A., Montillet, J.P. et al. (2005). Synchronization, TOA and position estimation for low-complexity LDR UWB devices. 2005 IEEE International Conference on Ultra-Wideband, Zurich, Switzerland, pp. 480–484. https://doi.org/10.1109/ICU.2005.1570035.

55. Palaa, S., Jayanb, S., and Kurupa, D.G. (2020). An accurate UWB based localization system using modified leading edge detection algorithm. Ad Hoc Networks 97: 102017. https://doi.org/10.1016/j.adhoc.2019.102017.

56. Cirulis, A. (2020). Large scale augmented reality for collaborative environments. In: Universal Access in Human-Computer Interaction. Design Approaches and Supporting Technologies, HCII 2020. Lecture Notes in Computer Science, vol. 12188 (eds. M. Antona and C. Stephanidis). Cham: Springer https://doi.org/10.1007/978-3-030-49282-3_23.

57. Hulka, K., Strniste, M., and Prycl, D. (2020). Accuracy and reliability of sage analytics tracking system based on UWB technology for indoor team sports. International Journal of Performance Analysis in Sport 20 (5): 800–807. https://doi.org/10.1080/24748668.2020.1788349.

58. Cheng, L., Chang, H., Wang, K., and Wu, Z. (2020). Real time indoor positioning system for smart grid based on UWB and artificial intelligence techniques. 2020 IEEE Conference on Technologies for Sustainability (SusTech), Santa Ana, CA, USA, pp. 1–7. https://doi.org/10.1109/SusTech47890.2020.9150486.

59. Di Pietra, V., Dabove, P., and Piras, M. (2020). Seamless navigation using UWB-based multisensor system. 2020 IEEE/ION Position, Location and Navigation Symposium (PLANS), Portland, OR, USA, 2020, pp. 1079–1084. https://doi.org/10.1109/PLANS46316.2020.9110146.

60. Kolakowski, M. and Djaja-Josko, V. (2016). TDoA-TWR based positioning algorithm for UWB localization system. *Proceedings of the 2016 21st International Conference on Microwave, Radar and Wireless Communications (MIKON)*, Krakow, Poland (9–11 May 2016).

61. Silva, B., Pang, Z., Åkerberg, J., and Neander, J. (2014). Experimental study of UWB-based high precision localization for industrial applications. *Proceedings of the 2014 IEEE International Conference on Ultra-WideBand (ICUWB)*, Paris, France (1–3 September 2014).

62. Rowe, N.C., Fathy, A.E., Kuhn, M.J., and Mahfouz, M.R. (2013). A UWB transmit-only based scheme for multi-tag support in a millimeter accuracy localization. *Proceedings of the IEEE Topical Conference on Wireless Sensors and Sensor Networks (WiSNet)*, Austin, TX, USA (20–23 January 2013).

63. Lo, A., Yarovoy, A., Bauge, T. et al. (2011). An ultra-wideband Ad Hoc sensor network for real time indoor localization of emergency responders. Delft University of Technology, Thales Research and Technology Limited and IMST GmbH, Delft, The Netherlands.

64. Gupta, A. and Mohapatra, P. (2007). A survey on ultra wide band medium access control schemes. Computer Networks 51: 2976–2993. https://doi.org/10.1016/j.comnet.2006.12.008.

65. Janssen, M., Busboom, A., Schoon, U. et al. (2012). A hybrid MAC layer for localization and data. *Proceedings of the IEEE 17th Conference on Emerging Technologies and Factory Automation (ETFA)*, Krakow, Poland (17–21 September 2012).

66. Kuhn, M.J., Mahfouz, M.R., Turnmire, J. et al. (2011). A multi-tag access scheme for indoor UWB localization systems used in medical environments. *Proceedings of the 2011 IEEE Topical Conference on Biomedical Wireless Technologies, Networks, and Sensing Systems (BioWireleSS)*, Phoenix, AZ, USA (16–19 January 2011).

67. Alcock, P., Roedig, U., and Hazas, M. (2009). Combining positioning and communication using UWB transceivers. *Proceedings of the 5th IEEE International Conference*, DCOSS 2009, Marina del Rey, CA, USA (8–10 June 2009).

68. Cardinali, R., De Nardis, L., Di Benedetto, M.-G., and Lombardo, P. (2006). UWB ranging accuracy in high-and low-data-rate applications. IEEE Transactions on Microwave Theory and Techniques 54 (4): 1865–1875.

69. Benedetto, M.G., De Nardis, L., Junk, M., and Giancola, G. (2005). (UWB)^2: Uncoordinated, wireless, baseborn medium access for UWB communication networks. Mobile Networks and Applications (MONET) 10 (5): 663–674. https://doi.org/10.1007/s11036-005-3361-z.

70. Subramanian, A. and Lim, J.G. (2005). A scalable UWB based scheme for localization in wireless networks. *Proceedings of the Conference Record of the Thirty-Ninth Asilomar Conference on Signals, Systems and Computers*, Pacific Grove, CA, USA (30 October–2 November 2005).

71. Guvenc, I., Gezici, S., and Sahinoglu, Z. (2008). Ultra-wideband range estimation: theoretical limits and practical algorithms in ultra-wideband. *Proceedings of the 2008 IEEE International Conference on Ultra-Wideband (ICUWB)*, Hannover, Germany (10–12 September 2008).

72. Ding, J., Zhao, L., Medidi, S.R., and Sivalingam, K.M. (2002). MAC protocols for ultra-wide-band (UWB) wireless networks: impact of channel acquisition time. *Proceedings of the SPIE 4869, Emerging Technologies for Future Generation Wireless Communications*, Boston, MA, USA (12 November 2002).

73. Ding, J., Zhao, L., Medidi, S.R., and Sivalingam, K.M. (2002). MAC protocols for ultra-wide-band (UWB) wireless networks: impact of channel acquisition time. *Proceedings SPIE ITCOM2002*, pp. 1953–54.

74. Ubisense Staff. Ubisense contact tracing. https://ubisense.com; also: https://youtu.be/VSgyZy-Xci8.

75. Zebra Technology, Lincolnshire, IL, www.zebra.com. Zebra is a global provider of enterprise mobile computing, data capture, barcode printing and radio frequency identification devices.

76. Turner, B. and Dorris, T. (2019). Methods and apparatus to generate site health information. US Patent 10,481,238, 19 November 2019, filed 15 February 2019. Uncopyrighted.

77. Lenares, A. (2020). UWB localization: time difference of arrival vs two-way ranging, 18 August 2020. https://www.inpixon.com/blog/uwb-localization-tdoa-vs-twr.

78. AIRTLS BV. Koxkampseweg 10, 5301KK Zaltbommel, The Netherlands. www.airtls.com.

79. Pozyx Staff. Pozyx UWB. www.pozyx.io.

8 RTLSs and Distance Tracking Using Wi-Fi, Bluetooth, and Cellular Technologies

Building on previous chapters, this chapter focuses on some of the other technical approaches that can be used for Office Social Distancing (OSD) and Office Dynamic Cluster Monitoring And Analysis (ODCMA): it covers Wi-Fi, Bluetooth® Low Energy (BLE), and also, briefly, cellular systems. See Figure 8.1 for symbology of the key active Real-Time Location Systems (RTLSs) technologies in commercial use.

8.1 OVERVIEW

As discussed in previous chapters, RTLSs estimate locations of mobile entities (people or objects) wearing or carrying tags within interior zones in buildings such as offices, factories, hospitals, and nursing homes, among others. Many existing RTLSs based on Radio Frequency (RF) signals, such as Wi-Fi or BLE, are designed to track mobile tags that transmit a radio message to a network of receiving devices such as "gateways," "bridges," sensors, emitters, or Wireless Access Points (WAPs), deployed in the zone of interest. See Figures 8.2 and 8.3 for illustrative graphical views of the environment.

Typically, the near-ubiquitous adoption of Wireless Local Area Networks (WLANs) as a common network infrastructure enables IT planners to deploy Wi-Fi-based localization with few incremental hardware costs. It is assumed in all the Wi-Fi methods that each occupant does carry a smartphone with Wi-Fi enabled, or he/she carries a Wi-Fi-based tag. The network of gateways customarily makes use of the received signal strength of the transmissions from a tag as a proxy for an estimate of the distance between the tag and each gateway; thereafter, a Positioning Engine (PE), also called a location engine, uses proximity or multilateration algorithms to estimate the locations of tags. These approaches, comprising of tags that transmit signals and of location engines that predominantly use multilateration, are common in the industry and provide location determinations that are acceptable for many use cases in office, industrial, manufacturing, and hospital environments. Although these methods fail to provide a highly accurate location determination for the environments, as is the case with Ultra-Wideband (UWB) systems, they may be good enough for OSD/ODCMA. They support position determination with accuracy in the 10- to 15-feet range.

As it should be clear at this juncture, radio technology for RTLSs may be divided into two main groups: (i) Wi-Fi-based wireless systems and (ii) non-Wi-Fi based wireless systems (e.g. RFID-based or UWB-based systems previously discussed). A main difference between these two main groups is that the Wi-Fi-based approach allows quick and relatively low entry cost into an RTLS-enabled service for a relatively small zone of interest and a small population of office users, given that the Wi-Fi infrastructure (e.g. Wi-Fi/WAPs) has almost always already been

High-Density and De-Densified Smart Campus Communications: Technologies, Integration, Implementation, and Applications, First Edition. Daniel Minoli and Jo-Anne Dressendofer.
© 2022 John Wiley & Sons, Inc. Published 2022 by John Wiley & Sons, Inc.

Active RFID   Wi-Fi tags   Bluetooth (BLE)

FIGURE 8.1 Symbology for various technologies.

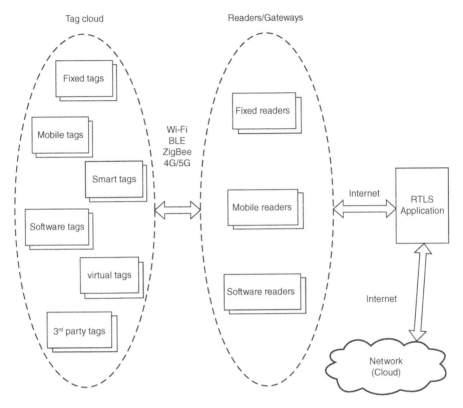

FIGURE 8.2 WLAN/cellular RTLS environment (loosely based on [1]).

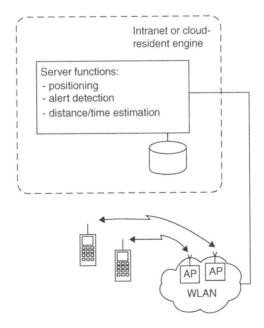

FIGURE 8.3 Basic WLAN/Wi-Fi RTLS architecture.

installed in institutions, factories, malls, hospitals, airports, and other venues. In contrast, the non-Wi-Fi based systems typically require their own communication network, at least at the "front-end," facing the tags. The main advantages of non-Wi-Fi based systems are that they can support a larger set of users, are more accurate, and can be cheaper on a per-user basis in large installations compared to a Wi-Fi-based RTLS. When using a large number of tags, the cost of a non-Wi-Fi based system is substantially lower than a Wi-Fi solution because the cost of a Wi-Fi-based tag is relatively high. Other advantages of non-Wi-Fi based system include the fact that (i) the tag size is smaller, (ii) the tag typically has a longer battery life because a non-Wi-Fi based system is designed to be much more energy efficient from the get-go, and (iii) it often achieves better localization accuracy [2]. Nonetheless, Wi-Fi-based RTLSs have a potential role to play in OSD/ODCMA.

In a Wi-Fi-based RTLS, the Wi-Fi tags transmit advertisements into the field of fixed receivers, these receivers often being in the form of sensors, gateways, bridges, readers, or emitters. The receivers attempt to locate tags by estimating a location on a floor plan, known as an "(x, y) location fix" for the map coordinates. Through the locating process of multilateration, one or more gateways or bridges measure the Received Signal Strength Indicator (RSSI) of the advertisement they hear from a tag and forward that RSSI information to a location engine. Signal strength can also be measured in dBm as Received Signal Strength (RSS). dBm is an absolute quantity representing power levels in mW (milliwatts); RSSI is a relative index used to measure the relative quality of a received signal to a client device, but it is not an absolute value. The IEEE 802.11 standards specify that RSSI can be on a scale of 0 to up to 255 and that each chipset manufacturer can define their own "RSSI_Max" value (some manufacturers, for example, use a 0–100 scale); the higher the RSSI value is, the better the signal is. The location engine uses the Received Signal Strength (RSS or RSSI) as an estimate of the distance between the tag and each reporting bridge, and the multilateration algorithm estimates the location of the tag on a floor plan by reporting the location as an (x, y) location on the floor plan.

The distance between the estimated (x, y) location of the tag and its true (x, y) location on the floor plan is called the "error." Current RTLS vendors measure their typical error (or "typical accuracy") in feet or meters. The typical error of an RTLS is defined by a statistical population distribution of a large number of sample location estimates and their "error" measurements. Hence, RTLS-equipment vendors often state their "typical error" or "typical accuracy" with phrases such as ". . . achieving 1-meter accuracy 90% of the time." In analytical sciences, precision and accuracy are two ways to think about errors. Accuracy refers to how close a measurement is to the true or accepted value; precision refers to how close measurements of the same parameter are to each other. Precision is independent of accuracy: it is possible to be very precise but not very accurate, and it is also possible to be accurate without being precise. The best quality observations are both accurate and precise [3]. An RTLS that uses only Wi-Fi radio signal strength will be able to locate each tag to within 1 m (approximately 3 feet, 1 yard) of its true location, but it cannot tell whether the asset is one-half meter to the left of a particular reference point, or one-half meter to the right of the reference point [4].

8.2 RF FINGERPRINTING METHODS

The RSSI-signature-based approach is particularly popular for localization in indoor environments. The idea here is to exploit the temporal stability in the RSSI received from a set of pre-deployed beacons at a certain location; this set of RSSI values is referred to as the RSSI signature or as the RF Fingerprint (RFF). RF fingerprinting methods are based on the principle that every location has a unique RF signature so that a location can, in theory, be identified by a unique set of values, including measurements from the constellation of infrastructure nodes receivers [5]. The number of hearable emitters (for example, cell towers in outdoor applications) has a significant impact on accuracy; for instance, for emergency localization over cellular systems, one needs to obtain readings from a multitude of towers to get a reasonable accuracy (say 60 m – but in suburban environments where there are fewer towers, the accuracy degrades to 100 m). Also, there is a significant variation of the estimated position depending on the handset antenna orientation.

A number of Wi-Fi-based RTLSs make use of the fingerprinting method. This method requires (i) the use of a database of appropriate scope, and (ii) a process training phase, followed by an operational tracking phase. The goal of these RSSI signature-based solutions is to exploit the mapping between a tag's location and the RSSI received from a set of pre-deployed beacons [6]. In the training phase, the area is surveyed to construct the reference RSSI signature for each sampled location – the set of RSSI signatures at various locations in the zone of interest is called the "'radio map."

One issue with this approach is that the database requires maintenance because the signature can age quickly in many environments as the environment changes (especially outdoors), and, as a result, the localization accuracy degrades as the RSSI signature ages. While there are several causes of the RF fingerprinting database instability, one of the major ones is the multipath: radio transmissions in busy indoor office or factory environments – which typically entail lots of stationary furniture and partitions, but also people movements, elevators movements, and possible equipment movement – will result in a complex multipath distribution, thus impacting the RF signature. In indoor environments, a small change in a physical location (in two or three dimensions) causes significant changes in the RF signature. These changes are the result of the combination of (i) the multipath phenomenon, which makes RF signature three-dimensional, and (ii) the short wavelengths at the radio frequencies of interest, which results in significant RF signature changes over distances of just one-fourth wavelength [7]. Nonetheless, studies [6] have shown that the effect of background obstacles, including movement of human beings and furniture in the building, as well as noises generated from the use of electronic devices, is relatively negligible; however, the effect of the foreground obstacles can be significant. People holding or walking by the receiving tag may change the average localization error from 1 to 3 m. Furthermore, the average localization error can differ by over 100% if the receiving tag is placed on the human wrist or held by hand (both preferable) instead of somewhere closer to the body: the position of the tag on the bodies impacts the accuracy because RF signals are affected when travelling through liquids or water, and, because human bodies are 60% water, they partially impede the signal; in fact, studies have shown that a tag worn on the wrist will be less impacted than a tag worn on the chest. In addition, the effect of antenna orientation is prominent: depending on the orientation of the antenna, the average localization error varies from 1 to 3 m, implying that if the receiving tag is not held at the same orientation, the error can be drastically different. These studies also show that the effect of beacon density varies depending on the location of the measuring point. For example, when there is a sufficient number of beacons nearby the measuring point, the average localization error may not degrade much. To tackle the RFF problem, solutions have focused on auto-calibration of the RSSI signature or focused on endeavoring to keep the RSSI signature fresh, but other approaches also exist [6, 8, 9].

8.3 Wi-Fi RTLS APPROACHES

8.3.1 Common Approach

A brief overview of this technology was provided in chapter 6. Almost invariably, workspaces are already configured with WLANs; hence, due to the near-ubiquitous deployment of Wi-Fi-based infrastructure in office buildings, warehouses, airports, schools, churches, and restaurants – to name just a few of the indoor environments of interest – Wi-Fi frameworks can be utilized for occupancy estimation, presence detection, and RTLS functions. In these environments, Wi-Fi-based location monitoring can be achieved by utilizing Wi-Fi transceivers (i.e. tags or smartphones – and, if/where additional sensing might be needed, one can adjoin the appropriate sensors to the tags; the tags transmit the sensor-measured parameters to a central server, which, in turn, distributes the measurements to upstream applications that need to monitor these auxiliary parameters).

WAPs (or hotspots) provide a means for determining the location of mobile tags or smartphones. One can often implement RTLS functions using existing Wi-Fi structure with some firmware changes and with the addition of a PE. Wi-Fi RTLSs use the tags or smartphones to transmit a Wi-Fi signal to multiple WAPs throughout the zone of interest; the receivers can locate the tag using Differential-Time of Arrival (DToA) methods and RSSI. Wi-Fi-based systems use these Time of Flight (ToF) measurements with a relatively wide bandwidth, giving an accurate location positioning within a few meters [10]. Dedicated Wi-Fi tags in principle work better than smartphones, but that approach entails additional hardware, cost, inconvenience, and tag-power management.

At a very coarse level, since smartphones are widely used by occupants, the number of smartphones can be treated as an indirect estimator for the number of occupants in a given space, and perhaps a "heat map" (as depicted in figure 6.4) [11]; however, this approach also has some limitations: for example, the occupants might have multiple smartphones, or they may not turn on Wi-Fi for their smartphones.

Early work in this arena was undertaken by Microsoft Research, who then proposed an RF-based indoor location tracking system by processing signal strength information at multiple base stations [12]. More recent research has focused on improving Wi-Fi-based RTLS by dealing with noisy signals, and newer solutions have achieved position estimation with reasonably high (>90%) accuracy within an error of 1 m. People's mobility and clustering, however, have been issues impacting accuracy: clustering can create interferences with surrounding signals, leading to significant degradation in location accuracy [12–15]. Differences in relative humidity levels in the zone of interest can also impact accuracy, and a separate radio map may need to be constructed for each humidity level. As noted earlier, traditional Wi-Fi-based location system utilizes wireless signal strength to estimate locations in two phases. First, the offline training phase collects RSSI from multiple WAPs at each sampled location. The results are saved in a radio map. Second, the online real-time estimation phase matches the RSSI from a target mobile device to each sampled location on the radio map. The coordinates of the target location can be estimated deterministically or probabilistically [13].

Wi-Fi RTLSs provide real-time locating of mobile devices by measuring the signal strength from the mobile device; specifically, a mobile device is tracked by continuously monitoring the signal strength as seen from various WAPs and reporting the signal strengths to a central PE. A signal strength of –75 dBm or better (higher) for client devices or tags is desirable because client devices with lower signal strength may be difficult to locate within the network. The central PE server determines the location of the mobile device and transmits the location back to the mobile device. This is referred to as position polling, or actively sampling the location of the mobile device. However, such continuous reporting between the mobile device and the central server can result in a reduction in overall network performance; possibly, only a few location updates received from the mobile device actually correspond to a real positional change because positional polling of the mobile device continues even if a person carrying the mobile device is stationary and places the mobile device on his or her desk. As such, positional polling not only imposes a performance penalty on the wireless aspect of the WLAN system, but it also consumes power from the mobile device reducing battery life.

Because standard Wi-Fi provides bidirectional communication, it requires a lot of transceiver power – high power consumption is inherent to Wi-Fi technology because Wi-Fi is intrinsically designed for high-bandwidth applications. Consequently, a battery-operated standard Wi-Fi

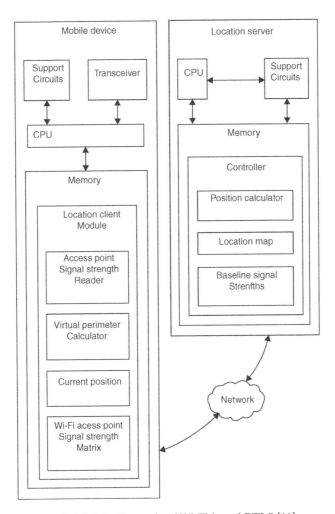

FIGURE 8.4   Example of Wi-Fi-based RTLS [16].

transceiver in a tag must use large and/or heavy batteries, or the batteries must be replaced relatively frequently.

Figure 8.4 depicts the basic elements of a Wi-Fi-based RTLS based on [16]. The tags transmit in one of three independent, non-overlapping Wi-Fi channels in the Industrial, Scientific and Medical (ISM) band at 2.4 GHz (e.g. Channel 1 at 2.401–2.423 GHz; Channel 6 at 2.426–2.448 GHz; and Channel 11 at 2.451–2.473 GHz; other channels in the ISM band can also be used if/as needed). The controller in the PE location server incorporates a position calculator for calculating a new position for the mobile device based on WAPs' signal strengths data received from the location client module. The location map in the location PE server is a graphical representation of a physical location, that is, a floor plan of the zone of interest containing the locations of all WAPs in the zone. The baseline signal strengths data element in the location server contain baseline signal strength readings of all WAPs represented in the location map generated when the location server is provisioned. The WAP signal strength associated with a particular WAP correlates with the distance of the mobile device from that WAP. Thus, based on the WAP signal strength data measured by the mobile device, the position calculator reads the information received from the location client module, determines which WAP the mobile device is measuring readings from and, based on the signal strength readings, compares these to the baseline measurements in the baseline signal strengths file, and determines where on the

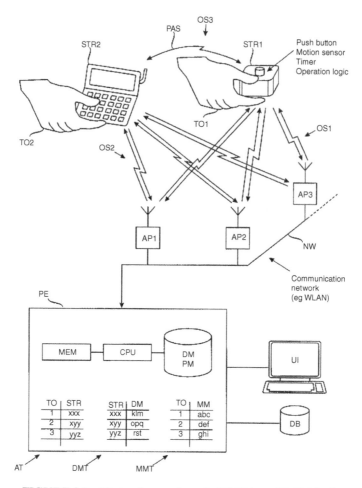

FIGURE 8.5   Mode of operation of a Wi-Fi-based RTLS [17].

location map the mobile device is located. The position calculator uses trilateralization techniques for calculating the position of the mobile device based on the WAP signal strength data. After calculation of the new position of the mobile device, the position calculator communicates the new position to the mobile device, and the new position is then stored as the current position.

Figure 8.5 further depicts one mode of operation of a Wi-Fi-based RTLS, providing some added details [17]. In the figure, two target objects (e.g. tags, entities) TO1 and TO2 are depicted as hands of persons carrying respective signal transceivers (signaling devices) STR1, STR2. The signal transceivers STR1 and STR2 are coupled to a communication network, such as a WLAN network, which comprises several WAPs shown as AP1, AP2, AP3, etc. The signal transceivers STR1 and STR2 observe signal strength from the access points AP1–AP3. The signal transceivers periodically, or when their movement stops, transmit observation sets OS1 and OS2 via the network to a centralized positioning server, the positioning engine. Alternatively, or additionally, the signal transceivers STR1 and STR2 can send signals for positioning, and these signals are observed by the access points AP1–AP3 and then relayed to the PE. The respective locations of the target objects influence the observation sets OS1 and OS2. The relation of location versus signal strength from the access points is modeled in a Data Model (DM) in a positioning engine PE. The DM may indicate a calibrated value for each of several locations in the environment. In some implementations, the data model is a probabilistic model that indicates a probability distribution for the signal strength (or other location-dependent physical quantities) for each of several locations, called sample points. A probability distribution provides more information than a single calibrated or expected value and is more robust in cases where the observations are ambiguous or contradictory. In the illustrated example, the first signal transceiver STR1 is a small positioning tag. To prolong battery life, the tag STR1 is designed to be active only intermittently. There are several ways to implement such an intermittent transmission of observations. For instance, the tag may be provided with an operation logic that puts the timer on sleep mode for most of the time and wakes up the tag in response to timer alerts or activity detections from a push button or motion sensor. When the tag wakes up, it observes the signal strength or other location-dependent physical quantities and sends the observation(s) via the network to the positioning engine. Along with the observed value of the location-dependent physical quantities, the tag may also indicate the activity that triggered the sending of the observation. For instance, the person depicted as target object TO1 can press the push button or otherwise indicate an alert condition, which is signaled via the network to the positioning engine. In an alternative implementation, the alert condition is signaled to another server cooperating with the positioning engine. Detection of the alert condition indicates that the tag and the person carrying it should be located as soon as possible. In some implementations, positioning of the tag STR1 is assisted by the second signal transceiver STR2, which is carried by another person depicted as target object TO2. The second signal transceiver STR2 is depicted as a smartphone or Personal Digital Assistant (PDA). Typical positioning systems rely on modeling of at least one location-dependent physical quantity, such as signal strength, at several locations in the environment, called sample points. Although signal strength of transmissions by the WAPs is a typical example of a location-dependent physical quantity, it is not necessary for the data model DM to know the locations of the WAPs, so long as the locations of the sample points are known. Some positioning systems may rely on physical quantities other than signal strength; yet other positioning systems are based on short-range transmissions, such as IR or Bluetooth. If communication or detection is possible between a short-range transmitter and sender, one of which is the signal transceiver to be located, the signal transceiver and its associated target object are within range of the short-range transmitter. Such techniques suffice to determine a room where the signal transceiver and its associated target object are located [17].

A variation of the Wi-Fi protocol that makes use of unidirectional communication from the tag to the WAPs, instead of bidirectional communication, is employed by some Wi-Fi infrastructure providers. This protocol is more power-efficient, given that it does not require association between the tags and the WAPs for communication. An example of such a protocol is the Compatible Extensions (CCX) protocol that, while using the Wi-Fi Physical layer (PHY) and Medium Access Control (MAC) structure, is not a conventional Wi-Fi/IP packet that would typically require two-way communication between the WAPs and the tags. CCX is a Cisco-developed mechanism that is, in fact, used in Cisco Wi-Fi-based RTLS.[1] A CCX-compatible tag transmits one-directional packets with a flexible format that is recognized by WAPs using a predefined header. The WAPs transmit the packet with other support information, such as an RSSI indicator, to a Cisco mobility services engine that uses the information from multiple

---

[1] CCX aims at enabling partners to provide WLAN devices and adapters in all relevant form factors and on all critical operating systems. The initial goal was to drive WLAN infrastructure adoption by defining and driving WLAN client innovations as required by Enterprise customers. Launched in 2002, the CCX program is a license-free technology initiative opened to all silicon and mobile device manufacturers (e.g. laptops, WLAN adapter, and silicon vendors). Along the way, CCX supported innovations that address enterprise customer requirements ahead of standards. Capabilities include: Location, Voice (call admission control, voice metrics), QoS, mobile device management, enhanced security (WAP 2), and enhanced RF performance [18].

WAPs to calculate location using triangulation. A CCX-based system has advantages over the standard bidirectional Wi-Fi communication because (i) the CCX-based system does not require for the tag to have an IP address; (ii) one-directional communication substantially reduces power consumption by the tag [19].

A number of new generations of RTLS utilize secondary technologies such as infrared (IR), ultrasound (US), or low-frequency (LF) RF. The secondary technologies are used to provide higher accuracy for positional resolution that cannot be provided by triangulation alone. Dual-mode tags with secondary technologies include at least two RF modules that can communicate with a network: a first RF module may provide a front end that is used to transmit and/or receive a Wi-Fi signal; a second RF module may provide a front end that is used to transmit and/or receive a non-Wi-Fi signal [1]. In a first operating mode, and in the absence of detection of a non-Wi-Fi system or wireless infrastructure, the dual-mode tag may operate only using Wi-Fi signals; the communication is one-directional from the dual-mode tag to WAP using the Cisco CCX protocol. In the first operating mode, the tags and the Wi-Fi infrastructure components are relatively unsynchronized with each other. Spatial resolution performance and tag battery life both may be poor because the tag uses Wi-Fi signals, which uses high power, while the dual-mode tag also scans the IR, US, and/or LF environment to detect predetermined transmissions from infrastructure nodes. Exciters may transmit a signal using technology such as IR, US, or LF. In a second operating mode, the dual-mode tag may operate as a non-Wi-Fi tag that communicates with the central monitoring system by use of a non-Wi-Fi communication system. The non-Wi-Fi communication system may operate in the 900 MHz communication band and provide an infrastructure and support network in order to communicate with dual-mode tags and nodes. The dual-mode tag can use the 900 MHz band for transmission, as well as synchronize timing and communication with the use of signals transmitted by exciters. Synchronized timing reduces positional uncertainty, provides high spatial resolution performance, and provides lower power consumption by the dual-mode tags, resulting in longer tag battery life. Refer to Reference [1] for additional details.

8.3.2 Design Considerations

An example of an implementation is described in [20] where it is noted that (at least) three aspects of the design are important: (i) placement of WAPs, (ii) distance between WAPs, and (iii) minimum bit rate (also see Table 8.1).

- Placement of WAPs. Proper placement of WAPs is recommended in order to fully leverage the benefits of location accuracy. WAPs should not be clustered inside the floor plan; instead, WAPs may be placed on the perimeter of the floor plan in order to provide consistent coverage throughout. Additional WAPs can be placed in the corners of the floor plan to enhance the location accuracy for client devices. These corner WAPs play a vital role in ensuring good location accuracy for clients that are inside the perimeter area.
- Distance between WAPs. Positioning of WAPs may impacts client location tracking as well as the wireless performance. WAPs should not be placed very close together or too distant from each other. The goal is to ensure that the Signal-to-Noise Ratio (SNR) value for a client does not drop below 20 dB as they roam between WAPs. Furthermore, a minimum of three access points (with four or more for better accuracy and precision) should be able to decode client frames at any given time for client location tracking.
- Minimum bit rate. Setting up a higher minimum bit rate value may help to balance client devices between all available WAPs in the network. This will also result in much accurate tracking since clients with lower RSSI values may roam to other WAP with a much better RSSI value. This may help to track the location of client devices more accurately.

TABLE 8.1   Basic best practices of overlay Wi-Fi-based RTLSs (summarized from [21])

Issue	Consideration
Minimum detected received signal thresholds	For mobile devices to be tracked properly, WAPs must report mobile device RSSI to their respective controllers at levels meeting or exceeding a minimum RSSI cutoff value (−75 dBm or better); mobile device's RSSI reported below this level may be discarded by the positioning engine. At least three WAPs (and preferably four or more for optimum accuracy) should be reporting this level of signal strength or better for any device being localized
Correct Access Point placement	WAP density has a significant effect on location tracking performance. Proper placement and density of access points is critical to achieving acceptable performance. In many office wireless LANs, WAPs are distributed throughout interior spaces just enough to provide coverage to the surrounding work areas. Although there is no single steadfast rule that yields the proper density in every environmental situation, a practical rule is to have an inter-WAP separation of 50–70 feet, which often results in one location-aware WAP being deployed approximately every 2500–4900 square feet
Minimizing excessive co-channel interference	In many cases, location-based services are added or retrofitted to an existing wireless design, some of which encompass wireless voice handheld devices. When designing a location-aware solution that will be used in conjunction with such latency-sensitive application devices, special care needs to be taken to ensure that excessive co-channel interference is not introduced into the environment. The needs of an optimal location-aware design must be balanced against the stringent QoS requirements of a properly designed wireless voice infrastructure
Multi-floor structures	In multi-floor structures, such as office buildings, proper location-aware design must be undertaken

8.3.3   Drawbacks and Limitations

It should be noted that Wi-Fi-based RTLS tags are the most power-hungry, the largest in size, and nearly the most expensive of all RTLS technology. At press time Wi-Fi tag cost in the $60–$100 range and the batteries only last days on one charge. Some other limitations include [22]:

- Installation can be challenging. For example, during deployment, one must survey the building by walking around with calibration devices to determine the WAPs. The design rules of the previous section should be given proper consideration.
- The organization will need more WAPs: one might need four to five times more than are actually required for the data load. Additionally, the WAPs that support a Wi-Fi RTLS are fairly expensive.
- Adding access might give rise to channel management problems, affecting the Wi-Fi systems performance (in practical terms, Wi-Fi tends to interfere with itself when there are more than three access points in close proximity).
- To enable an indoor positioning system using Wi-Fi RTLS, one needs infrastructure that supports Time of Arrival (ToA) algorithms; the vendor set is somewhat limited.
- The organization will need a license for the location software that run on the location engine, which can increase the overall cost of the system.

8.3.4   Potential Enhancements

Numerous proposed enhancements to the basic Wi-Fi RTLS configuration have emerged in the published literature. A few proposals follow (this list is not intended to be exhaustive).

- As discussed above, a location server calculates the position of a mobile device based on the strength of multiple WAP signals measured by the mobile device and transmits that location to the mobile device. However, to improve performance, in Wi-Fi-based RTLSs, it is desirable to limit redundant positioning polling from a mobile device. In a proposed approach [16], the mobile device establishes a virtual perimeter around its position and monitors the signal strengths of WAPs located within the virtual perimeter until changes in signal strengths indicate a change in location of the mobile device. If a change in signal strengths does occur, the mobile device has moved to a new position, that is, a position outside the virtual perimeter, and the mobile device communicates the new signal strengths from the WAPs in range to the location server. The location PE server then calculates and communicates the new position to the mobile device. The mobile device establishes a new virtual perimeter around its new location, and the method iterates.

- It was mentioned above that number of new generations of RTLS utilize secondary technologies to provide higher accuracy. Although a CCX-based system is adequate to support position location using triangulation methods, it is not adequate by itself to support position location methods having greater positional accuracy, such as those used in new generations of RTLS (e.g. with UWB methods) [19]. In spite of the fact that usage of one-directional CCX tag transmission helps reduce the power consumption, with CCX there is no back channel from the server to the Wi-Fi tags, precluding the realization of benefits that could arise from being able to communicate bidirectionally with the tag for network-level information that would allow the tag better utilize its resources or allow possible configuration updates. We noted that new generations of RTLS utilize secondary technologies; however, the secondary technologies consume additional power from the tags, thereby further burdening conventional Wi-Fi tags that already suffer from relatively poor power consumption. One way to make the secondary technologies on the tag consume less power is through the use of synchronization. Improved synchronization allows a tag to activate certain circuitry only when it may be needed. In order to synchronize the end devices (i.e. tags and infrastructure nodes – aka Exciters), there must be a return link back from the system (via a WAP) to the end devices. CCX protocol is incapable of supporting the return link because the CCX protocol is one-directional. Therefore, a need exists to reduce the power consumption of Wi-Fi-based tags and to help improve the system positional accuracy, in particular, the system positional accuracy when used in conjunction with secondary technologies such as IR, LF, and/or US. Reference [19] describes one possible improvement. The reference suggests an Enhanced CCX (ECCX) protocol to address the shortcomings of the existing system. The ECCX protocol is based upon the CCX protocol as supplemented by a message field in the data message (e.g. a field in a header or elsewhere in a message packet) that sends a request for an acknowledgment signal (ACK) from at least one Wi-Fi WAPs. The WAPs are configured to receive and act upon these data messages. The WAP is designed to transmit an ACK upon receiving an ECCX message.

- Another enhancement is possible: considering Figure 8.5 above, in some future implementations, positioning could be enhanced by utilizing information that is based on observations from one or more Positioning-Assisting Signals (PASs), one of which is denoted by reference signal PAS in the figure, as advanced in reference [17]. The positioning-assisting signal PAS is sent by one signal transceiver, e.g. STR2, and received by another one, e.g. STR1. Neither the origin nor the destination of the positioning-assisting signal PAS is indicated by the data model or otherwise known a priori. Positioning on the basis of signals between two mobile signal transceivers is based on the fact that although neither the origin nor the destination of the positioning-assisting signal is known a priori, an observation of the positioning-assisting signal nevertheless provides useful information. This is depicted as the third observation set OS3, which is based on the positioning-assisting signal PAS and is an example of a positioning-assisting observation set. To that end, the positioning engine may be operatively coupled to a signal propagation model that indicates a signal value probability distribution as a function of a distance travelled by the signal. The signal propagation model may also consider obstacles between the signal's originating and terminating locations in the environment. This information can be used to derive an additional location probability distribution, which can be used to resolve ambiguities regarding either target object's location.

A number of other variations or improvements have been discussed in the literature, in addition to the ones discussed above. Specifically, there is ongoing research aimed at making more accurate systems by using various mathematical methods to filter out the inaccurate input data, which tend to be intrinsically noisy in the Wi-Fi environment. The short list that follows highlights some elaborations of the basic Wi-Fi approach to localization that have been documented [11] (these approaches, however, have seen very limited, if any, commercial deployment):

- A Wi-Fi-based system that utilizes a coarse-grained localization approach based on the connected access point with metadata information and occupancy patterns [23].
- A Wi-Fi-based system that captures a snapshot occupancy by matching the RSS measurements with the measurements of anchors in different zones. Thereafter, the estimation performance is improved using temporal correlations of successive snapshots of occupancy [24].
- A Wi-Fi-based system where, instead of collecting Wi-Fi RSS from smartphones, the system uses routers to scan Wi-Fi enabled-smartphones along with a localization algorithm based on On-Line Sequential Extreme Learning Machines (OS-ELMs) [25]. OS-ELM, a sequential learning algorithm, can learn the training data, not only one-by-one but also chunk-by-chunk (with fixed or varying length), and discards the data for the training that has already been done. At any time, only the newly arrived single or chunk of observations (instead of the entire past data) are seen and learned. A single or a chunk of training observations is discarded as soon as the learning procedure for that particular (single or chunk of) observation(s) is completed [26–29]. Compared with other gradient–descent-based learning algorithms, ELM provides better performance at higher learning speed and the learning phase in many applications is completed in seconds.
- A Wi-Fi-based system that directly estimates occupancy based on Wi-Fi power measurements, since Wi-Fi RSS management between Wi-Fi transmitters and receivers is influenced by occupants between them [30]. Wi-Fi signals have the tendency of getting disturbed by the motion of occupants and other movements in a zone. If one measures the level of this variation, the variation can represent the human activity in that zone [31].
- A Wi-Fi-based system based on Channel State Information (CSI), a fine-grained value derived from the physical layer, refers to known channel properties of a communication link. This information describes how a signal propagates from the transmitter to the receiver and represents the combined effect of, for example, scattering, fading, and power decay with distance. It consists of the attenuation and phase shift experienced by each spatial stream on every subcarrier in the frequency domain. It follows that CSI is more sensitive to environmental variance created by a moving object. Often, one can easily acquire CSI information from many off-the-shelf 802.11-based devices [32].
- A Wi-Fi-based system that models occupancy dynamics as a Markov Chain process [33].

8.3.5 Illustrative Examples

A handful of exemplary products are highlighted in this section to give weight to the availability of the technology (no product is explicitly sanctioned herewith).

Cisco Systems Traditional RTLS   Cisco Systems, Inc. has traditionally offered an RTLS solution with emphasis on (i) being able to quickly and efficiently locate valuable assets and key personnel, (ii) improving productivity via effective asset and personnel allocation, (ii) reducing loss because of unauthorized removal of assets from company premises, and (iv) coordinating Wi-Fi device location with security policy enforcement.

Cisco's system uses the RFF approach to provide a Wi-Fi-based RTLS without the need for specialized time-based receivers or other specialized hardware that must be mounted alongside each access point. Standard WLAN clients or Wi-Fi 802.11 active RFID tags can be tracked, in addition to other entities (e.g. rogue clients). RFF technology offers the key advantages of an indoor location patterning solution but with significantly less effort required for system calibration. While both approaches support onsite calibration, the Cisco RFF approach requires less frequent recalibration and can operate with larger inter-access point spacing. RFF can also share RF models among similar types of environments and includes prepackaged calibration models that can facilitate rapid deployment in typical indoor office environments. The Cisco RFF approach yields better performance than solutions employing pure triangulation or signal strength lateration techniques – these techniques typically do not account for effects of attenuation in the environment, making them susceptible to reductions in performance. Cisco RFF begins with a better understanding of RF propagation as it relates specifically to the environment in question. Except for the calibration phase in location patterning approaches, none of the traditional lateration or angulation techniques take environmental considerations directly into account in this manner. RFF then goes a step further and applies statistical analysis techniques to the set of collected calibration data. When properly deployed, the accuracy and precision of the Cisco RTLS solution in indoor deployments are represented in two ways, as follows: (i) accuracy of less than or equal to 10 m, with 90% precision; (ii) accuracy of less than or equal to 5 m, with 50% precision. In other words, given proper design and deployment of the system, the error distance between the reported device location and the actual location should, in 90% of all reporting instances, be within 10 m. In the remaining 10% of all reporting instances, the error distance may be expected to exceed 10 m [21].

WAPs forward received signal strength information to WLAN controllers with regard to the observed signal strength of any Wi-Fi clients, 802.11 active RFID tags (also rogue access points or rogue clients). In normal operation, WAPs focus their collection activities for this information on their primary channel of operation, going off-channel and scanning the other channels in the regulatory channel set of the access point periodically. The collected information is forwarded to the WLAN controller to which the access point is currently registered. Each controller manages and aggregates all such signal strength information coming from its access points. The location appliance uses Simple Network Management Protocol (SNMP) to poll each controller for the latest information regarding each tracked category of devices. The location appliance synchronizes with each controller containing access points participating in location tracking during controller synchronization. Synchronization occurs either on-demand or as a scheduled task.

The Cisco Location Appliance (the locating engine/PE) is added to support location and statistics history and serves as a centralized positioning engine for the simultaneous tracking of up to 2500 devices per appliance. No proprietary client hardware or software required: the solution is implemented as a network-side model and not client-side; thus, it can provide location tracking for most Wi-Fi clients without the need to load proprietary client tracking software or wireless drivers in each client; for example, there is no dependency on proprietary software being resident in RFID asset tags. Cisco's RTLS interoperates with active RFID asset tags from popular vendors

Cisco Meraki and Meraki Marketplace   Cisco Systems, Inc. acquired Meraki in 2012. The division focuses on cloud-controlled Wi-Fi, routing and security; all managed from a centralized dashboard. Cisco Meraki WAPs can track location of client devices independently, using the signal strength of each client device. This helps to locate client devices that are either stationary or moving inside the intended area. Using this approach, network administrators can easily determine the location of any desired client device within the network perimeter. Cisco Meraki WAPs also integrate with RTLS software from AiRISTA (originally Ekahau, which was acquired by AiRISTA in 2016) and Stanley Healthcare (previously known as AeroScout). These RTLS vendors offer the ability to track clients or Active RFID tags with great accuracy in real-time, graphical formats. Cisco Meraki WAPs support AiRISTA/Ekahau and Stanley/AeroScout tags in both unassociated "blink mode" and associated "connected mode." Blink mode is used in order to conserve the battery of the Wi-Fi tags as they are not connected to the AP full time [20].

The Meraki Marketplace is a catalog of Technology Partners that showcases applications developed on top of the Meraki platform, allowing customers and partners to view, demo, and deploy commercial solutions. A press time example of an RTLS available at the Marketplace is Phunware Inc. Multiscreen-as-a-Service (MaaS) Location Based Services (LBS). Key features of the Phunware MaaS LBS app portfolio include [34]:

- Indoor routing and navigation: proximity and real-time blue dot mapping, navigation and wayfinding using BLE beacons, Wi-Fi, Li-Fi, and Global Positioning System (GPS).
- Off-route notifications: real-time notifications for route modifications and adjustments.
- Location sharing: share, view, and route to app users' locations.
- Customizable Points Of Interest (POIs): create, configure, and display unique POIs.
- Simple real-time map management: edit, manage, and update maps.
- Additional RTLS services: conduct asset tracking, issue proximity alerts, monitor traffic patterns.
- Native map experience: multilayer vector maps, stitched into Apple Maps and Google Maps, following the latest technology and design patterns including pinch zoom, panning and rotation.
- Landmark routing: developer landmark configuration options for more user-friendly, turn-by-turn directions and routing.

8.4   BLE

8.4.1   Bluetooth and BLE Background

Bluetooth is a wireless technology standard targeted at Personal Area Networks (PANs). The standard was originally developed by the IEEE 802.15.1 group; it is now maintained by the Bluetooth Special Interest Group (SIG), which guides the development of the specification and manages the product qualification program – the SIG had approximately 36 000 member companies in 2019. The standard has evolved considerably since the early versions published in 2002, as summarized in Table 8.2.

BLE was designed as an elaboration of classic Bluetooth for power-efficient burst, point-to-point communication, as well as for broadcast data in one-to-many environments. BLE is very energy-efficient compared to classical Bluetooth technologies and compared to Wi-Fi. BLE 4.x supports a symbol rate of 1 mega-symbols per second (Msps) – thus, BLE 4.x is faster (four times faster to be exact) than other mesh technologies, for example, based on IEEE 802.15.4, which runs at 0.250 Msps. Bluetooth hardware (e.g. beacons, access points, transceivers) has been on the market for several years and, besides other applications, the hardware has been used in the context of proximity applications [35, 36].

There were 4.6 billion Bluetooth device shipments in 2020 and, as seen in Figure 8.6 [37], annual Bluetooth shipments are expected to reach 6.2 billion devices by 2024. BLE is expected to be used by 1.8 billion smartphones by 2024 and the *cumulative* BLE single-mode devices

TABLE 8.2   Bluetooth Protocol Versions

Version	Key features
Bluetooth 1.0 and 1.0B	Early "buggy" version, late 1990s/early 2000s
Bluetooth 1.1	IEEE Standard 802.15.1–2002
Bluetooth 1.2	IEEE Standard 802.15.1–2005; higher transmission speeds in v1.1, >700 kbps; faster discovery and connection
Bluetooth 2.0 + EDR	Added Enhanced Data Rate (EDR) for faster data transfer; bit rate >2 Mbps
Bluetooth 2.1 + EDR	Adopted by the Bluetooth SIG in July 2007; improved security
Bluetooth 3.0 + HS	Adopted by the Bluetooth SIG in April 2009. Bluetooth v3.0 + HS (high speed) supports data transfer speeds of up to 24 Mbps via a Bluetooth-negotiated/colocated 802.11 link. It also added a number of other features such as: Enhanced Modes, Alternative MAC/PHY, Unicast Connectionless Data, and Enhanced Power Control
Bluetooth 4.0 (also called Bluetooth Smart)	Adopted in June 2010. It defines the concepts of Classic Bluetooth, Bluetooth High Speed and Bluetooth Low Energy (BLE) protocols. Bluetooth High Speed is based on Wi-Fi protocols, and Classic Bluetooth supports legacy Bluetooth protocols BLE is a subset of Bluetooth v4.0; it encompasses a new protocol stack aimed at rapid establishment of simple links and at very low power applications powered by a coin cell
Bluetooth 4.1	Adopted in December 2013 as an incremental software update to Bluetooth v4.0 (not a hardware update). It also adds new features that improve consumer usability. These include increased co-existence support for LTE, bulk data exchange rates, and support of simultaneous multiple roles
Bluetooth 4.2	Adopted in December 2014, focusing on the Internet of Things. Improvements include: Low Energy Secure Connection with Data Packet Length Extension, Link Layer Privacy, and IPv6 Mesh over BLE using the Internet Protocol Support Profile (IPSP) version 6
Bluetooth 5	Adopted in December 2016; the new features are mainly focused on new Internet of Things support. It includes BLE options that can double the speed (2 Mbps burst) by trading off range, or up to fourfold the range by trading off data rate. Version 5 also adds functionality for connectionless services
Bluetooth 5.1	Adopted in January 2019. Some of the enhancements include Angle of Arrival (AoA)/Angle of Departure (AoD) mechanisms that are used for location and tracking of devices, and mesh-based model hierarchy
Bluetooth 5.2	Adopted in December 2019, focusing mostly on audio features (audio runs on BLE lowering battery consumption, and allow the protocol to carry sound, e.g. for headphones and hearing aids)

planned to be shipped over the 2020–2024 period is expected to be 7.5 billion. By 2024, 100% of all new platform devices (phones, tablets, and laptops) will support both Bluetooth radio versions, the Bluetooth classic, and the BLE, so that the dual-mode Bluetooth will be the standard configuration for platform devices. Due to its ubiquity, BLE-based approaches may become increasingly important in localization applications; in fact, the Bluetooth SIG expects the number of shipments for location services (item finding, asset tracking, wayfinding, and access control) to increase from 186 million units shipped in 2020 to over 500 million by 2024 (at a CAGR of 32%), observing: "[expecting a] *4x growth in annual shipments of Bluetooth location services devices by 2024: location services is still the fastest growing solution area – continued demand for location services is powering a surge in wayfinding, asset tracking, item finding, and access control solutions that incorporate Bluetooth technology*" [37].

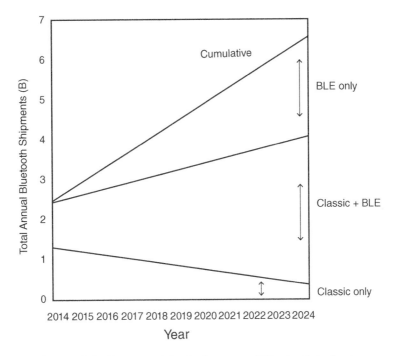

FIGURE 8.6   Forecast of the deployment of Bluetooth technology.

Besides BLE, other recent advancements in Bluetooth technology facilitate the deployment of scalable, smart building applications in commercial and industrial environments, with an emphasis on reliability, performance, security, and multi-vendor interoperability. For example, the specifications describing Bluetooth Mesh networking were published in 2017, specifically V1.0. This new Bluetooth capability is designed for applications such as smart buildings, smart factories, commercial lighting, and smart industry, including connected lighting, building automation, and sensors. Bluetooth Mesh expands the capabilities of Bluetooth, complementing other Bluetooth systems such as BLE and Bluetooth Basic Rate/Enhanced Data Rate (BR/EDR), each of which has intrinsic strengths and application foci (Bluetooth BR/EDR was developed for the transmission of a predictable or isochronous stream of data over point-to-point connections between two devices).

8.4.2   RTLS Applications

As just noted, BLE is energy-efficient compared to baseline (classical) Bluetooth technologies and Wi-Fi: BLE provides low power consumption and low cost while maintaining similar communication range of classical Bluetooth. BLE is optimized for burst data communications, for example, to send identifiers or short status messages. BLE can be utilized for occupancy detection and indoor localization; other BLE-based applications targeted to building occupants include indoor navigation, activity recognition, and remote healthcare monitoring [38]. Recently, contact tracing apps based on the BLE technology found in smartphones have been deployed in some venues (more on this below).

Generally, basic BLE-based RTLS solutions are considered "proximity-based systems": they can pinpoint locations within only about 100 square feet, making them ideal for use cases that do not require very exact localization. Proximity-based RTLS are less expensive than precision

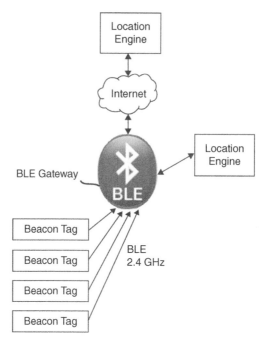

FIGURE 8.7   Simple BLE RTLS setup.

systems, they utilize less expensive tags, and generally, they require less infrastructure [39]. As noted, BLE is widely available in smartphones, obviating the need for a discrete tag in some applications. These solutions, however, may not be totally adequate for OSD/ODCMA applications.

At a broad level, one can use smartphone-based BLE to estimate the cumulative number of indoor occupants in a specific zone (e.g. [38, 40, 41]); more elaborate systems support more accurate proximity determination (within only 100 square feet). Parenthetically, an early example of the use of classical Bluetooth indoor positioning via RSSI, prior to BLE, is described in reference [42].

A basic RTLS setup using BLE beacons is shown in Figure 8.7; a more complex arrangement is depicted in Figure 8.8. BLE-based RTLS (generally) uses tags to send out a transmission to a reader (also called "bridge"); the reader then transmits the location data to the PE in the intranet or in the cloud. This method can detect the presence of an asset inside a mid-size room; for example, it could help a materials management team reduce the search zone for a machine or tool from an entire building to just a few rooms.

Bluetooth applications beyond point-to-point connectivity accelerated after Apple announced iBeacon in 2013, a BLE-based technology that enables a device (specifically, a beacon) to send push notifications to nearby iOS devices. Each BLE device that advertises information using the iBeacon protocol is identified by the following values [40]:

- Proximity UUID (Universally Unique IDentifier): a 128-bit value that identifies a beacon region (that can be composed of many beacons).
- Major value: a 16-bit unsigned integer that can be used to group related beacons that have the same proximity UUID.
- Minor value: a 16-bit unsigned integer that differentiates beacons with the same proximity UUID and major value.

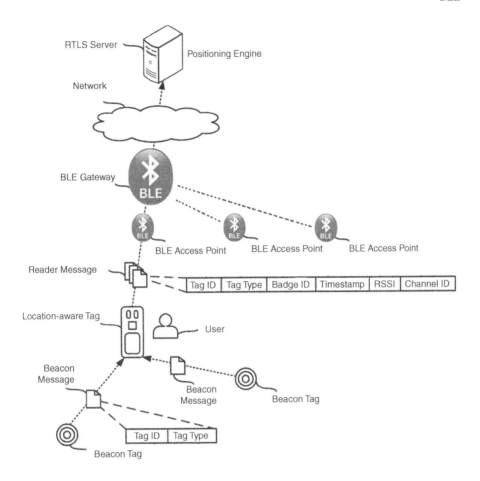

FIGURE 8.8   More complex BLE RTLS setup (loosely based on [43]).

Typically, a complex BLE proximity environment has four types of elements [39]:

- Reference-point beacon tags: simple, inexpensive BLE tags that are strategically placed in multiple places in the zone of interest (for example, one per room). Beacons are able to generate real-time information.
- Location-aware tags: the same tag hardware that is used by normal BLE beacons can be reprogrammed to instead listen for reference-points tags. The tag then processes its own location algorithm and connects to a Bluetooth/BLE Access Point (AP) to relay this information.
- Access points (also referred to as readers or bridges): devices that receive the (encrypted) location (or sensor) data from the tags and sends that data to the PE server via the gateway. Bluetooth/BLE APs are typically spaced about every 100 feet in a zone of interest and are, thus, less dense than reference points.
- Gateway: a device that forwards data to the cloud where a PE may be located or directly to a local PE. A single gateway can connect to a number of APs throughout a large zone of interest. The connection to the server can be achieved via several networking technologies such as Wi-Fi-based Internet access, cellular/IoT means (e.g. LTE-M), or a LoRa (long range)-based wireless network.

- Positioning engine: a server (and storage) that uses the data received from the gateway to determine the location of tags using one or more algorithms (e.g. multilateration, proximity, RFF, and so on). The PE can also support more advanced functions, for example, dwell times in different areas, heat maps, visualization tools, storage, and trend analysis. As noted, the PE can be local or cloud-based.

Figure 8.9 is another diagram illustrating typical components used in a BLE-based RTLS, as described in [3]. This system includes (i) one or more fixed (in-zone) beacon transmitters that generate transmissions containing a report of the motion-status of objects or entities in the beacon's zone, as determined by a motion sensor in the beacon; (ii) a set of signal transmissions – the set of transmissions is received by a fixed infrastructure of bridges and relayed to a central location PE server; and (iii) a database identifying the in-zone beacons and storing the reported motion status patterns for the immediate proximity of each beacon. The motion status reported in the beacon's advertisement is at least one bit that toggles to represent "I see motion (in my zone)" or "I see no motion (in my zone)" and may also be comprised of several bits that include a description of how much motion is seen, along with indicators of recent history of motion-state transitions. The history of motion status may indicate that there was no motion one minute ago, but there is motion now. One or more tags transmit a radio signal containing the tag's motion status to one or more bridges in a fixed infrastructure. For a tag, the motion status is either (i) a bit that says it is moving (or not), or (ii) numerical readings from its onboard accelerometer, or (iii) an increased transmission rate that implies that the tag is in motion. The bridge retransmits the received signal strength of the tag's message, and the tag's motion status, via some wired or wireless network to a central PE location server. The central location PE server may employ trilateration algorithms on the signal strength reports it receives from multiple bridges to in order to form an estimate of the location of the tag. The central PE location server also processes the content of the tag's motion-status message, comparing it to the coincident motion status reported by the fixed beacons. The location server may also analyze patterns of beacon-reported motion status over time, determining which patterns appear to be people walking along walking paths. The location PE server may also analyze signal-strength-based tag-location estimates over time, determining which patterns appear to be tags that are traveling along walking paths with those people. All of this information, including (i) signal

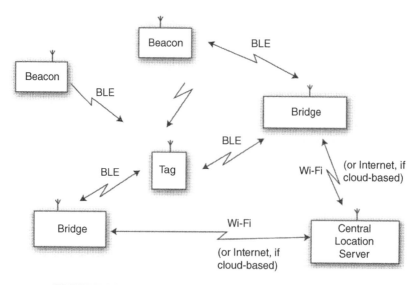

FIGURE 8.9 Another conceptual view of a BLE RTLS [3].

strengths of tag transmissions heard at the bridges, (ii) coincident motion reports from the tags and fixed beacons, and (iii) coincidence of human movement and tag movement, is factored into the location algorithm at the central PE location server. The central PE location server produces a "location estimate" for the tag [3, 44–46]. Radio signals sent from the tag to the bridges provide data that the PE may use to derive an estimate of the location; however, this estimate may or may not be good enough for desk-level-accurate positioning, e.g. for OSD applications.

A typical system (as described in [43]) includes one or more beacon tags affixed to fixed reference points (or assets) within the environment and that transmit (e.g. periodically, aperiodically, and/or episodically) beacon messages. The beacon messages are received by a (mobile) reader badge (also referred to simply as "readers"), that listens for beacon messages transmitted in the environment. The reader badges process the received beacon messages and communicate information obtained from the beacon messages to one or more RTLS PE servers via a communication infrastructure. For example, reader badges/bridges may aggregate and communicate a batch of beacon messages (e.g. a threshold number of beacon messages, a threshold interval of time, e.g. a window of interest), to the RTLS server via a Wi-Fi infrastructure.

- Beacon tags are installed throughout a zone of interest or building. Beacon tags are low-cost, low-power transmitters of beacon messages. A beacon message (sometimes referred to a "beacon") includes information about the beacon tag such as a unique identifier (e.g. a tag identifier such as a MAC address) and a tag type identifier (e.g. whether the beacon tag is affixed to a fixed-location asset or to a mobile asset). In some examples, the beacon tags broadcast beacon messages (e.g. advertise, communicate, transmit) at pre-set frequencies.
- A reader badge is a mobile wireless bridge that facilitates mobile tracking by "listening" and receiving beacon messages broadcast by beacon tags. Bridges are often stationary devices. The reader badge includes a BLE controller (and/or other low-power, short-range radio frequency wireless controller) used to receive connection-less beacon messages broadcast by beacon tags. The reader badge may also include a Wi-Fi controller (or other technology) to establish a connection with the RTLS server. In some examples, the reader badge/bridge collects a number of beacon messages or waits a period prior to communicating the reader messages. The reader badge generates and communicates a reader message when a beacon message from a beacon tag is received. A reader message includes information received from the beacon message, such as a unique identifier of the source beacon tag and a spatial location of the source beacon tag. The reader badge may include a timestamp identifying when the beacon message was received by the reader badge in the reader message. In some examples, the reader badge includes an RSSI value.
- The PE determines which beacon tags are proximate (e.g. near or closely located) to the reader badge. For example, the PE can compare the RSSI strength of a beacon message to a threshold and if the RSSI strength satisfies the threshold (e.g. the RSSI strength is greater than a threshold), the PE identifies the source beacon tag as proximate to the reader badge.

The low energy requirements of BLE enables the RTLS to work more efficiently and offer more capabilities for a smaller cost; a BLE beacon's battery can last up to two years or more. RFID tags have a very short range (typically 2 m), BLE has a range of about 100 m. Wi-Fi and Bluetooth work on the same transmission principles, but Bluetooth is designed to be as low energy as possible. Figure 8.10 depicts the typical components of a beacon and Figure 8.11 depicts some commercial examples of tags. In general, BLE RTLSs use the RSSI strength. BLE uses different frequencies and frames than Wi-Fi but, from a functional perspective, they are similar.

The benefits of a proximity RTLS solution is that it is generally inexpensive and can use open-source tags. For example, iBeacons (tags) range from $2 to $10 depending on size, battery

Asset Beacon

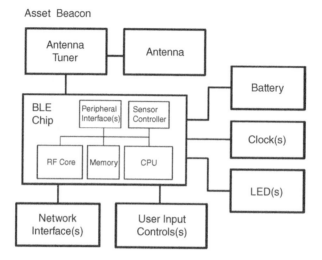

FIGURE 8.10   Typical beacon elements.

iBeacon

- Robust
- Easy to implement
- Functional with any app
- Great with iOS

Eddystone

- Flexible
- Able to broadcast additional data
- Customizable
- Great with Android

Card Beacon

- People monitoring
- Employee management
- Package tracking

FIGURE 8.11   Examples of BLE beacons.

size, and enclosure materials, and are available in a large number of form factors, making application customization straightforward. Card-based beacons are thin, mobile devices that are typically used for people monitoring or asset tracking; for example, a hospital might attach beacons to medical equipment.

8.4.3   BLE-Based Contact Tracing

Digital contact tracing, which can automatically notify an individual if they have crossed paths with someone who tested positive for COVID-19, has been proposed as a way to augment manual contact tracing [47]. Lessons learned in the COVID-19 pandemic will clearly be applicable and applied to future pandemics. In principle, digital contact tracing can be used to contact large numbers of people at a fraction of the cost of manual contact tracing. BLE is one of the technologies that has been proposed for that purpose. BLE was designed as a PAN communication technology, not for determining the distance between objects or smartphones, but the technology has become, of necessity, a de facto option for contact tracing during the pandemic. Some have labeled this application as a "highly improvised . . . Bluetooth-based solution" [48]. This has to be seen in the context of statements from the Bluetooth SIG such as "*Bluetooth technology in smartphones drives location services – 1.8 B Bluetooth enabled handsets will be actively engaged in location services by 2024: from indoor navigation and item finding to point of interest information solutions, more than 1.8 billion actively engaged handsets by 2024 will continue to make Bluetooth location services an integral part of the smartphone experience*" [37].

In 2020, the U.S. National Institute of Standards and Technology (NIST) organized an event named "Too Close for Too Long (TC4TL) Challenge" that leveraged the talents of AI researchers around the world to help evaluate and potentially improve the baseline BLE performance in detecting when smartphone users are standing too close to one another. Most contact tracing apps generally try to collect some form of information about a smartphone user's encounters with other people and notify those users if they were potentially exposed to a confirmed COVID-19 carrier; however, each app has its own approach to privacy and can differ in whether it collects more specific location data based on GPS or merely just records close encounters with other smartphones based on BLE interactions [48, 49]. The NIST challenge attracted several AI research teams from around the world who demonstrated how machine learning might help boost proximity detection by analyzing the patterns in BLE signals and data from other phone sensors. The teams' testing results, presented during a final evaluation workshop held on 28 August 2020, showed that BLE's capability alone in detecting nearby smartphones is, if anything, imperfect. Preliminary results also show that when people hold phones in their hand, one can get relatively acceptable proximity detection, but when they carry the phones in their pocket or in their purse, the performance of this proximity detection technology appears to degrade.

NIST's effort focused on evaluating whether vendors' machine learning models could improve the process of detecting a smartphone's proximity based on the combination of BLE signal information and data from other common smartphone sensors such as accelerometers, gyroscopes, and magnetometers. They also assessed scenarios such as both people holding the phones in their hands, as well as one or both people having the phones in their pockets. The latter is important, given how BLE signals can be weakened or deflected by a number of different materials. The NIST results were presented as a Normalized Decision Cost Function (NDCF) that represents proximity detection performance when accounting for the combination of false negatives (failing to detect a nearby phone) and false positives (falsely stating that a nearby phone has been detected). The fact that the baseline performance of BLE signal detection for detecting nearby smartphones was found to be somewhat weak may not bode well for the extant digital contact tracing efforts using BLE-based apps – especially given the much higher error rates for situations when one or both phones is/are in someone's pocket or purse. The conclusion was that many future challenges remain to be addressed before researchers can deliver enhanced BLE-based proximity detection and a possible performance boost from machine learning. For example, BLE-based proximity detection could likely become more accurate if phones spent more time listening for BLE chirps from nearby phones, but tech companies such as Google and Apple (initially) limited that listening time period in the interest of preserving phone battery life [48].

Contact tracing work has also been undertaken by MIT-led Private Automated Contact Tracing (PACT) project, pioneering the Bluetooth-based privacy protocol at the heart of Apple and Google's solution. The PACT system seeks to automate contact tracing by detecting and logging proximity between phones using BLE signals, or "chirps," from phones within an approximate 6-foot radius and picked up for a particular duration of time. Unlike early digital contact tracing efforts from the World Health Organization and MIT's SafePaths which relied on GPS data, the PACT system does not collect specific location data. Instead, the system relies on phones sending out anonymous Bluetooth "chirps" – random, rotating numbers which do not reveal from where or whom they were sent. Then, if a person tests positive for the new coronavirus, they can upload all the chirps their phone has sent out in the last two weeks to a database. If any of those chirps match ones picked up by someone else's phone, a notification will inform that person of a possible exposure [47]. Some of the data are being collected with robots: by moving robots equipped with smartphones around a room under various conditions, the MIT team is gathering data on the signal strength of the chirps for various distances and amounts of times.

It has been well reported that in the Spring of 2020, in a rare act of cooperation, Google and Apple released specifications for software developers to build digital contact tracing apps for Apple and Google mobile operating systems, which jointly encompass the majority of smartphones around the world [49]. The vendors published the Google Apple Exposure Notification (GAEN), which has been used in exposure notification or contact tracing apps. The firms stated that apps using their contact tracing Application Programming Interfaces (APIs) must be made by or for the use of government health authorities; gathered information is only for use for COVID-19 exposure information; and users must be opt-in only and shall consent before sharing a positive test result [47]. The firms also pledged to discontinue the use of the system once the crisis has passed.

8.4.4 Illustrative Examples

An exemplary BLE-based product is highlighted in this section to give weight to the availability of the technology (no product is explicitly sanctioned herewith).

AiRISTA Flow AiRISTA develops and manufactures identification-and-track-and-trace products using passive, active, and semi-active RFID; GPS; BLE; and other technologies. AiRISTA's basic product, *AiRISTA Flow*, makes use of location-aware active RFID badges that operate over standard Wi-Fi networks (the fairly well-known Ekahau RTLS business was acquired by AiRISTA Flow in 2016). AiRISTA Flow solutions use existing WLANs to gain real-time visibility into the assets, people, and workflows that drive success with our innovative RTLS platform. The analytics system – Unified Vision Solution (UVS) software platform – is written in the AI-centric language, "Go," and performs AI processing of data consumed by machine learning models; machine learning algorithms and federated data bus mechanisms are utilized. The stream-based architecture can scale to support over 1.8 million records per second. Applications can subscribe to topics within a stream as part of a workflow orchestration that spans an entire enterprise. UVS deploys brokers as endpoint and data stream relationships are made and dissolved. The software is designed so that the platform can combine information from assets company-wide, delivering insight across the delivery chain. The software platform manages data in three tiers: in-memory, locally stored, and long term. The hardware consists of asset tags, location-aware active RFID personnel badges, embedded sensors, and the wireless infrastructure. Use cases include healthcare (for personnel safety, asset tracking, patient flow, temp monitoring, hand hygiene, wander management, and social distancing and contact tracing); industrial (for personnel safety, asset tracking, inventory management, condition and environmental monitoring, wireless ID, workflow management, and social distancing and contact tracing); and hospitality (to motivate, track, or monitor compliance for hand washing), among others [50].

In 2020, the firm offered a version of their product specially targeted to OSD/ODCMA (see Figure 8.12). The tags use a unique feature of BLE that detects other BLE devices in the area. The AiRISTA Flow implementation of BLE not only provides beaconing of an outbound signal, but it also scans (listens) for other BLE devices. The stronger the signal received from another tag, the closer the tag. Once the power threshold representing 6 feet is met, we alert the individual and record the ID of the other tag. The tags can be configured over the air; for example, if safe distancing recommendations change, the tags can be remotely configured to the new distance settings. The tags communicate with each other autonomously and alert the employee when within range of another tag. Data collected by the tags are communicated to the cloud-based software platform,

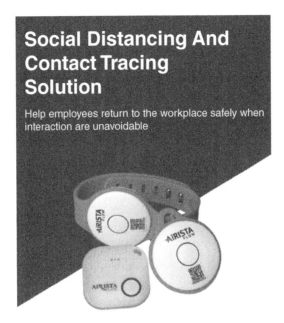

FIGURE 8.12 Example of RTLS tags for OSD/ODCMA (Courtesy AiRISTA).

which analyzes contact information and manages the tags. When people come within 6 feet of each other for a period of time, the device makes an audible chirp. A record of the contact is made on the tag and uploaded to the AiRISTA Flow software system (triggers for tag distances and event duration are configurable over the air). Tags communicate with the software platform using existing Wi-Fi systems, employees' smartphones, or small gateways provided by AiRISTA Flow. The system supports BLE and BLE+Wi-Fi tags. BLE tags are low cost and offer long battery life. BLE+Wi-Fi tags are badge-style tags and require recharging every couple of months (depending on tag configuration). AiRISTA Flow also offers an application that makes a smartphone act like a tag. To speed deployment and keep costs down, an existing Wi-Fi network is typically used to communicate to the cloud-based software platform. The BLE+Wi-Fi tags communicate directly with Wi-Fi access points. BLE tags can communicate with access points that support BLE. The location of the tag is tracked. The tag location is collected and made available via a table in the cloud portal. A location "snapshot" is made each time the tag uploads its contacts to the cloud software via an access point or gateway – the unique ID the access point or gateway, and the time are recorded. The system administrator associates that access point or gateway with a physical location in the building such as break room, lobby, and so on [50].

- Reference [51] noted:

    Until there is a vaccine for the coronavirus, social distancing will be required as we get back to work and try to return to some level of normalcy. And for those not yet immune contact tracing is needed to help prevent flare ups. The Arista Flow Social Distancing And Contact Tracing solution uses a wireless device warned by employees to help enforce CDC guidelines for social distancing and automate contact tracing. The tag is worn as a wrist strap pendant or key fob and detects proximity of other tags. When people come within six feet of each other, for a period of time, the device alerts the wearer, and a record of the contact is made in the Arista Flow software system. [. . .] The tags use a unique feature of Bluetooth Low Energy, which detects other BLE devices in the area as tags come near to each other, they register the stronger signal at a distance of six feet. For example, the tag alerts the employee with a chirp and at the time of the event is recorded along with the tag IDs. The Arista Flow cloud-based software platform can be deployed quickly to record employees contacts, provide contact tracing, and generate reports for trend analysis and problem detection. Tags communicate with the software platform using existing wireless infrastructure, employees' smartphones, and tablets, or small gateways provided by Arista Flow. People are eager to return to the workplace but want to do so knowing they are taking the necessary precautions. The Arista Flow Social Distancing And Contact Tracing solution will reinforce best practices and avoid virus flare ups to help keep employees safe. Other modules are available for the Arista Flow platforms such as asset tracking, duress alerting, environmental monitoring, and many more."

Mist (Juniper Networks) Virtual Bluetooth Beacon Technology   Mist offers enterprise-grade systems that simultaneously support Wi-Fi, BLE, and IoT that support (among other functionality) personalized location services, such as wayfinding, proximity notifications, and asset location. Mist utilizes virtual BLE (vBLE) technology (virtual beacons replace physical hardware), such that no battery beacons or manual calibration are required (machine learning eliminates manual configuration). Real-time wayfinding helps employees, guests, and customers get around with turn-by-turn directions with accuracy of up to 1 m (3.3 feet) with sub-second latency. Real-time proximity notification and alerts are available to greet patients, clients, or customers as they arrive onsite. The system uses a Dynamic vBLE-16 antenna array in the APs to cover an entire room with Bluetooth LE signal. Figure 8.13 [52] illustrates Mist's approach.

- Reference [53] noted:

  Indoor location services provide a huge opportunity to engage with employees, guests, and customers, while optimizing resources to save time and money. But widespread adoption of these services has been limited by physical beacons and costly site surveys: they just don't scale. This all

FIGURE 8.13   Mist (Juniper Networks) vBLE concept (courtesy).

changes with Mist. With the invention of scalable enterprise-grade BLE we have flipped the model by virtualizing the indoor location experience. Here is how. Our patented BLE directional antenna array enables all Mist endpoint to blanket an entire location with BLE. Location is computed in the Mist intelligent cloud. This lets you go beyond just listening to a tag to interacting with an entire space. With this new model beacons are completely virtual. Create as many as you want, wherever you want and change them as often as you want without the hassle and cost of physical beacons. With machine learning manual calibration and site surveys are a thing of the past. The Mist platform also includes comprehensive zone analytics like visits and dwell times providing rich insight into personnel and resource utilization. The world is ready for new indoor location experiences.

8.5   CELLULAR APPROACHES

The question is if in-building localization can be achieved without having to deploy a dedicated RTLS infrastructure. Approaches based on the GPS suffer from certain inaccuracies when locating objects in closed (i.e. indoor) environments; it may not even work at all. As an alternative, cellular systems can be utilized for (some) in-building localization situations. In fact, 3GPP Release 9 standardized cellular positioning techniques for LTE (Long Term Evolution) systems, for supporting Location Application for Emergency Services (LAES) applications, as follows:

- Primary method: A-GNSS (Assisted Global Navigation Satellite System) or A-GPS (Assisted Global Positioning System).
- Fall-back method: Enhanced Cell-ID (E-CID) and OTDOA (Observed Time Difference of Arrival), including DL-OTDOA (Downlink OTDOA).

Positioning Reference Signals (PRSs) were added in the Release 9 to be used by the User Equipment (UE) for OTDOA positioning (a type of multilateration). The Release 9 specification also required neighboring base stations (specifically, Evolved Nodes B [eNBs]) to be synchronized. In the OTDOA technique, the Time of Arrival of the signal coming from several neighboring eNB base stations is calculated. The UE position can be estimated in the handset (UE-based method) or in the network (NT-based, UE-assisted method) once the signals from three base stations are received. The measured signal is the Common Pilot Channel (CPICH); this channel is a downlink channel broadcast by eNBs where the signal has constant power and conforms to a known bit pattern. The propagation time of signals is correlated with a locally generated replica. The peak of correlation indicates the observed time of propagation of the measured signal. Time difference of arrival values between two base stations determines a hyperbola. At least three reference points are needed to define two hyperbolas. The location of the UE is in the intersection of these two hyperbolas [54], see Figure 8.14.

While these methods might satisfy the mandatory FCC E911 emergency location requirements, the accuracy, reliability, and availability of these location methods fall short of the needs of LBS or RTLS users, who require highly accurate positioning/locating within buildings, shopping malls, urban corridors, and so on. Furthermore, although cellular wireless communication systems provide good coverage in urban and some indoor environments (e.g. when Distributed Antenna Systems [DASs] are used), the position accuracy of these systems is limited by self-interference, multipath and non-line-of-sight propagation. Furthermore, newer FCC 911 requirements are more stringent than the existing ones, and while the accuracy of A-GNSS/A-GPS may be sufficient in open spaces, this methodology is generally unreliable in indoor environments [54].

The indoor and outdoor location inaccuracies of cellular systems are due mainly to the physics of RF propagation, in particular, due to losses/attenuation of the RF signals, signal scattering,

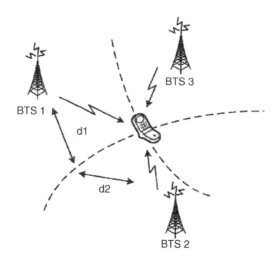

FIGURE 8.14 OTDOA approach for LTE [54].

and reflections. The losses/attenuation and scattering issues can be solved by employing narrow-band ranging signal(s) and operating at relatively low RFs, for example, at Very High Frequency (VHF) range or lower. Although at VHF and lower frequencies the multipath phenomena (e.g. RF energy reflections) are less severe than at Ultra High Frequency (UHF) and higher frequencies, the impact of the multipath phenomena on location-finding accuracy makes location determination less reliable and precise than required by many applications, including for OSD/ODCMA. As a rule, conventional RF-based identification and location-finding systems mitigate multipath by employing wide bandwidth ranging signals. For example, for 60-feet accuracy, one needs about 40 MHz of bandwidth; for 30-feet accuracy, one needs about 80 MHz of bandwidth; and for 20-feet accuracy one needs about 110 MHz [55, 56]. Spatial diversity and/ or antenna diversity techniques are used in some cases; however, the spatial diversity may not be an option in many tracking-location applications because it leads to an increase in the required infrastructure – furthermore, the antenna diversity has a limited value, because at lower operating frequencies, for example, VHF, the physical size of antenna subsystem becomes too large. An antenna array could, in theory, also be used to mitigate the multipath, but at VHF and lower frequencies, the size of the antenna array will significantly impact device portability. The issue remains that, because of a very limited spectrum, the narrow bandwidth ranging signal does not lend itself to the multipath mitigation techniques that are currently used by conventional RF-based identification and location-finding systems; the reason is that the ranging signal distortion that is induced by the multipath is too small for reliable detection/processing in presence of noise.

As noted in earlier chapters, DASs are commonly employed indoors; however, each antenna does not have a unique ID; therefore, the UE equipment cannot differentiate between the multiple antennas. This phenomenon prevents the usage of the multilateration method, employed in the Release 9 and Up Link OTDOA (UL-OTDOA) systems. To address this predicament, one needs to add hardware and new network signals to the existing indoor wireless network systems [56]; however, these modifications for indoor wireless network antenna systems are not always possible because upgrading the existing systems would require effort and cost. Furthermore, in the case of an active DAS, the best accuracy is limited to about 50 m, and in practice, this accuracy can actually be lower because of RF propagation phenomena, including multipath (signals that are produced by multiple antennas will appear as reflections, e.g. multipath).

Now, if all antennas' locations are deterministically known, it is possible to provide a location fix in DAS environment without the additional hardware and/or new network signals (as suggested in [56]), if the signals' paths from individual antennas can be resolved, such as by using multilateration and location consistency algorithms. Proposals have been made in the recent past for multipath mitigation methods for object/people identification and location-finding that utilize narrow bandwidth ranging signal(s) and operate in VHF or lower frequencies. Reference [54] describes a possible enhancement for a system for cellular-based RTLS, that employs a narrow bandwidth ranging signal operating in the UHF, VHF, HF, LF, and VLF bands. Employing a multipath mitigation processor increases the accuracy of object/device/entity tracking and locating. The proposal includes small, highly portable base units that allow users to track, locate, and monitor multiple persons and objects. Each unit broadcasts an RF signal with its ID, and each unit is able to send back a return signal, which can include its ID as well as voice, data, and additional information. Each unit processes the returned signals from the other units and, depending on the triangulation or trilateration and/or other methods used, continuously determines their relative and/or actual locations. The concept can also be integrated with products such as GPS devices, smartphones, two-way radios, and PDAs. In this environment, the infrastructure consists of (i) one or more wireless Network Signals Acquisition Units (NSAUs) and (ii) one or more Locate Server Units (LSUs) that collect data from NSAU(s) and analyze it, determine range and locations, and convert it into a table e.g. of phone/UEs IDs and locations at an instant of time. The LSU interfaces to the LTE/cellular network. Multiple NSAU/LSU units could be deployed in various locations in a large infrastructure. If NSAU(s) have coherent timing, the system will provide better accuracy; the coherent timing can be derived from the GPS clock and/or other stable clock sources. The NSAU communicates with LSU via a Local Area Network (LAN), a Metro Area Network (MAN), and/or the Internet. In some installations or instances, the NSAU and LSU could be combined/integrated into a single unit. Figure 8.15 depicts some aspects of the proposal, particularly in a DAS context.

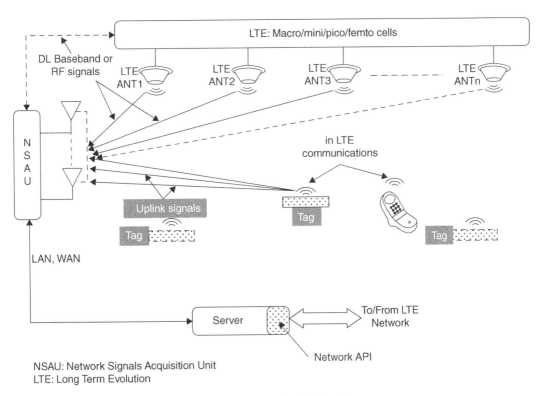

FIGURE 8.15   Example of DAS-based system [54].

8.6 SUMMARY

The following points and Table 8.3 (based on reference [57]) serve to summarize the discussion about the key RTLS technologies discussed in this chapter (with RFID systems also included for comparison):

- Wi-Fi-based RTLSs that utilize Wi-Fi tags are able to leverage existing Wi-Fi ecosystems, making them popular in industries such as healthcare. These tags act almost the same as Bluetooth beacons with one key difference: energy usage. Wi-Fi was designed to be highly efficient and therefore uses a large amount of bandwidth; the result is higher energy usage (Wi-Fi tags often must be charged every 24–48 hours) and higher associated costs. Wi-Fi tags may be applicable to OSD/ODCMA.
- BLE is designed to use a lot less energy than other PANs. It supports tags that can run for several years on coin-cell operated batteries (beacons can last several years). Due largely to the simplicity of the technology, Bluetooth-based beacons can be created quickly and for a lower cost than alternative solutions (one-fifth the price of competing technologies). Bluetooth version 5 enjoys four times the range and double the speed of its predecessor. It also has an 800% increased capacity for advertisements. BLE-based systems may be applicable to OSD/ODCMA.
- Passive RFID-based systems are well-established solutions in a wide range of applications for the detection and identification of items. Passive RFID tags are simple and inexpensive (and can be easily disposed along with product packaging when used for logistics). However, in order for a passive RFID tag to be successfully read, it must receive sufficient RF power for its internal logic to be activated and transmit back to the reader with sufficient SNR; this requirement sets limits on the maximum tag range to about 10 m. Additionally, RFID tags are typically difficult to locate accurately because of multipath fading, and this can significantly restrict their use in applications where accurate location sensing is important (unless choke points are used, as discussed in Chapter 6). Passive RFID system operating at UHF promises to offer read ranges of the order of 10 m using tags that do not require their own power source [58]. Passive RFID may have some limited application OSD/ODCMA.
- Active RFID-based systems use small battery-powered tags to broadcast signals. Unlike passive tags, these provide a long "read range" (up to 100 meters) and include a power source for broadcasting. These systems, however, are more expensive than their passive counterparts. Active RFID may be applicable to OSD/ODCMA.

TABLE 8.3   Additional comparison of technologies [57]

RTLS type	Cost	Accuracy/reading distance	Battery life	Upfront hardware costs
BLE (Kontakt.io)	Gateway; $89 Tag: $17	4 m (room level)	7 mo for card beacons 2+ yr for tough beacon	$25 900
Wi-Fi	Tag $120	3–4 m	24 h for personal tags 3–4 yr (low frequency)	$120 000
Active RFID	Reader $1000–$5 k Tag: $50–$100	3–5 m	3–10 yr	$150 000

TABLE 8.4 Overall comparison of sensor technologies for RTLSs

Presence/count technology	OSD applicability	Presence/ detection accuracy	Count accuracy	Popularity	Existing infrastructure	Cost	Privacy issue	Issues, considerations, possible limitations
Wi-Fi	Possibly high	Medium to high	Medium to high	Medium[a]	Yes	Low	Some	Tag cost relatively high; not too accurate; scalability issues
Active RFID	Possibly high	High	High	High	Sometimes yes	Medium	Some	Inconvenient; might miss visitors; cost of tags (for large groups)
BLE	Possibly high	Medium to high	High	Medium	No	Medium	Some	Need to turn on Bluetooth on devices or use tags
BLE (smartphone)	Medium to high	Medium	Medium to high	Medium	Partially	Low to medium	Some	Need to turn on Bluetooth on smartphone; privacy issues
UWR	Possibly high	High	Medium to high	Medium	No	Medium to high	Some	Scalability is a recent breakthrough; might miss visitors cost of tags (for large groups)
Zigbee	Some	High	High	Medium	No	Medium	Some	Inconvenient; might miss visitors
PIR	Some	High	Low	Medium	No	Low	No	Misses static occupants
Videocamera	Some	High	High	Low[b]	Yes	Medium	Yes	Illumination requirements; Line-of-Sight-(LOS)
Chair sensor	Limited	High	High	Low	No	Medium	No	Deployment complexity; relatively high cost for larger offices, misses standing people
Vibration	Limited	High	Medium	Low	No	High	No	Scalability; limited product set
LED	Limited	Medium to high	Medium	Low	Yes	Medium	No	Limited product set

[a] Medium for OSD-like applications, high for intranet/hotspot usage.
[b] Low for real-time OSD-like applications, relatively high for security applications.

Table 8.4 synthetized from a variety of sources (including among others [10, 11, 59, 60]) provides a comparison of sensor technologies for RTLSs discussed in the last three chapters.

REFERENCES

1. Bassan-Eskenazi, A., Sharir, N., and Friedman, O. (2016). Location measurements using a mesh of wireless tags. US Patent 9,390,302, 12 July 2016, filed 21 November 2013. Uncopyrighted.
2. Amir, I., Annamalai, K., and Naim, A. (2016). System and method for multimode Wi-Fi based RTLS. US Patent 9,341,700, 17 May 2016, filed 1 February 2013. Uncopyrighted.
3. Staff of Exploring Our Fluid Earth Practices of Science. (2020). Precision vs. Accuracy. https://manoa. hawaii.edu/exploringourfluidearth/about-site.
4. Swart, J.A. (2019). Real-time location system (RTLS) having tags, beacons and bridges, that uses a combination of motion detection and RSSI measurements to determine room-location of the tags. US Patent 10,390,182, 20 August 2019, filed 10 December 2018. Uncopyrighted.
5. Martin, A., Ionut, C., and Romit, R.C. (2009). Surroundsense: mobile phone localization via ambience fingerprinting. *Proceedings of the 15th Annual International Conference on Mobile Computing and Networking*, pp. 261–272.
6. Lina, T., Nga, I., Laua, S.Y. et al. (2008). A microscopic examination of an RSSI-signature-based indoor localization system. HotEmNets'08 (2–3 June 2008). Charlottesville, Virginia, USA: ACM.
7. Prevatt, T. (2018). Angle of arrival (AOA) positioning method and system for positional finding and tracking objects using reduced attenuation RF technology. US Patent 10,091,616; 2 October 2018, filed 9 October 2017. Uncopyrighted.
8. Haeberlen, A., Flannery, E., Ladd, A.M. et al. (2004). Practical robust localization over large-scale 802.11 wireless networks. *Proceedings of ACM MOBICOM* (2004).
9. Lim, H., Kung, L.C., Hou, J.C., and Luo, H. (2006). Zero configuration, robust indoor localization: theory and experimentation. *Proceedings of IEEE INFOCOM* (2006).
10. Ridolfi, M., Van de Velde, S., Steendam, H., and De Poorter, E. (2018). Analysis of the scalability of UWB indoor localization solutions for high user densities. Sensors (Basel) 18 (6): 1875. https://doi. org/10.3390/s18061875.
11. Chen, Z., Jiang, C., and Xie, L. (2018). Building occupancy estimation and detection: a review. Energy and Buildings 169: 260–270.
12. Bahl, P. and Padmanabhan, V.N. (2000). RADAR: an in-building RF-based user location and tracking system. *Proceedings IEEE INFOCOM 2000. Conference on Computer Communications. Nineteenth Annual Joint Conference of the IEEE Computer and Communications Societies (Cat. No.00CH37064)*, vol. 2, 775–784, Tel Aviv, Israel. doi:10.1109/INFCOM.2000.832252s.
13. Chan, L., Chiang, J., Chen, Y. et al. (2006). Collaborative localization: enhancing WiFi-based position estimation with neighborhood links in clusters. In: Pervasive Computing, Lecture Notes in Computer Science, vol. 3968 (eds. K.P. Fishkin, B. Schiele, P. Nixon and A. Quigley). Berlin, Heidelberg: Springer https://doi.org/10.1007/11748625_4.
14. Yie, J., Yang, Q., and Ni, L. (2005). Adaptive temporal radio maps for indoor location estimation. International Conference on Pervasive Computing (8–12 March 2005). Kauai Island, HI, USA. https:// doi.org/10.1109/PERCOM.2005.7.
15. Chen, Y.C., Chiang, J.R., Chu, H. et al. (2005). Sensor-assisted Wi-Fi indoor location system for adapting to environmental dynamics. *Proceedings of the 8th ACM International Symposium On Modeling, Analysis And Simulation Of Wireless And Mobile Systems* (October 2005), pp. 118–125. https://doi. org/10.1145/1089444.1089466.
16. Young, S. (2014). Method and apparatus for limiting redundant positioning polling from a mobile device in a real-time location system (RTLS). US Patent 8,922,432, 30 December 2014, filed 20 December 2011.
17. De Lorenzo, R.F. (2020). Processing alert signals from positioning devices. US Patent 20200258373, 13 August 2020.
18. Cisco Staff. (2006). Cisco compatible extension overview. Cisco Systems, Inc. https://www.cisco.com/ web/partners/downloads/765/ccx/Comp_Ext_Cust_Preso.pdf.

19. Amir, I. (2016). System and method of enhanced RTLS for improved performance in wireless networks. US Patent 9,298,958, 29 March 2016, filed 2 May 2013. Uncopyrighted.

20. Cisco Staff. (2020). Real-time location services (RTLS). https://documentation.meraki.com/MR/Monitoring_and_Reporting/Real-Time_Location_Services_(RTLS). Article ID: 2152.

21. Cisco Staff. (2020). Cisco unified wireless location-based services. https://www.cisco.com/en/US/docs/solutions/Enterprise/Mobility/emob30dg/Locatn.html.

22. Ray, B. (2017). The shortcomings of WiFi RTLS. https://www.airfinder.com/blog/rtls-technologies/shortcomings-of-wifi-rtls.

23. Balaji, B., Xu, J., Nwokafor, A. et al. (2013). Sentinel: occupancy based HVAC actuation using existing WiFi infrastructure within commercial buildings. *Proceedings of the 11th ACM Conference on Embedded Networked Sensor Systems*, p. 17. https://doi.org/10.1145/2517351.2517370.

24. Lu, X., Wen, H., Zou, H. et al. (2016). Robust occupancy inference with commodity WiFi. IEEE International Conference on Wireless and Mobile Computing, Networking and Communications (WiMob), pp. 1–8.

25. Zou, H., Jiang, H., Yang, J. et al. (2017). Non-intrusive occupancy sensing in commercial buildings. Energy and Buildings 154: 633–643.

26. Liang, N.Y., Huang, G.B., Saratchandran, P., and Sundararajan, N. (2006). A fast and accurate online sequential learning algorithm for feedforward networks. IEEE Transactions on Neural Networks 17 (6): 1411–1423. https://doi.org/10.1109/TNN.2006.880583.

27. Huang, G.B., Zhu, Q.Y., and Siew, C.K. (2004) Extreme learning machine: a new learning scheme of feedforward neural networks. *Proceedings of the International Joint Conference on Neural Network (IJCNN2004)*, vol. 2, Budapest, Hungary (25–29 July 2004), pp. 985–990.

28. Huang, G.B. and Siew, C.K. Extreme learning machine: RBF network case. *Proceedings of the 8th International Conference on Control, Automation, Robotics and Vision (ICARCV 2004)*, Vol. 2, Kunming, China (6–9 December 2004), pp. 1029–1036.

29. Huang, G.B., Zhu, Q.Y., Mao, K.Z. et al. (2006). Can threshold networks be trained directly? IEEE Transactions on Circuits and Systems II: Express Briefs 53 (3): 187–191.

30. Depatla, S., Muralidharan, A., and Mostofi, Y. (2015). Occupancy estimation using only WiFi power measurements. IEEE Journal on Selected Areas in Communications 33 (7): 1381–1393. https://doi.org/10.1109/JSAC.2015.2430272.

31. Azam, M., Blayo, M., Venne, J.S., and Allegue-Martinez, M. (2019). Occupancy estimation using Wifi motion detection via supervised machine learning algorithms. IEEE Global Conference on Signal and Information Processing (GlobalSIP), Ottawa, ON, Canada, pp. 1–5. https://doi.org/10.1109/GlobalSIP45357.2019.8969297.

32. Xi, W., Zhao, J., Li, X.Y. et al. (2014). Electronic frog eye: counting crowd using WiFi. IEEE INFOCOM, pp. 361–369.

33. Wang, W., Chen, J., and Song, X. (2017). Modeling and predicting occupancy profile in office space with a WiFi probe-based dynamic markov time-window inference approach. Building and Environment 124: 130–142.

34. Staff. Cisco Meraki launches Phunware location based services in Meraki marketplace. Businesswire (25 February 2020). https://www.businesswire.com/news/home/20200225005341/en/Cisco-Meraki-Launches-Phunware-Location-Based-Services.

35. M-Way Solutions Staff. Smart Beacon Management with BlueRange, Version 1.1 – Status 01/2018, M-Way Solutions GmbH, Stresemannstr. 79, 70191 Stuttgart, Deutschland. https://www.bluerange.io/downloads/en_bluerange_guide_2018.pdf.

36. Minoli, D. and Occhiogrosso, B. (2020). IoT-driven advances in commercial and industrial building lighting and in street lighting. In: Industrial IoT: Challenges, Design Principles, Applications, and Security (ed. I. Butun). Springer.

37. Bluetooth SIG. Bluetooth Market Update 2020. https://www.bluetooth.com/wp-content/uploads/2020/03/2020_Market_Update-EN.pdf.

38. Filippoupolitis, A., Oliff, W., and Loukas, G. (2016). Occupancy detection for building emergency management using BLE beacons. International Symposium on Computer and Information Sciences, 233–240. https://doi.org/10.1007/978-3-319-47217-1_25.

39. Airfinder Staff. (2020). Asset location technologies & the selection process. www.airfinder.com

40. Conte, G., De Marchi, M., Nacci, A.A. et al. (2014). Bluesentinel: a first approach using iBeacon for an energy efficient occupancy detection system. ACM Conference on Embedded Systems for Energy-Efficient Buildings, BuildSys'14, 5–6 November 2014, Memphis, TN, USA, pp. 11–19. https://doi.org/10.1145/2676061.2674078.

41. Corna, A., Fontana, L., Nacci, A.A., and Sciuto, D. (2015). Occupancy detection via iBeacon on android devices for smart building management. Design, Automation & Test in Europe (DATE) Conference & Exhibition, 9–13 March 2015, Grenoble, France, pp. 629–632. https://doi.org/10.7873/DATE.2015.0753.

42. Pei, L., Chen, R., Liu, J. et al. (2010). Inquiry-based bluetooth indoor positioning via RSSI probability distributions. *Proceedings of the 2010 Second International Conference on Advances in Satellite and Space Communications*, SPACOMM '10, Athens, Greece (13–19 June 2010), pp. 151–156. https://doi.org/10.1109/SPACOMM.2010.18.

43. Cannell, M.J., Nguyen, D., Geiger, E. et al. (2018). Healthcare beacon device configuration systems and methods. US Patent 10,311,352, 4 June 2019, filed 4 April 2018. Uncopyrighted.

44. Swart, J.A. (2017). Bluetooth low energy (BLE) Real-time location system (RTLS) having tags that harvest energy, bridges that instruct tags to toggle beacon modes on and off, beacons and bridges that self-report location changes, and optional use of a single beacon channel. US Patent 10,028,105, 17 July 2018, filed 31 May 2017. Uncopyrighted.

45. Swart, J.A. (2018). Bluetooth low energy (BLE) Real-time location system (RTLS) having tags, beacons and bridges, that use a combination of motion detection and RSSI measurements to determine room-location of the tags. US Patent 10,251,020, 2 April 2019, filed 18 June 2018. Uncopyrighted.

46. Swart, J.A. (2019). Bluetooth low energy (BLE) Real-time location system (RTLS) having simple transmitting tags, beacons and bridges, that use a combination of motion detection and RSSI measurements to determine room-location of the tags. US Patent 10,231,078, 12 March 2019, filed 18 June 2018. Uncopyrighted.

47. Scudellari, M. (2020). COVID-19 digital contact tracing: apple and google work together as MIT tests validity - developers are building and testing an opt-in automated system to slow the spread of the coronavirus. But will anyone use it?. IEEE Spectrum, 13 May 2020.

48. Hsu, J. (2020). Can AI make bluetooth contact tracing better? Machine learning shows some promise in boosting contact tracing technologies meant for detecting nearby phones. IEEE Spectrum, 8 September 2020.

49. Hsu, J. (2020). Survey finds Americans skeptical of contact tracing apps mistrust and misunderstanding pose dual challenges to contact tracing apps in the United States. IEEE Spectrum, 7 July 2020.

50. AiRISTA Website. https://www.airistaflow.com.

51. AiRISTA Flow video. https://youtu.be/fFK2C_0M4Yg.

52. Mist (Juniper Networks). https://www.mist.com/ble-indoor-location-services/

53. Mist video. https://youtu.be/b3m_j9oyv2o

54. Prevatt, T. and Buynak, M.J. (2019). Angle of Arrival (AOA) positioning method and system for positional finding and tracking objects using reduced attenuation RF technology. US Patent 10,440,512, 8 October 2019, filed 20 September 2018. Uncopyrighted.

55. Salous, S. (1997). Indoor and outdoor UHF measurements with a 90 MHz bandwidth. IEEE Colloquium on Propagation Characteristics and Related System Techniques for Beyond Line-of-Sight Radio, pp. 8/1–8/6.

56. Chan, B.H., Lee, J.A., and MacDonald, W.M. (2011). Method and apparatus for UE positioning in LTE networks. US Patent 20110124347, 26 May 2011, filed 15 September 2010. Uncopyrighted.

57. Kontakt Staff. (2020). Comparing RTLS infrastructure: weighing bluetooth and competitors. www.kontakt.io. Uncopyrighted.

58. Sabesan, S., Crisp, M., Penty, R., and White, I. (2016). RFID tag location systems. US Patent 9,367,785, 14 June 2016, filed 18 April 2011. Uncopyrighted.

59. Koyuncu, H. and Yang, S.H. (2010). A survey of indoor positioning and object locating systems. IJCSNS International Journal of Computer Science and Network Security 10 (5): 121–128.

60. Labeodan, T., Zeiler, W., Boxem, G., and Zhao, Y. (2015). Occupancy measurement in commercial office buildings for demand-driven control applications: a survey and detection system evaluation. Energy and Buildings 93: 303–314. https://doi.org/10.1016/j.enbuild.2015.02.028.

9 Case Study of an Implementation and Rollout of a High-Density High-Impact Network

This chapter[1] provides an up-to-date case study of the implementation and rollout of a high-density, high-impact network that embodies many of the concepts, approaches, and technologies discussed in the previous chapters. It discusses the revamping and the modernization of the passenger-touch points for both the general operations and business/concession-facing wireless network infrastructure at Baltimore/Washington International Thurgood Marshall Airport (BWI) undertaken by Slice Wireless Solutions, Inc. (SliceWiFi) during the years 2020 and 2021. Such a synthesis comprises what can be called a Wireless SuperNetwork (WiSNET)™, the more general essence and scope of which is described in detail in the next chapter. This new network is inclusive of airport-wide Distributed Antenna System (DAS), Wi-Fi system, and all other wireless technology in use by concession and operations that include customer facing services.

Airports are the ultimate case study of environments, where high-density, high-impact networks are practically compulsory, given the broad functional and operational requirements of the supporting customer-facing infrastructure, in view of the fact that there will be thousands of users within a confined area at any one time – 24 hours a day, 365 days a year – having multiple devices and time on their hands, and with service expectation that is immediately responsive and will be reliable, multifaceted, easy to use, secure, free (or for a very low fee), and providing low latency and high throughput. High-density networks at airports must ensure that users are able to have sufficient throughput regardless of the thousands of other users within close proximity. Airports operate like mini cities with multidimensional, multi-client communication service demands that are ever-changing.

In a legacy airport environment, there are a relatively high number of independent networks numbering into the thousands, servicing passengers, concessions, airlines, and maintenance functions. Each business, for example, will typically require several of these networks to support their computer systems; their retail needs such as cash registers and inventory processes; their vending and autopay kiosks; their employee, asset, and equipment tracking; their Internet access; and their external communications, such as cellular usage. Each such network typically has its own unique operating parameters driven by multiple telecommunications carriers and Internet Service Providers (ISPs) having multiple wireline and wireless infrastructures, including twisted pair voice and data lines, and fiber campus backbones, all with complex, often "unmanaged" cabling plants. Due to the multi-generational nature of

---

[1]The authors would like to thank Mr. Ed Wright for important contributions to the writing of this chapter. Mr. Wright was responsible for the design and rollout of the new BWI Thurgood Marshall Airport network.

High-Density and De-Densified Smart Campus Communications: Technologies, Integration, Implementation, and Applications, First Edition. Daniel Minoli and Jo-Anne Dressendofer.
© 2022 John Wiley & Sons, Inc. Published 2022 by John Wiley & Sons, Inc.

the technology deployments, network documentation is often missing, incomplete, or out-of-date, and the airport administrators use mostly, filed permits with dated information to the extent of what is actually installed at their site and where. Layered on top of these disparate infrastructures are groups of hardware-, software-, and security-vendors and service providers that often do not cooperate for the greater "good" of the airport stakeholders. It is tautological to state that this orbit of providers is not managed by or as a single entity; traditional "high-density" networks in an airport typically entail thousands of independent networks not working together within one environment – for example, a panoply of 1500 rogue networks and counting in BWI – and there is no sharing of infrastructure, or consistency of design, grade-of-service, quality-of-information, or security or operational processes. It is worth noting that airports can serve as proxy for a number of other public/private environments that have mixed-use high-density public access with private/enterprise communications, along with Real-Time Location Systems (RTLSs) and Internet of Things (IoT) requirements.

Traditionally, when developing high-density networks, there are several key issues that impact infrastructure design:

- Ability to support the breadth of services required by all groups of distinct stakeholders, particularly in the context of multimedia/multiservice, Quality of Service (QoS), throughput, latency, security, and usability.
- Ability to support the number of expected users within a certain area (density) at the peak hour and over the 24×365 operational horizon.
- Ability to support the physical topology (coverage area) of the airport, including terminals, baggage areas, garages, parking lots, and cargo and maintenance buildings.
- Ability to adapt grow/reduce and change with little to no disruption due to constant change due to users' types, technology requirements, physical airport improvements, and security and health requirements.

It is challenging to meet these requirements when the legacy infrastructure is comprised of a combination of (i) an inconsistent set of dated overlay networks, and (ii) constrained resources provided by traditional third-party telecommunication carriers. In particular, the concern has been – and has been proven in specific situations – that current airport network wireless infrastructure is not able to handle the traffic spikes, the new technology and customer demands, and the long-term growth in the customer base. New designs, approaches, technologies, and network administration are needed to meet evolving airport requirements.

This chapter addresses the following: (i) the current environment and problems with the existing wireless network at the BWI Thurgood Marshall Airport; (ii) the Request For Proposal (RFP) Requirements for a new network at the BWI Airport; and (iii) the newly deployed #MyBWI-FI communications infrastructure.

9.1   THURGOOD MARSHALL BWI AIRPORT DESIGN REQUIREMENTS

In 2019, the Maryland Department of Transportation (MDOT) and the Maryland Aviation Association (MAA) issued an RFP for BWI Thurgood Marshall Airport with an objective to: "establish, operate, and maintain a next-generation DAS and airport-wide Wi-Fi system." The Airport administration's goal was to enhance the customer experience for the traveling public utilizing BWI Airport. Airport wireless and telecommunications services are provided for the passengers, the concessions (stores), the employees, the vendors, and the contractors within the airport real estate. What was unusual and noteworthy about this RFP and why it was selected is that in most cases the wireless communications are driven by the carriers and cellular communication needs and in this case Wi-Fi and wireless services were put in the forefront. What follows is a summary of the main components of the public RFP and the award-winning team's observations and responses to the RFP.

9.1.1 Broad Motivation

The administration owns and operates the BWI Airport, which is comprised of approximately 3600 acres of property. The airport comprises a total of 2.423 million square feet; 5 concourses (4 domestic, 1 international); 73 jet gates for 36 airlines including commuter, charter, and cargo. The indoor space including the concourses and the ticketing areas is approximately 1 072 000 square feet; not to mention the rest of the outdoor areas including the parking garage and lots, and rental car facility. The administration's primary responsibility is the operation and management of the airport, a multimillion dollar enterprise serving roughly 75 000 passengers a day. In 2018–2019, total passenger traffic at BWI Airport exceeded 27 million. BWI Airport experienced over 2.3 million passengers in December 2019. Southwest Airlines remains the largest carrier; Spirit Airlines, American Airlines, Delta Air Lines, Alaska Airlines, British Airways, and Condor Airlines had passenger increases in 2019. In 2019–2020, BWI Airport was the 22nd busiest airport in the United States and maintains its position as the busiest in the Washington–Baltimore Region. BWI Airport usage has been growing at record rates year after year. The Airport averages 320 daily nonstop departures to 90 destinations. BWI Airport's economic impacts include $9.3 billion in total economic, more than 100 000 jobs, over $4 billion in total earnings, and approximately $600 million in total state and local revenue. While some of the traffic volumes were impacted during 2020 (passenger traffic at Baltimore/Washington International Thurgood Marshall Airport fell 65% for the year), the distribution and acceptance of the vaccines at press time were driving the economy toward business normalization.

It was determined by the airport administration that the increase in the airport's passenger traffic, passengers' use of mobile devices, and the airport's growing evolution in form and function as a leading regional gateway required a premium reliable wireless network to support a large group of stakeholders, not the least of which are the flying public. Airport tenants and employees also need to stay seamlessly connected. As a secondary factor, the passenger and cargo service are relied upon by many residents and businesses in the region, as a regional economic engine and a catalyst for local growth. From passenger airlines to taxi services, to retail operators in the terminal and security services, as well as a host of other businesses, these activities require information networks that connect to personnel who work at the airport and contribute to the regional economy. Figure 9.1 is an aerial view showing the terminal and concourses of the airport.

A lot of new construction has taken place in recent years, including a new security checkpoint to serve domestic and international travelers, a new secure connector between Concourse D and Concourse E, and configured airline gates to support additional international flights. Currently, the administration is planning for a Concourse C and D secure connector, as well as other projects such as automated parking, new hotel, gas station, smart bathrooms and robotic delivery and customer service and automated wayfinding applications are underway or planned and construction is expected to meet airline expansion and passenger growth. Several major capacity-enhancing projects are underway or planned soon after the Covid-19 normalization. Foremost is the Concourse A and B Connector redesign to expand the upper-level outward another 60 feet. and construction of a new baggage handling area underneath it on the lower level.

The growing relevance of wireless communications (supported by high-capacity backbone core networks and wired networks) along with the recent and expected longer-term growth at BWI Airport has driven the growing need for advancement of the in situ legacy network architecture and services to enable a large number of passengers a higher grade of service. Each new airline and service provider built their own infrastructure, and their wireless networks were developed

FIGURE 9.1   Aerial view showing the terminal and concourses of the airport.

independently of each other. Clearly, a fully integrated solution is highly desirable especially from a use of spectrum capacity, an administration, service, security, and consistency perspective. It is important, however, that these networks offer the same private Virtual Local Area Network (VLAN) and benefits enjoyed that are not normally found in a shared network environment.

9.1.2   Status Quo Challenges

There were critical problems with the existing BWI environment. The new infrastructure is aimed at solving these issues.

First and foremost, BWI did not control its wireless networks. The airport's concession developer entered into separate and shared revenue sublease agreements for both Wi-Fi and DAS services, in 2009 and 2013, respectively. The Wi-Fi service evolved from coverage for a single food court and later focused on the public, with the expansion of service in other common areas of the airport. The concession developer further subcontracted a company to manage both DAS and Wi-Fi service at BWI Airport. Neither BWI nor the concession developer received direct revenue or managed the very networks that connected their concession or travel customers to the airport. The only networks that were managed were the DAS and the Wi-Fi networks and these networks were managed from the subcontractor's perspective, not the airport's. All other concession, operation, or business networks were independently requested, procured, paid for, and installed, focusing on the immediate need and not the future. This scenario is typical for most airports.

The existing DAS was based on an end-of-life modular multiservice access router (Cisco 2600), providing flexible Local Area Network (LAN) and Wide Area Network (WAN) configurations, multiple security options, voice, and data integration, but only covered selected areas within the airport. All four major carriers occupied the DAS and paid monthly rent to the subcontracted company. This system moved revenue away from the airport and development activities were not transparent but more importantly no plans for upgrading or increasing revenue generation were required. This process enabled outdated DAS and Wi-Fi systems to prevail to an ever-increasingly disgruntled user community and to reduce traffic from repair situations due to failing, aged equipment. The new goal was to provide a next-generation neutral host system capable of supporting all wireless carriers and their respective frequencies and technologies as well as the operational 800 MHz trunked radio systems and advanced technologies like

5G. The modernization had the goal to increase revenue, so that state-of-the art systems can be maintained and the airport might partake in both increased customer satisfaction and rental fees. In addition, apron and ramp areas also had to be covered that extended to the fully deployed ends of each of the airport's jet bridges. Gaps in DAS coverage included passenger curbside pickup and drop-off areas, the hourly parking garage, and the airport's consolidated rental car facility.

The existing Wi-Fi also had major issues. The public Wi-Fi service was developed in 2012 with an airport concession (food, beverage, and retail) location focus and later expanded, but never fully developed to provide Wi-Fi access to all public areas and tenant areas within the airport. The Wi-Fi service Access Points (APs) at BWI Marshall supported 802.11ac Wave 1 protocol features operating in the 2.4 and 5 GHz frequency bands. The bandwidth provided to the user was 2 Mbps. There were approximately 190 APs (Cisco Aironet 3702i series) within the airport terminal building. BWI Airport's ASQ (airport service quality) rating had consistently been held back by poor Wi-Fi service. The service provided very slow speed, very limited coverage area, and built-in drops in service every 20 minutes unless the user paid a fee. Given that Wi-Fi is one of the critical services that customers depend on, this unacceptable service level was noted as a major customer complaint for BWI users. BWI had to do something dramatic to turn the situation around. Their RFP declared that all Wi-Fi services would be high-speed and would be free. The demand for Wi-Fi at the airport led the management to conclude that the entire system would need to be removed and replaced. The new proposed Wi-Fi system would need to offer convenient and user-friendly, free Wi-Fi for the traveling public, visitors and employees including a high-density, high-speed Wi-Fi experience in the public spaces of the airport, including all levels and all concourses of the terminal building, the baggage claim areas, car parking, and car rental facilities. It was determined that the Wi-Fi service would initially comply with IEEE 802.11ac (Second Wave, Wave 2) and support 80 MHz channels minimally.

In addition to improving both the DAS and Wi-Fi services at BWI Airport, the administration wished to enhance passenger in-person experiences by connecting with customers through their mobile devices. The administration was looking to facilitate a seamless mobile journey from home to gate for its international and domestic customers. One goal was to establish an airport beacon system throughout the airport terminal building to better comprehend the customer experience, deploy intelligent management solutions, and to inform passengers of concession sales and other purchasing incentives. The administration envisioned establishing an airport beacon system for airlines and concession tenants of the airport for a fee to provide a platform for such tenants to receive incentives offers and delivery options to passengers as they travel through the airport. The airport also planned to create a mobile app for customers to engage with service providers at the airport and the surrounding region, which may be used in conjunction with the beacon technology. The necessary protocols on the airport's app to activate the beacon systems and function had to be configured. The beacon system should also provide analytics (real-time statistics and location, geo-fencing, traffic patterning, etc.), including RTLS and Location Based Services (LBSs). Other services were envisioned to use the Wi-Fi and beacon systems to implement a connected Digital Directory and Smart Restrooms touchless technologies. Asset tracking was also a part of the requirement.

9.1.3   RFP Requirements

The airport had a need but not a budget so the entire contract would have to be funded by the winner through the fees made from selling DAS and Wi-Fi-related services. The administration wished to deploy next generation of wireless services throughout the airport, with the primary goal to enhance the passenger's experience, while improving the operation of its internal customers (airlines, concessions, etc.). The MAA asked respondents to propose the best business model, for achieving innovative Wi-Fi, DAS, and other wireless systems in and around the airport. Specifically, MDOT/MAA articulated the following objectives for the wireless technology service at BWI Marshall Airport:

- To identify and implement industry best practices to maximize project success.
- To provide a high-speed free Wi-Fi experience throughout the airport at no cost to both the airport and traveling public.
- To include Wi-Fi, DAS, and other wireless technology elements that are "future-ready" to the greatest extent possible; considering technological advances and system upgrades needed to maintain reliable service.

The administration noted that current and future technology trends necessitated ubiquitous wireless connectivity to support the mobile economy, which now ranges from tablets to smartphones, from wearables to IoT sensors, from social media platform to Over the Top (OTT) streaming video, from Cloud Computing to Big Data, and from Artificial Intelligence (AI) to Deep Learning (DL) systems supporting smart buildings and smart campuses, to list just a few. MDOT/MAA identified these trends as a significant opportunity to innovate and integrate the Wi-Fi, DAS, and other wireless systems in and around the BWI Marshall Airport to provide a complete and comprehensive solution with abilities to finance, install, establish, and provide all required services and materials associated with its operations – all with the ultimate goal of delivering an optimal user experience and creating additional revenue for the airport while at the same time improving the operation of its internal processes. MDOT/MAA wanted a solution to be designed, installed, owned, operated, marketed, maintained, upgraded, and managed by a single service provider. Some of the points that came forward by the team who implemented the new design when identifying the problem and defining the problem statement were as follows:

1. The problem
   - Dense areas have many concessions and many rouge networks
   - Multiple types of networks identified that included:
     ◦ Wi-Fi
     ◦ BLE (Bluetooth® Low Energy)
     ◦ DAS
     ◦ Private Long Term Evolution (LTE)
     ◦ IoT
   - No proper monitoring processes to know what is going on in the airport
2. The required solution
   Remove or takeover all the networks and rebuild networking environments without disrupting service in the interim. The solution would have to be:
   - multi-network
   - multi-industry
   - multi-applications
3. The Multi-focused users
   - Consumers
   - Employees
   - Vendors
   - External companies seeking access

4. The expected customer experience

The customer experience at the BWI Marshall Airport was expected to be more interactive along with better control over the activities by the airport management. Some of the suggested solutions to enhance customer experience were customer portals, digital kiosks, applications, and so on. It was imperative that all solutions maintained strict privacy of the user, especially the collection of Personally Identifiable Information (PII).

5. Strategies to generating income

An internal and external strategy also was to be defined to generate income through the enhanced solution. As more network services are consolidated onto the shared platform, additional revenue can be extracted rather than going to other ISP, Cable, and Telco companies.

6. Future plan – smart city

The bigger picture for the wireless solution was to expand the network services to include Private LTE, IoT, and beacons in order to offer enhanced services to airlines, suppliers, contractors, and concession operators, eventually linking to the greater Baltimore area, restaurants, hotels, malls. Convention Center public venues, and transportation systems to implement the concept of a true smart city that BWI would be in the center.

As a result, some of the conceptual requirements for the project are listed below (technical requirements are identified further below):

Integration:

- Independently accommodate multiple wireless service providers and technologies
- Provide and install all additional infrastructure elements required, such as Cat6A cables, Singlemode (SM) fiber, backbone pathway, fiber panels, racks, RJ45 panels, Uninterruptible Power Supply (UPS), and Power Distribution Units (PDUs) as required.
- Provide systems and network equipment with data outlets, connectors, adapters, and terminating equipment necessary to interconnect system equipment.
- Provide cutover plans, post-operational and system test plans, subject to administration approval, and perform testing as specified.
- Adhere to current industry best policies for network configuration and comply with all administration network security policies.
- Continually ensure, provide, and maintain current industry standards for cybersecurity for wireless services.
- 24/7/365 on-call/on-premise technical support with quick restoration including repair and/or replacement.

Implementation:

- Install and configure software to ensure fully operational systems.
- Provide system reliability, redundancy/resiliency, and adequate capacity to accommodate sustained Very High Throughput (VHT) operations to meet the future data demands and technologies of wireless services and customers.
- Monitor performance, analyze data, and produce timely reports in a form that allows administration to review monthly and annual performance results to determine if minimum service levels are being met.
- Provide adequate capacity and expansion capability to meet the future data demands and technologies of wireless services and customers.
- Provide system reliability and redundancy/resiliency.

Innovation:

- Design, engineer, install (build), own, operate, market, maintain, upgrade, and manage all requested wireless services.
- Accommodate independently as many wireless service providers and technologies as possible.
- Implement a free Wi-Fi system that provides one-click access to high-speed Internet without requiring any personal identity information such as email addresses or credit card numbers.
- Implement strategic advertising features such as splash pages, video, targeted advertising, and so on.
- Provide a fair market return to the administration.

9.2   OVERVIEW OF THE FINAL DESIGN

The key elements of the system deployed at BWI consists of the following:

- Carrier-neutral cellular service coverage was expanded across the entire airport footprint, from the existing public areas to the rental car facility, firehouse, and parking garages. Expanding the cellular and Wi-Fi coverage to the furthest corners of the airport campus enabled future smart city extensions with interconnected locations such as the train station, mall, convention center, and bus terminals. A physical upgrade to a 5G-ready infrastructure is also planned.
- Free, uninterrupted, and high-speed Wi-Fi for passengers to enjoy capable of exceeding today's demand for video chat, streaming, and downloading, including access to entertainment services such as short films, premium movie channels, and local and regional content.
- Indoor wayfinding and location-based messaging enabled via a network of Virtual Bluetooth Beacons that assist passengers in navigating the airport, locating the nearest restrooms, and informing them of retail and food promotions as they pass by, including internal and external concierge delivery. Brands, businesses, and application developers interested in integrating with BWI-FI's network to reach BWI's target market are supported.
- The Wi-Fi solution can be remotely updated, configured, and maintained with minimum or no service disruption. The Wi-Fi system solution interfaces with tenant existing systems (i.e. Ethernet networks, applications, and databases) and supports thousands of concurrent users with automatic RF Transmit Power Control (TPC), Configurable Acceptable Client Received Signal Strength Indicator (RSSI), Dynamic Frequency Selection (DFS), optimized AP roaming (load balancing), client mobility with seamless hand-off, Wireless QoS, and wireless security extensions including WPA2, AES, 802.11x EAP.
- The security solutions including wireless intrusion protection and malware detection systems must meet industry standards for cybersecurity for public Wi-Fi systems and services, including (at minimum) Wi-Fi Security Guidelines NISTSP800-153 and National Institute of Standards and Technology (NIST) framework for improving Critical Infrastructure Cybersecurity.
- The design provides an open access model that supports IP roaming across the entire airport terminal building, jet bridges, and ramp areas and must be have proper placement with RF signals.
- The solution does not cause any interference with existing communications or data systems. The Wi-Fi landing page automatically connects to the user without additional clicks. While the Wi-Fi service at BWI Marshall is free to passengers, airport tenants (airlines, concessionaires) have specialty Wi-Fi and advertising services for a fee (e.g. push promotional content to users – advertisements, coupons, etc).

- The Wi-Fi portal was also allowed to have advertising.
- Asset tracking of physical resources via Bluetooth sensor tags helps to allocate airport resources, including the 10 000 employees who keep BWI running on a daily basis, to maintain cleaner restrooms and faster location of luggage and passenger carts and wheelchairs, reducing wait times and increasing passenger comfort.
- Future-ready technology that will also support AI-assisted autonomous vehicles and robotic utilization as those services become available.

From a service perspective, the roadmap described in Table 9.1 needed to be implemented; namely, supportive subsystems had to be designed, procured, installed, and activated (in parallel with the existing infrastructure and services, not to have any service gaps), with an eye to having

TABLE 9.1   MyBWI-FI technology plan

Technology segment (maximally integrated by networking, operations, administration)	Services	Utilizations
DAS	For all major US wireless service providers	4G LTE, migratable to 5G
Broadband	Internet, VoWi-Fi/VoIP, entertainment OTT	High-density high throughput Wi-Fi 5 (802.11ac Wave 2) and seamlessly migratable to Wi-Fi 6 (802.11ax)
Private Wi-Fi	Kiosks, mobile apps, LTE roaming	Enables Smart Restrooms operations including custodian location tracking, and custodian dispatching, consumables resupply, and restroom cleaning when usage thresholds are met Supplements connectivity gaps in DAS coverage Provides lower-cost connectivity for kiosks and information/advertising displays
BLE	Wayfinding, proximity, indoor tracking	Enables concession proximity marketing Enables operator asset tracking such as luggage carts and wheelchairs
CBRS/LTE (private LTE via CBRS)	Kiosks, mobile apps, outdoor tracking; Future cellular (in roaming but not out roaming)	Enables Smart Restroom connectivity, which includes people counting, stall occupancy detection Enables airport operations including baggage scanning, airplane flight preparation and maintenance, airport vehicle location tracking and maintenance Supplements connectivity gaps in Wi-Fi coverage
IoT/LoRaWAN	Sensors, lighting, building management (all airport buildings) management, campus management, environment management	Indoor technology complemented by LoRa WAN gateways Enables low-bandwidth wide-area sensor data collection Includes Smart Restrooms, consumables monitoring, plumbing leak detection, temperature, humidity, and air quality monitoring
Mobile APP management	Tablet, phones, laptops, watches, wearables	Supports design, development, and deployment wireless integrations for beacon integration, map/directory synchronization, proximity messaging, blue dot wayfinding and asset tracking

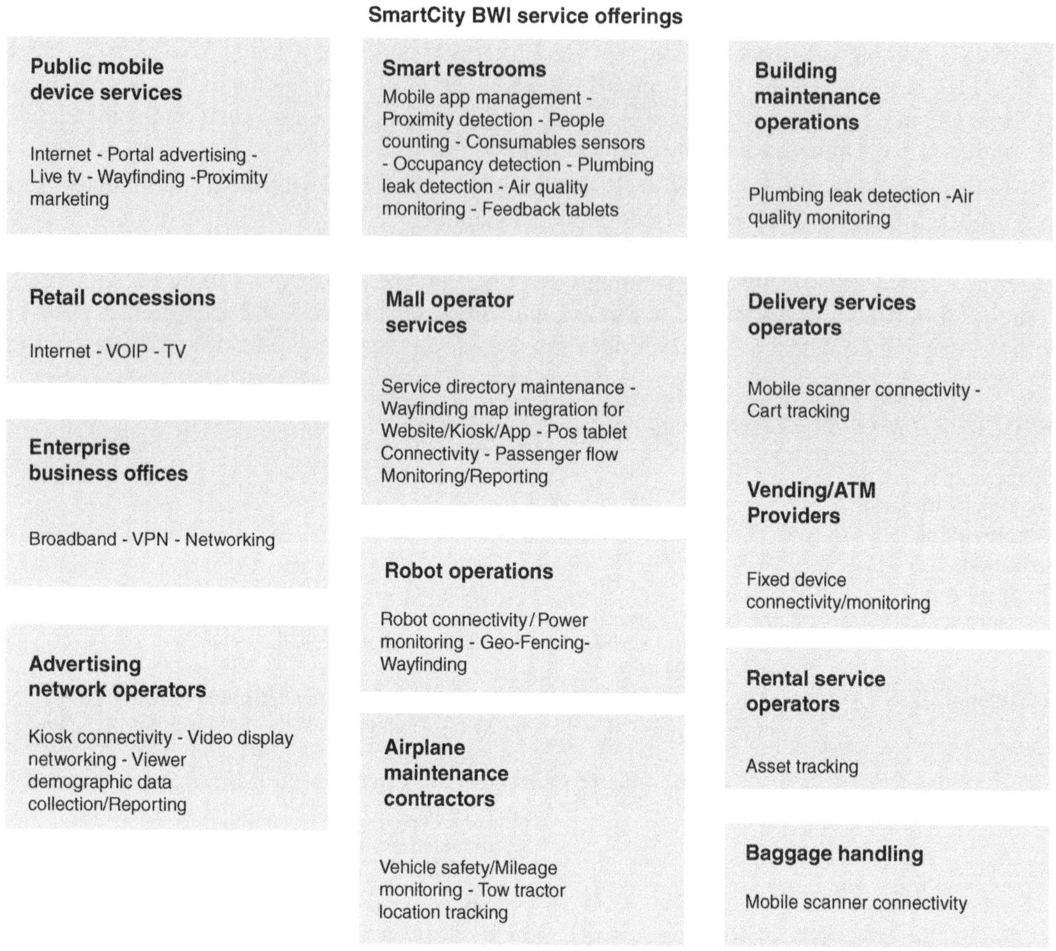

SmartCity BWI service offerings

Public mobile device services

Internet - Portal advertising - Live tv - Wayfinding -Proximity marketing

Smart restrooms

Mobile app management - Proximity detection - People counting - Consumables sensors - Occupancy detection - Plumbing leak detection - Air quality monitoring - Feedback tablets

Building maintenance operations

Plumbing leak detection -Air quality monitoring

Retail concessions

Internet - VOIP - TV

Mall operator services

Service directory maintenance - Wayfinding map integration for Website/Kiosk/App - Pos tablet Connectivity - Passenger flow Monitoring/Reporting

Delivery services operators

Mobile scanner connectivity - Cart tracking

Enterprise business offices

Broadband - VPN - Networking

Robot operations

Robot connectivity / Power monitoring - Geo-Fencing- Wayfinding

Vending/ATM Providers

Fixed device connectivity/monitoring

Advertising network operators

Kiosk connectivity - Video display networking - Viewer demographic data collection/Reporting

Airplane maintenance contractors

Vehicle safety/Mileage monitoring - Tow tractor location tracking

Rental service operators

Asset tracking

Baggage handling

Mobile scanner connectivity

FIGURE 9.2 BWI services obtainable over the new network.

a maximally integrated panoply of technologies by way of common networking, operations, and administration.

Figures 9.2 and 9.3 depict the services that are provided by the new network infrastructure as well as a high-level view of that infrastructure.

9.2.1 DAS Solutions

Carrier-neutral cellular service coverage will be expanded across the entire airport footprint, from the existing public areas to the rental car facility, firehouse, and parking garages. Expanding the cellular and Wi-Fi coverage to the furthest corners of the airport campus will enable future smart city extensions with interconnected locations like the train station, mall, convention center, and bus terminals. A physical upgrade to a 5G-ready infrastructure is also planned. The expectation is set from the ground up physically building and expanding the cellular DAS.

The neutral host DAS will provide in-building coverage enhancement for cellular services inside the airport, hourly parking garage, and the rental car facility. The DAS solution needs to support frequencies used by national Wireless Service Providers (WSPs) such as AT&T Mobility, T-Mobile, and Verizon Wireless. The solution supports all active technologies in the 700 MHz, 850 MHz, 1900 MHz (PCS), and 1700 MHz/2100 MHz (AWS) frequency bands. The DAS is

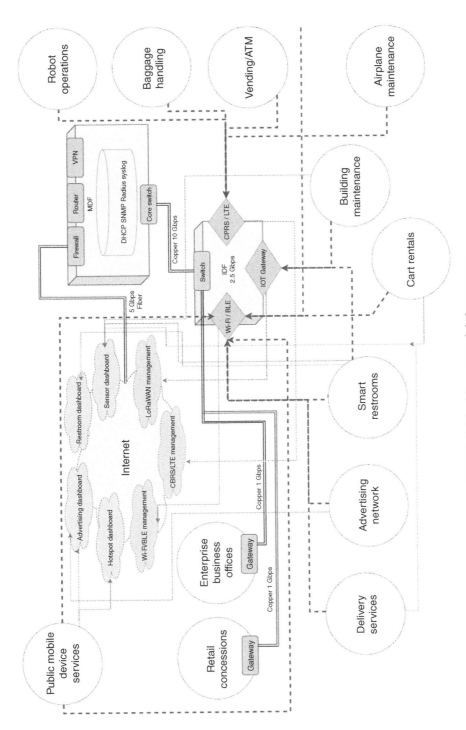

FIGURE 9.3 New network infrastructure.

TABLE 9.2 DAS Requirements for the BWI Airport

Targeted coverage areas		
Building	Area of coverage	
BWI Thurgood Marshall Airport	Baltimore, MD	
Indoor neutral host DAS coverage areas	• Concourse A, AB, C, D, DX, DY, E and the ticketing terminal • 95% of all areas in scope • Approximately 1 072 000 square feet	
Outdoor neutral host DAS coverage areas	• Hourly parking garage and rental car facility • 95% of all building areas • Approximately 2 000 000 square feet	
Total coverage area	Approximately 3.072 000 square feet	
Design specification overview		
Services	Uplink, MHz	Downlink, MHz
700 commercial band	698–716 and 776–787	716–728 and 728–757
Cellular band	824–849	869–894
PCS band	1850–1915	1930–1995
AWS band	1710–1755	2110–2155
High-density coverage criteria	Dominance or minimal – 85 dBm: LTE/VoLTE/HD Audio/ SISO/EVDO/UMTS/HSPA in 95% of the building	
Normal density coverage criteria	Dominance or minimal – 85 dBm: ITE/VoLTE/EVDO/ UMTS/HSPA in 95% of tire building	
Wireless service providers	AT&T Mobility, Sprint, T-Mobile, and Verizon Wireless	

being installed at BWI Airport, with the DAS head-end being installed in the main distribution frame (MDF) Telecommunications Rooms (TRs) (MDF TR). The DAS head-end provides distribution of the carrier signal(s) that are connected over single-mode plenum-rated fiber between the head-end location and the distributed TRs. The neutral host DAS provides in-building coverage enhancement for the commercial wireless services. The solution must meet the stringent requirements of the WSPs to ensure that the installed system will be approved to allow the legal rebroadcasting of their licensed spectrum. The DAS system supports multiple services (voice/data/video) in a modular architecture, so that services can be added or removed without disturbing existing services. The architecture enables a "pay as you grow" strategy for incrementally adding service-specific modules when new services are required. This solution encompasses the design, installation, operation, and maintenance for the DAS.

The DAS solution consists of a DAS head-end that is flexible, modular, and scalable that can be expanded if needed in the future. The DAS is designed to support 3 WSPs and 11 bands on Day One, with coverage requirements of – 85 dBm for CDMA over 95% and – 95 dBm for LTE over 95%; RSCP with 8 dB dominance over the macro for 95% of the coverage area. Table 9.2 shows the DAS requirements for the BWI Airport.

The scope of work was as follows:

PHASE 1: Requirements Gathering
    WSP Engagement and Coordination
    Technical Requirements
    DAS Signal Source
    Site Survey (RF Survey)

PHASE 2: Detailed Solution Design (Final Concept Design)
PHASE 3: Installation and System Commissioning
PHASE 4: System Testing and Verification (System Operability)
    System Verification and Acceptance Test Plan
    System Inspection
    Cable Test Results
    Signal Coverage Testing
PHASE 5: Close-out & As-Built Documentation (Documentation and Training).

In consideration of the fact that the Wireless LAN (WLAN) system is being upgraded and expanded a DAS with similar architecture to a traditional WLAN network was sought with single-mode fiber providing connectivity from the head-end to the closets and CAT6A cabling in the horizontal ceiling connecting the equipment in the IDFs to the Universal Access Points (UAPs) in the ceilings. The DAS head-end equipment is co-located with the WSP RF signal source equipment at the head-end location within the airport.

In conjunction with the equipment plan, there was a need to contact the WSPs and coordinate participation and confirm their requirements. At the appropriate time during the installation, coordinate with the airport administration to schedule the onsite commissioning and testing with each of the WSPs. After equipment selection, a detailed RF (radio frequency) site survey is needed to assess the facility and verify assumptions made while creating our initial design. Information to be collected during this survey includes, but is not limited to, the following:

- Building construction related to the system installation
  ◦ Cable, pathways, and antenna locations
  ◦ Space, electrical, and environmental controls for active equipment locations
- Building construction related to RF propagation characteristics
  ◦ Dense partitioning vs. open floor spaces
  ◦ Wall construction such as sheetrock vs. cinderblock
- Ambient RF environment
  ◦ Macro signal strength bleeding into the building
  ◦ Serving microsites
  ◦ Spectrum snapshots

Based upon the site survey and requirements analysis, a final, detailed system design was generated, clearly being compatible with the technical specifications as required by the WSPs. The detailed design included:

- Design drawings to include floor plans with equipment locations identified.
- Equipment specifications including power, space, electrical, and HVAC requirements.
- WSP channel loading requirements.
- Cable pathways and antenna locations.
- Predictive heat maps highlighting design and coverage criteria are being met.
- Interconnection diagrams.
- Bill of materials highlighting manufacturer, model number, quantity, and warranty.
- Link budget analysis.
- WSP RF signal source information.

For the physical installation, a DAS subcontractor was engaged to install the antenna equipment and the cabling infrastructure. This includes but is not limited to minimum bend radius, tensile loading, and vertical and horizontal support of the cable throughout the entire pathway. Installation standards included but not were limited to:

- Institute of Electrical and Electronics Engineers (IEEE)
  - IEEE C2 – National Electrical Safety Code
  - IEEE STD 100 – The Authoritative Dictionary of IEEE Standards Terms
- National Fire Protection Association (NFPA)
  - NFPA72 – National Fire Alarm and Signaling code
  - NFPA 70 – National Electrical Code
  - NFPA 70E – Electrical Safety in the Workplace
- ANSI/TIA/EIA Standards
  - TIA-EIA-568-B Commercial Building Wiring Standard
  - TIA-EIA-569-A, Commercial Building Standards for Telecommunications Spaces and Pathways
  - TIA-EIA-606-A, Administration Standard for Commercial Telecommunications Infrastructure
  - J-STD-607-A, Commercial Bldg. Grounding and Bonding Requirements for Telecommunications

Signal coverage testing is required after installation. Walks throughout the facility were undertaken to measure and record the resultant signal strength coverage, signal quality, and other pertinent WSP performance parameters. Results providing "heat map" coverage showing signal strength readings were then provided as part of the project close-out package. The following diagram shows the type of test and results that were measured and reported as part of the acceptance test criteria. The heat map (e.g. see Figure 9.4) provides a summary of received signal strength readings throughout the surveyed area in the various frequency bands. Testing in this manner records continuous signal strength readings throughout the facility, providing a comprehensive record of the coverage performance of our neutral host DAS solution for all frequency bands. In addition to the walk test measuring RF parameters, voice and data testing were done to verify that both the uplink and downlink transmissions are working correctly, ensuring the proper operation of the overall system.

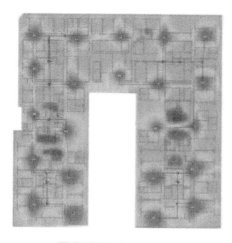

FIGURE 9.4 Exemplary plot for LTE.

9.2.2   Broadband, BLE, IoT

This portion of the work entailed installing a high-density high throughput Wi-Fi 5 system (802.11ac Wave 2), and seamlessly migratable to Wi-Fi 6 (802.11ax) when the standard is completely finalized. The existing Wi-Fi network had physical service coverage gaps that needed to be filled, as well as major grade-of-service gaps in terms of throughput, latency, and availability. The process to design, upgrade, and install the new Wi-Fi infrastructure included the methodology that follows (in addition to other project management steps).

- Calculated the expected number of clients that will be served: This number was used throughout the planning process. One approach is to use the maximum passenger capacity for a given zone and assume a certain number (often 0.5–2) of devices per person.
- Determined the number of APs needed: Although the APs do not have a hard client limit (they are limited by bandwidth, not number of clients), as a practical matter, 50 client sessions is a safe limit and is convenient for planning. To avoid too many active clients, enable power reduction, band steering, and ensure there are enough APs installed in the environment to support the required load.
- Calculated the backhaul required: Network performance is often plagued by limited backhaul. To calculate the backhaul requirement, multiply the bandwidth limit by the expected number of clients. While it is unlikely that all devices will use up to their full bandwidth limit, this conservative calculation will minimize the odds that the backhaul is insufficient.
- Maximized the number of APs that are connected to the wired network: This allows an AP to use the full bandwidth of its wired connection, rather than having to go through a neighboring AP via a mesh link. If possible, do not use any mesh APs.
- Used multi-radio APs: With multi-radio APs, the wireless network can make optimal decisions about channel assignment, band steering, and mesh networking. This maximizes throughput and minimizes channel interference for clients. As planners, one has no control of the client technology used: some passengers may have newer clients; other passengers may have older clients. Higher throughput 802.11n clients can operate on the 5 GHz band without being slowed down by older 802.11b/g clients, which remain on the 2.4 GHz band. Moreover, if mesh links are necessary, they can be provided on multiple radios, significantly improving the performance of the network across multiple mesh hops.
- Mapped the APs: Name the APs and place them on the map appropriately. The network decides the best mesh route based partially on the locations of APs on the map.
- Ensured signal strength: The signal strength between a client and an AP should be at least 20 dB for optimal stability and performance (anything less than 10 dB is unusable.
- Budgeted for spare hardware: Spare hardware should be readily available in case of failures. For example, have an extra switch, APs, cables, and associated power supplies.

Some AP configuration settings that have been taken into consideration include the following:

- Enabled bandwidth limits: When bandwidth limits are not enabled, a small number of clients can quickly saturate a channel. Higher limits (25–100 Mbps) will enable higher-bandwidth applications such as OTT; however, this requires that there be enough local and wide-area bandwidth available to support all users at this limit. Application traffic shaping can also be used; this allows the administrators to block applications that might be considered abusive, such as P2P file-sharing applications.
- Used Distributed Dynamic Host Configuration Protocol (DHCP) Scopes: Centralized DHCP servers often fail or become slow when hundreds or thousands of clients request an IP address in a short time. Using the appropriate NAT strategy, one can spread the DHCP load among all the switches.

- Enabled auto channel assignment (when appropriate): Auto channel assignment allows the Cloud Controller to assign channels to APs in the network using RF information that the APs constantly send up to the Cloud Controller. Unlike traditional wireless solutions, in which channel assignment decisions are made by each AP in a localized manner, the Cloud Controller ensures that channel assignments make sense locally as well as globally, relative to the rest of the network.

- Enabled channel spreading: Channel spreading enables APs in the same vicinity (e.g. in the same zone) to broadcast on different channels, so that channel utilization on each channel is minimized. This maximizes throughput and minimizes interference in the network.

- Enable band steering: Band steering forces 5 GHz-capable wireless devices (e.g. most 802.11n clients) to migrate away from the 2.4 GHz band. This opens up radio spectrum for legacy wireless devices (e.g. 802.11b/g clients). This is highly beneficial since there are many more 5 GHz channels than 2.4 GHz channels.

- Reduced transmit power: The reduction of transmit power enables administrators to create "microcells," such that a user associates only with the nearest AP in a room containing multiple APs. This guarantees an even distribution of users across the APs deployed in a physical space. This configuration allows for a greater number of individual channels, which might be desired in a high-density setup. This change is only applied to 5 GHz radios since 2.4 GHz radios already use 20 MHz channels by default.

Regarding proximity messaging and wayfinding applications, the use of virtual BLE (vBLE) technology enables one to deploy and move virtual beacons with the simple click of a mouse (or via APIs), eliminating the need for physical beacons.

Juniper Mist was the first vendor to bring Enterprise-grade Wi-Fi, BLE, and IoT together to deliver personalized, location-based wireless services without requiring battery-powered beacons. Mist Wi-Fi APs include a BLE antenna array in each WAP and they offer a Bluetooth-only AP (BLE AP) that can be daisy-chained in pairs from their WAP to improve the location and accuracy of the beacon system without requiring additional power or full-length cable run. This technology was briefly described in Chapter 8.

- Virtual beacons should be located 20–25 feet in open areas or hallways.
- Virtual beacons should also be located in front of each retail establishment, restroom, gate area, ticket counter, entrance, exit, and passageway.
- BLE APs should be a minimum of 50 feet apart in open areas.
- BLE APs need only be at the ends of long hallways or passageways to provide accurate wayfinding.
- A beacon network should be divided and managed by zone.
- A beacon network can provide wayfinding and push notifications if embedded in a mobile app.

Given the design principles just enumerated, a set of appropriate network elements were selected for the rollout, as follows:

- Network Infrastructure:
  - Core Ethernet Switches
  - Access Layer Ethernet Switches
  - WLAN Controller Solution
    a. Mist Intelligent Cloud
    b. Samsung Axis SDN Edge Gateway

- ◦ APs
    - a. Indoor APs
    - b. Outdoor APs
- • Cable Plant
    - ◦ Optical Fiber Backbone (Existing)
    - ◦ Equipment Racks/Cabinets
    - ◦ Category 6A Horizontal Cable (New)

Mist Wi-Fi and Bluetooth Access Points   In addition to delivering Wi-Fi 802.11ac Wave 2 range and performance, Mist APs have a dynamic vBLE 16 antenna array for the industry's most accurate location services. Additionally, all Mist APs have extra radios for collecting data and enforcing policies in conjunction with the Mist Cloud, which is critical when doing analytics, machine learning, location services, and event correlation. This solution allows a single, enterprise-grade platform for Wi-Fi, BLE, and IoT (Mist APs also have an expansion port for IoT functionality). This selection makes Mist APs a single, enterprise-class platform for all wireless connectivity options.

The new switching gear is Juniper Networks® EX4300 line of Ethernet switches, which supports Layer 2 and Layer 3 switching. Both 1 GbE access and multigigabit switch options are available. The EX4300 enables a variety of deployments, including campus, branch, and data center access (e.g. see Figure 9.5). The EX4300 switches can be interconnected over multiple 40 GbE Quad Small Form-factor Pluggable Plus (QSFP+) transceiver ports to form a 320 gigabit per second (Gbps) backplane. A flexible uplink module that supports both 1 and 10 GbE options is also available, enabling high-speed connectivity to aggregation- or core-layer switches that connect multiple floors or buildings. All EX4300 switches include High Availability (HA) features such as redundant, hot-swappable internal power supplies and field-replaceable fans to ensure maximum uptime. In addition, Power over Ethernet (PoE)-enabled EX4300 switch models offer standards-based 802.3at PoE+ for delivering up to 30 W on all ports to support high-density IP telephony and APs. Additionally, a multigigabit model, the EX4300-48MP, supports IEEE 802.3bz-compliant 100 Mbps, 1 Gbps, 2.5 Gbps, 5 Gbps, and 10 Gbps speeds on access ports. This enables 802.11ac Wave 2 access points, which require higher bandwidth,

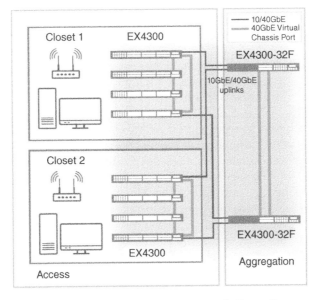

FIGURE 9.5   Access/aggregation design using Cluster Network Controller over proprietary software.

to connect to the switch. The EX4300 multigigabit switch also supports up to 95 W of power on any of the access ports, enabling PoE++ devices requiring more than 30 W to connect to and draw power from the switch. The EX4300 multigigabit switch enables higher levels of Medium Access Control Security (MACsec) AES256 encryption on all access and uplink ports, protecting customer traffic from unauthorized access.

Of the million square feet of terminal/concourse space, the RFP only required to provide coverage for the public areas – the tenant areas will be built out on an as-needed basis; however, these areas comprise a relatively small subset of the total. In heuristic terms, one Wi-Fi AP covers an area that is 50×50 feet (obviously the exact coverage depends a lot on the physicality of the environment, and so the cited heuristic is just that, a heuristic, a method of making use of readily-available, general information about a system to draw analytical conclusions, in lieu of using more complex algorithms, computation, or richer data). That heuristic would imply that 400 APs are needed. In fact, detailed analysis demonstrated that providing complete coverage for all the public areas for both Wi-Fi and BLE with enough density to support "Blue Dot wayfinding" and asset tracking required 380 Wi-Fi/BLE APs (BT43) and 760 BLE-only APs (BT11). In addition, the design required the installation of three (3) Samsung Axis Servers – 3-Node Clusters with one (1) Hot Spare for load balancing and failover and with 4000 SUL (Simultaneous User Licenses) (the Axis Gateway servers are discussed below). As we saw in Chapter 6, wayfinding is the process by which people navigate through a space, to find their trajectory from one location to another; in the context of a large complexes or smart buildings, this means navigating inside or between buildings to particular rooms. For example, facility managers may opt to identify emergency exits for accessibility compliance and building safety codes. In simplest terms, wayfinding can be achieved with landmark-based text directions, stationary kiosks, or directional cues built into a space, or communicated via signage or better yet digital signage; in some implementations, there could be QR codes at stationary kiosks or other fixed structures to enable users to capture their turn-by-turn wayfinding cues to their smartphone. This basic solution does not entail RTLSs. Blue Dot, as industry term, describes mechanisms to fully enable wayfinding: it describes an automated wayfinding method, that is, an Indoor Positioning System (IPS) technology. Blue Dot is a navigation tool, typically implemented in a smartphone app, mimicking an outdoor GPS capability that positions a person (or object) relative to a map, with a little "blue dot" following the person around and providing real-time location and navigation cues for use, as one moves. Effectively, it is a RTLS solution. Blue Dot allows users to accurately pinpoint/visualize their location (or location of assets inside of buildings using smart phones/mobile devices or other sensors). To make Blue Dot navigation work indoors, one must leverage indoor maps and a positioning/tracking system together. Users will obviously have to download an appropriate app on their smartphones.

The new gateway is the *Samsung Axis*, an SDN Edge Internet Gateway. The device is a software-defined cloud orchestrated edge gateway for multi-tenant networks. It offers instant provisioning and on-boarding; provides private, secure, ubiquitous, and high bandwidth experience to end-users; and enables secure, connected experiences. This solution creates a Personal Area Network (PAN) for each user by using unique micro-segmenting capabilities that provide private connectivity to every user; end-users can access their PAN from anywhere in the building/venue/campus, in a fast, reliable, and seamless way, without worrying about network security and bandwidth challenges. The secured gateway prevents inadvertent actions from adversely affecting the wireless networks and reduces the instances of network hogging that creates a sub-optimal user experience. Basic functionality includes:

- Per-user traffic shaping: Enables real-time policy-driven restrictions and prioritization on a per-user/per-device basis and usage trends. This allows a guaranteed bandwidth rate, and at the same time, creates an opportunity for the network owner to upsell service tiers.
- Zero operator intervention end-user multitier self-provisioning billing system: Provisions end-users with a captive portal to process various forms of payments and upgrades, avail of

special promotions and offers, while extending a no-cost service support – all without the intervention of the network owner.

- Fully integrated unified threat management system: Stateful firewall and intrusion protection security that integrates seamlessly with the billing engine to enable premium service offerings to support revenue generation.
- Advanced client-side link control and routing: Aggregating several uplinks to provide high throughput of a large connectivity link.
- Software Defined (SD) network management: Increases network agility, performance, and optimization through SD-WAN capabilities, through policies defined by the network managers.
- Web experience manipulation: Using redirection of end-user web requests to specifically designed advertising templates, inject advertising and insert banners using webpage rewriting, thereby augmenting revenue potential for network owners.

The Gateway Network Controller solution can generate reports on transactions by users, devices, and other groups based on the policies set by the network operator at specified regular intervals, ranging from every minute to every month, and delivered via email or shared through a GUI. The Gateway Network Controller solution enables network operators to have complete cognizance over the HTTP activity of the end-user population. The granularity of analytics data that can be generated on how customers use bandwidth can help network operators make real-time and agile decisions on capacity and network planning while augmenting user experience. Network managers can also use the analytics generated to engage in marketing activities and promotions, which can be pushed instantaneously to capitalize on unique opportunities.

Software included the following:

Mist Cloud with Marvis AI   The software optimizes the wireless user experience with Marvis AI, which uses machine learning algorithms to adapt in real time to changes in user, device, and application behavior. This ensures predictable and reliable Wi-Fi and enables accurate BLE location-based services. The software is built on modern cloud elements: (i) micro-services for SaaS agility (i.e. fast, risk-free changes); containers ensure portability and fault tolerance; (ii) Big Data collection/analysis from every wireless user; (iii) built for speed and reliability using modern cloud elements such as Storm, Spark, and Kafka; (iv) SaaS agility for upgrades and patches; and (v) proactive AI-Driven IT operations, using AI to monitor network trends in real time and send alerts when service levels degrade below defined thresholds; recommendations are provided for troubleshooting and/or proactive configuration changes (see Figure 9.6 depicting Mist's Proactive Analytics and Correlation Engine [PACE]). Mist collects over 100 pre- and post-connection states in near real time from every wireless device: Monitor visits and dwell times, with detailed drill-down into zone traffic patterns and congestion points. Reports can cover any standard or customized time period (see Figure 9.7).

Security   On the switches, the dynamic VLANs feature was enabled; this will make sure each user is issued a single VLAN per device and eliminate the possibility of users sniffing the network data once connected to the network internally. A network intrusion detection system was installed along with the firewall and the RADIUS server on-premises. The Wi-Fi portal has an open network connection with the portal, and the hotspot portal is associated with the RADIUS server. The RADIUS server checks that information is correct using authentication schemes such as PAF, CHAP, EAP. Once the user has logged into the hotspot portal and the RADIUS server connection has been successful, the user is forwarded onto a secure wireless network where all data between the user and the access point is encrypted. This eliminates the possibility of a man-in-the-middle attack. As noted earlier, the Network Intrusion Detection system undertakes deep packet inspection to prevent the user from downloading torrents or use of any other

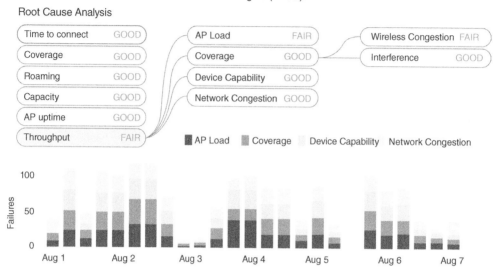

FIGURE 9.6 Machine learning analyzes data and provides insight, such as root cause identification and remediation.

unwanted Layer 7 applications. The firewall is used to block incoming network connections from outside, making a more secure network environment. The firewall also does bandwidth/traffic shaping to make sure users do not abuse the network by uploading/downloading too much data. Industry-standard encryption is utilized for data communications between network admins, infrastructure hardware/software, end-users and the Mist Cloud, and stored data are block encrypted. Mist encrypts and secures data in the following methods:

- AP to Mist Cloud: Communication between the Mist Cloud and the AP uses HTTPS/TLS with AES-128 encryption, and mutual authentication is provided by a combination of digital certificate and per-AP shared key created during manufacturing.
- UI or API Access: API communication (including UI access) uses HTTPS/TLS and is encrypted with AES-384.
- Internal to Cloud: Data within the cloud are stored using AES-256 encryption. They are hosted in a SOC2 type 2 compliant datacenter across multiple availability zones/regions.
- Management Console: Accessed over HTTPS connection, using 2048-bit RSA key.

Once the hardware and software were selected, the deployment process entailed the following steps:

1. Overall design review: go over the design in detail with the design team and all parties involved to make sure all aspects meet the expectations of each party involved.
2. Equipment review: go over main equipment to be installed at the facility to make sure all questions are answered, and any issues can be resolved early in the process.
3. Manage the overall schedule: layout the initial overall schedule and review it with all parties involved to make sure everyone's expectations are met. Additionally, if there are any hurdles or challenges within the schedule, they can be brought up early in the process to be resolved.

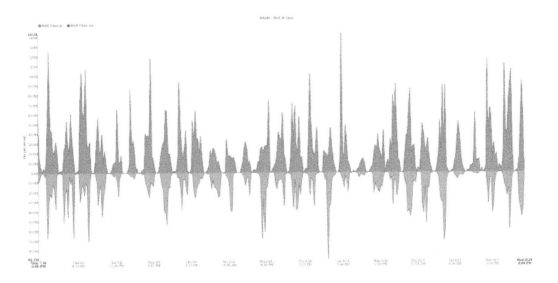

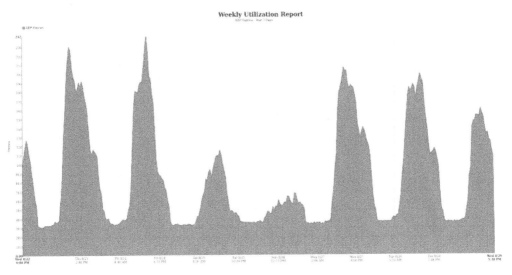

Weekly Utilization Report

FIGURE 9.7   Sample reports.

4. Installation: develop installation plans acceptable to all parties and supervise the phased installation.

5. Installation validation: ascertain that once installation of components is complete, each party will have the opportunity to review placement and aesthetics.

6. Commissioning: once all installation is complete, the entire system is commissioned and tested multiple times.

7. Optimization: work with the installation crew, along with carriers to optimize each section/zone along with the entire facility.

8. Testing: testing on individual components along with zones and the entire system throughout the entire project. Stress testing was to be provided by the actual users. Speed performance tests were taken both at quiet times and busy times to indicate total performance.

9. Activation and turnover.

The entire process was initiated during the 2Q2020 through 4Q2021 with installation completion of all technology within all designated airport areas by 2022.

10 The Age of Wi-Fi and Rise of the Wireless SuperNetwork (WiSNET)™

The previous chapters discussed several key technical building blocks of a modern network. There is, in fact, an end-user expectation for a transparent, easy-to-use, service-oriented, multi-media-based secure network that transcends locality and navigates along wherever the user is, sojourns, or transits. However, it is still rare to find all the elements discussed earlier in this text being efficiently integrated into one cohesive technical, service, and administrative platform, if at all, and it is practically impossible to meet all or the majority of the simple service criteria just cited.

Yet, the opportunity now exists to go beyond a mere assemblage of disparate technologies into a complex network, but to a denouement, to a state where the network is endogenously architected on a dynamically elastic infrastructure that is based on one fundamental building block to support all services in an elemental but scalable manner: *Wi-Fi technology*. It turns out that, indeed, Wi-Fi technology allows one to build and deploy a new generation multitier network, a Wireless SuperNetwork (WiSNET), that delivers all the service desiderata, yet is easy to use, service-rich, cost-effective, easy to maintain and administer, scalable, and secure.

At press time, the Wi-Fi ecosystem was already valued at $3 trillion, and it was expected to reach $4 trillion by 2023 (that equates to an intrinsic value of more than $500 for every human being on the planet) [1]. The mobile Internet is becoming an economic human necessity on par with electricity and clean water: according to Cisco's annual internet report, 66% of the world's population will be using the Internet by 2023; over 70% of this Internet access connectivity will be over a wireless network [2].

This chapter discusses the intrinsic elements of the WiSNET, whose genesis and impetus were the new all-inclusive networking infrastructure recently engineered by the Slice Wireless team at Baltimore/Washington International Thurgood Marshall Airport (BWI) discussed in Chapter 9 [3, 4].

10.1 WHAT PRECEDED THE WiSNET

During the past 50 years, the industry has seen major advancements in Information Technology (IT). Fifty years ago, the world was still mostly analog, voice based, low bandwidth, hardware-enslaved, monopolistic, and carrier-dominated. A transition to a digital world took off, never to stop, in the 1970s as digital switches and fiber optics started to make their presence. Data networks, first based on dedicated links (but later based on packet technologies heralded under the ARPAnet) started to become more broadly available in the 1980s, although generally expensive. The Internet Protocol (IP) was born early in the decade from a pregenital version.

High-Density and De-Densified Smart Campus Communications: Technologies, Integration, Implementation, and Applications, First Edition. Daniel Minoli and Jo-Anne Dressendofer.
© 2022 John Wiley & Sons, Inc. Published 2022 by John Wiley & Sons, Inc.

Local Area Networks (LANs) were standardized and timidly deployed also in the 1980s as were first-generation wireless cellular services.

Then came the Internet and the standardization of the World Wide Web, with major rollouts starting in the early 1990s. The IP at the networking layer won out other competitors and became the de facto standards for data transmission in the LAN, in the Wide Area Network (WAN), and in the Internet. Also came newer video-encoding schemes, such as Moving Picture Experts Group (MPEG) standards that leapfrogged the earlier "Consultative Committee for International Telephony and Telegraphy" (CCITT, now ITU) schemes.

Packetization of voice had its modest beginning in the 1970s, although voice compression, also known as vocoding, was as foreign to carriers as the term vocoding,[1] defining the science, sounds. The 1990s saw the beginning of the deployment of Voice over IP (VoIP), especially as the Session Initiation Protocol (SIP) came of age at the end of that decade.

Major advances in semiconductor technology, following Moore's Law, continued furiously through the five decades referenced in Section 10.1, continuing to the present, making digital signal processing for voice, video, and other applications practical and cost-effective, as well as supporting a migration to second-, third-, fourth-, and fifth-generation cellular. The Internet of Things (IoT) concept took form, systematizing earlier disjoined disciplines, with the goal of embedding intelligence into every imaginable object used in modern life.

While Wireless Local Area Networks (WLANs) saw near-ubiquitous deployment in companies' intranet, Wi-Fi also started to be deployed in public hotspots and people's homes in the 2000s. At the same time – while the previous core network technologies migrated away from private lines and even from the Asynchronous Transfer Mode view to an all-IP-based system at Layer 3 (possibly under a MultiProtocol Label Switching [MPLS] paradigm) – Ethernet-based WAN core services started to make their presence in the carrier's world, including Cable TV and Internet access networks, offering an advanced mechanism either as a replacement of Synchronous Optical Network/Optical Transport Network (SONET/OTN) or at least as a Layer 2 vehicle running on and/or "managing" the SONET physical network.

So, by the 2010s, the voice world effectively became IP-based, video (such as IP Television [IPTV]) and streaming became IP-based, the Internet IP-based, and the IoT is all IoT-based. People say that we "used to design and deploy voice networks and struggled to make them carry data – now we design and deploy data networks and struggle (in a manner of speech) to make them carry voice."

The sum total of this retrospective is that a major migration to IP-for-all-services has taken hold and VoIP and packet-based video are king. What comes next?

10.2   WHAT COMES NEXT

We advance the thesis, herewith, that a major additional migration will take hold in this decade. Namely, that (except for geographically-dispersed in-transit regional situations) *Wi-Fi will become the prevalent method to deliver all services in all stationary venues, be it the home, office, airport, stadium, plane, and the inside of an automobile*, to list just a few. Wi-Fi will be the near-universal "last quarter mile." Carriers will need to deliver Wi-Fi as a direct service to the customers, not a plethora of other non-interworking services. Naturally, the core backbone will consist of other technologies, particularly fiberoptic-based facilities.

The lessons from VoIP need to be appreciated and projected forward: gone are the days where specialized, closed, vendor-specific hardware provided a fixed, inflexible, slow-to-enhancements,

---

[1] "A vocoder (a portmanteau of voice and encoder) is a category of voice codec that analyzes and synthesizes the human voice signal for audio data compression, multiplexing, voice encryption or voice transformation." See https://en.wikipedia.org/wiki/Vocoder

uni-threaded, undifferentiated, integrated service. If a user needed some additional service, a different overlay network had to be provided. IP has taken over, and VoIP and Over The Top (OTT) are powerful integrative technologies. But now, Wi-Fi will extend that progression to the delivery layer, making the service homogeneous, ubiquitous, economical, dynamic, and quickly upgradeable. Wi-Fi will be the last mile. Carriers will need to adapt and adopt Wi-Fi as the last mile or see severe diminutions in revenues. There will be applications for 5G, particularly for people in transit, on highways, or in remote open areas (say, at the beach or while enjoying a fishing outing on a boat close to the shore), and for dispersed sensors that might make use of Narrow Band IoT (NB-IoT) (or successors), but as discussed in previous chapters, to achieve true high bandwidth, millimeter wave (mmWave) frequencies are needed; however, these frequencies have transmission limitations. Some services will emerge at the new sub-6 MHz bands, but they will remain relatively expensive, will be tethered to a given carrier, and will be at low bandwidth by current standards.

At the same time, as mobile networks become more complex, cloud and Artificial Intelligence (AI) technologies are becoming more important in moving the management of these networks from one of managing network elements to one of managing the end-to-end user experience. Wireless networks of the future and the teams that manage them will need an AI "network assistants" on their team to help (B. Friday, personal communication).

10.3 THE SUPER-INTEGRATION CONCEPT OF A WIRELESS SUPERNETWORK (WiSNET)

This book has discussed several technical solutions, but these are often deployed as stand-alone systems. In some cases, the use of multiple solutions from multiple providers is useful, practical, even inevitable; such an approach, however, requires the separate administration, engineering, provisioning, operations, managing, sparing, vendor-stewardship, and billing of the various systems and service providers. The movement toward an integrated solution that is consistent both at the technology level and at the administration level (including engineering, provisioning, operations, managing, sparing, vendor-stewardship, and billing) is achievable and desirable. While 5G solutions may bring a degree of conversion at the wide-area level, it is likely that bona fide LAN technologies (such as 802.11ax) will continue to serve the needs of institutions at the local level for the foreseeable future, possibly forever. A major advancement toward integrated solutions took place when a large number of siloed technologies, including voice, entertainment video, OTT video, OTT voice, videoconferencing, video security, situational awareness, and all web-based activity has migrated to TCP/IP.

Instead of a multitude of overlay networks using a multitude of protocols, a disjointed operational support infrastructure, and a set of fragmented cybersecurity mechanisms (if even present), the opportunity now presents itself to develop a WiSNET that is based on an open-network platform and can afford a "service disruption" in the sense of revamping the revenue model. It should be noted that while the wireless equipment revenue has been growing in the United States in recent years at around 10% a year, the service revenues have remained relatively stagnant. The goal of the WiSNET is taking multiple network technologies and allowing them to plug and play into one open wireless network serving not only large venues but also possibly corporate intranets in sprawling corporate headquarter buildings. Such architecture can bring down the cost and speed of implementation by eliminating traditional red tape (i.e., engineering, permitting, and installation).

We noted in Chapter 1 that institutional (corporate/campus) networks (INETs) have evolved considerably in the past 60 years, which spans the era of data communications, also as highlighted in Section 1.4. Figure 1.9 (replicated below as Figure 10.1) depicts a view of this evolution. As the decade of the 2020s dawned, one can assert that we have reached a stage where "super-integrated," feature-rich INETs have emerged. Such "super-integration" in the

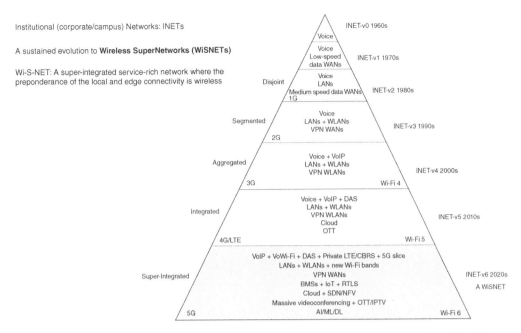

FIGURE 10.1   (Also shown as Figure 1.9): INET-v6: A Wireless SuperNetwork (WiSNET).

corporate/campus now covers (i) voice services in the form of VoIP, VoWi-Fi, Distributed Antenna Systems (DASs), and private (virtual) cellular networks (in addition to the public cellular networks migrating to 5G); (ii) extreme reliance of Wi-Fi access in the form of Wi-Fi 6 and also new usable bands; (iii) Virtual Private Networks (VPNs) as the wide-area connectivity of choice; (iv) integration of Building Management Systems (BMS) to support smart building/smart campus functionality along with RTLS and more general IoT functionality; (v) cloud-based services and analytics, also based on the concepts of Software Defined Networks (SDNs) and Network Function Virtualization (NFV); (vi) massive use of videoconferencing, as well as (corporate reception where/as needed) of OTT/IPTV video feeds (e.g., business TV); and (vii) the widespread use of AI, Machine Learning (ML), and Deep Learning (DL). The concept of a WiSNET was briefly introduced in Chapter 1 as being a super-integrated service-rich INET network where the preponderance of the local and edge connectivity is wireless; such a WiSNET, in addition to the underlying technologies just cited, enjoys and, in fact, requires a unified, highly flexible, cost-effective management and administration apparatus. Figure 10.2 depicts graphically the earlier present mode of operation (left hand side) and the evolving future mode of operation, the WiSNET, on the right hand side.

The WiSNET embodies a unified, comprehensive, scalable architecture for high-density, high-throughput multimedia communications, supporting open, scalable, and inexpensive technology for secure, Quality of Service (QoS)-enabled, high-mobility services for a plethora of users having a variety of connectivity and access requirements that span multiple-use cases. Integration can occur at the technology level, as well as at the administration and service provisioning/service monitoring level, where one provider bundles all requite systems by doing the design, the feature selection, the deployment, the grade-of-service monitoring, the intrinsic multi-vendor management, the technology tracking and refreshment, the overall security tracking, and the cost and billing support. Public venues such as airports, stadiums, convention centers, and so on benefit from installing such integrated solutions.

A deeper question is if all Layer 2 technology will migrate to Ethernet, both at the local level – for wired and wireless access – and at the metro and regional level, according to the thesis we

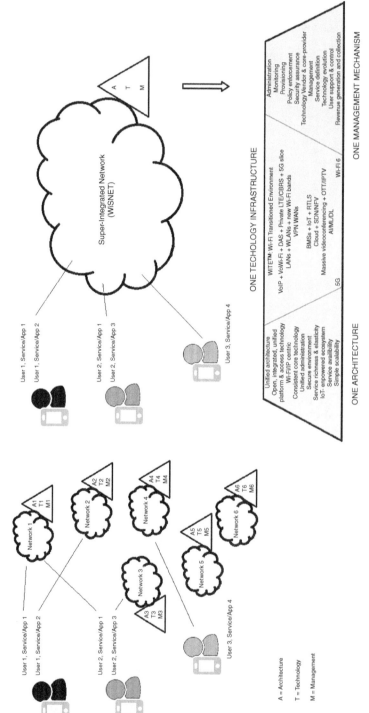

FIGURE 10.2 Present mode of operation (left hand side) and evolving future mode of operation, the WiSNET (right hand side).

postulated in Section 10.4, can all services, such as voice, entertainment video, OTT video, OTT voice, videoconferencing, video security, situational awareness, and ubiquitous IoT-sensing (or a large majority thereof), be delivered using Wi-Fi? We believe that the answer is yes.

Long-Term Evolution (LTE) was termed to describe the progression in cellular services from 3G to 4G. We define here the term **WiTE**™: Wi-Fi Transitioned Environment, an environment powered by a ubiquitous Wi-Fi infrastructure for the delivery of all services, including voice, video, multimedia, situational awareness and navigation, and IoT-based sensing and actuation. With the WiSNET transitioned environment, Wi-Fi is no longer a stepchild of the telecom industry.

10.4   THE MULTIDIMENSIONALITY OF A SUPERNETWORK (WiSNET)

An advanced network in this evolved context typically has a multidimensional formulation in terms of services, technologies, and design and operations (e.g., as implied in Figure 5.30).

- Services that are QoS-rich, reliable, usable, and secure span at least these classes: (i) Web and apps access; (ii) OTT and multimedia; (iii) voice and VoWi-Fi; and (iv) navigation and broad IoT support.
- Technologies spans at least the following domains: (i) advanced Wi-Fi services; (ii) advanced DAS (including LTE, private LTE, 5G, and 5G slicing); (iii) advanced wayfinding/RTLS; and (iv) advanced IoT sensing and asset tracking.
- Design and operations encompass at least these features: (i) unified, scalable architecture for high density and high throughput; (ii) standard/open inexpensive technology for high mobility; (iii) secure and QoS-enabled ecosystem for high service availability; and (iv) unified, cost-effective management/administration.

As depicted in Figure 10.3, these characteristics can be summarized as follows:

1. Architectural aspects of a WiSNET.
2. Technology aspects of a WiSNET.
3. Management aspects of a WiSNET.

These features are further discussed in Section 10.6, after providing a glimpse of what gave rise to the concept of the Wireless SuperNetwork.

10.5   THE GENESIS OF THE WiSNET CONCEPT DEFINED IN THIS TEXT

The experience of the Slice Wireless team acquired while revamping and modernizing the passenger- and business-facing network infrastructure at BWI airport discussed in Chapter 9 has enabled us to define the concept of the WiSNET as described in the sections that follow. A quick genesis of the motivational *tour-de-force* that gestated the WiSNET follows here as a bridge to the architectural, technology, and management formulation of a generic WiSNET to follow.

Traditional "high-density" networks typically entail a multitude of independent networks not working together within one environment; almost invariably, there is no sharing of infrastructure or revenue (by revenue arrangements). Furthermore, there is a distinct lack of forward-looking providers who think beyond the mundane collection of fractionalized Wi-Fi revenues. Typically, in an airport, every operator ran through the facility in a roughshod manner: each provider simply built, installed, and managed their own network. Further, anyone was able

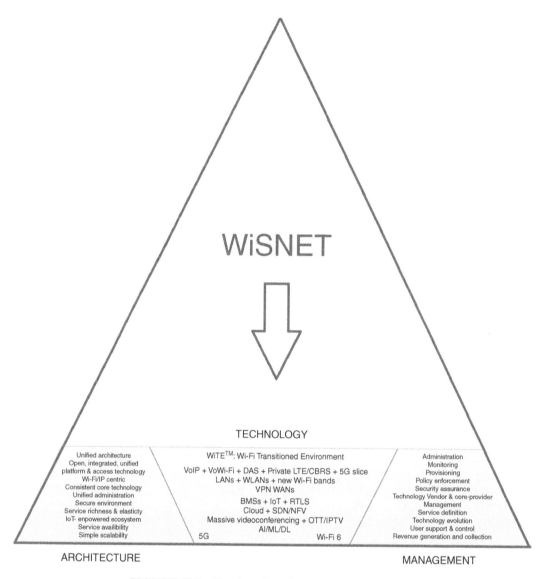

FIGURE 10.3   The three key dimensions of a WiSNET.

to install their own "rogue" Wi-Fi Access Points (APs). Some operators charged their customers for Wi-Fi access while still others offered "free" Wi-Fi access but usually with a side condition (such as becoming a subscriber to their network). Thus, one finds a situation where independently deployed or operated APs are turned up all the time. Someone who owns a store in the airport buys an AP and then uses it for their store operations or (worse) allows users have "free" Wi-Fi as a small inducement to get customers in the door. When one thinks about this kind of setup from the perspective of the customer (passenger in an airport) and from the perspective of the airport, this approach is less than optimal. Importantly, how does the owner of the venue (the airport administration) manage all these disparate Wi-Fi networks? As a result of all of this, airline fees have been static or in decline when compared with passenger numbers, while revenue from retailing and other commercial activities (products and services provided to passengers) has grown significantly. Even during the COVID-19 pandemic, the travel population in the United States exceeded one million for the 2020 Thanksgiving and Christmas holidays. It is anticipated that ongoing investment in airport retail, leisure, rideshare, and dining facilities will generate substantial revenues for airports, transforming landside and airside space into a hub of diverse activities.

The BWI airport saw the first deployment of the WiSNET (banded there as MyBWI-Fi – Figure 10.4); other airports may soon follow (Sarasota, FL [SGR] is in fact following suit). In an airport such as BWI, there are many different wireless providers. BWI was able to take charge of their wireless assets rather than running a *laissez-faire* operation; Table 10.1 identifies the wireless and partners who are all placed into a WiSNET run by Slice Wireless. As a result, the different and disparate assets are managed under one WiSNET. Instead of airports looking at wireless operations as a cost center, they now have control over all of these assets and can generate income and provide passengers with a better experience. Even off-site vendors, whether

FIGURE 10.4   MyBWI-FI SuperNetwork at the Baltimore, MD airport.

TABLE 10.1   Plethora of Players at BWI – In a WiSNET, All of These Are Managed by One Administrator

Account	Category	Service
AT&T	Carrier	DAS
T-Mobile	Carrier	DAS
Verizon	Carrier	DAS
Spirit Airlines	Tenant	VPN
Fraport	Tenant	Wi-Fi
Enterprise	Tenant	Wi-Fi
Clear Channel	Media Partner	Wi-Fi
Minute Suites	Tenant	Internet & Phone
Prime Flight	Tenant	Internet

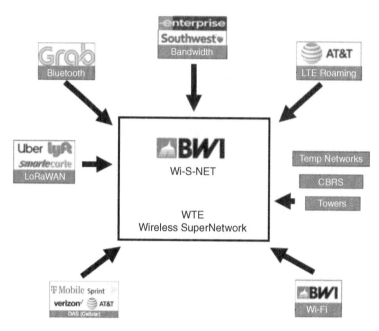

FIGURE 10.5 The digital transformation of a high-density network – The BWI WiSNET.

requested by travelers or by a venue-based establishment with an app that wants to link to this network, can now reach the million of people who pass through an airport such as BWI each year. At BWI alone, their potential reach two million captive travelers/consumers a month but will combine networks from multiple airport platforms into a seamless service ecosystem.

From a service perspective, the WiSNET has aimed at, and has delivered, "superior user experience." Figure 10.5 depicts a totally different environment brought about by rethinking the entire airport network: with a coordinated effort, the passenger experience is clean and simple as implied in the figure. Passengers can login to a secure network without having many different wireless providers from which to choose. A WiSNET turns the table on airport connectivity from disjointed and money losing to a transparent, coordinated, secure, user-friendly, and money-making operation.

The WiSNET is able to make airports and other densely populated public areas a part of the sharing economy. Just as ridesharing, housing and office-sharing, and shared food-delivery have changed their industries by opening up infrastructure and giving more power to the user, WiSNET is designed to garner the benefits of "we" to do the same by unlocking the Wi-Fi network for more affordability, convenience, and efficiency. Such a design allows the redirection of independently deployed wireless and telecom technologies to one technology-integrated and one service-integrated network platform that USERS can access, not just telecommunication companies.

10.6 THE DEFINITION AND CHARACTERIZATION OF A WiSNET

A WiSNET aims at supporting high-density large venues, as well as other environments. The network makes pervasive use of Wi-Fi as the local access and distribution apparatus for the so-called last quarter mile. The network is characterized by a well-defined set of architectural, technology, and management features. Architectural features of a WiSNET include (i) a unified architecture; (ii) an open, integrated, unified platform and access technology; (iii) a Wi-Fi/IP centric protocol suite; (iv) having a consistent core technology; (v) utilizing a unified administration; (vi) establishing a secure environment; (vii) providing service richness and elasticity;

(viii) supporting an IoT-empowered ecosystem; (ix) having high service availability; and (x) enabling simple scalability. Technology features of a WiSNET include (i) support for voice-related services; (ii) support for data-related services; (iii) support for WAN-related services; (iv) support for building management-related services; (v) support for cloud-supported services; (vi) support for video-related services; and (vii) support for artificial-related services. Management feature of a WiSNET include (i) administration capabilities; (ii) monitoring capabilities; (iii) provisioning capabilities; (iv) policy enforcement capabilities; (v) security assurance capabilities; (vi) service definition capabilities; (vii) user support and control capabilities; (viii) technology vendor and core provider management tools; (ix) revenue generation and collection tools; and (x) technology evolution mechanisms.

10.6.1 Architectural Aspects of a WiSNET

This section discussed architectural aspects of a generic WiSNET; sections that follow discuss the technological platforms and the management elements of a generic WiSNET. The innovative, all-encompassing, bleeding-edge design/engineering/deployment of the BWI airport developed by Slice Wireless in 2020/2021 brought into focus the need for a prescriptive system architecture. Each of these subsections describes important definitional aspects of the WiSNET.

The importance and value of defining a system architecture and ultimately standardizing on a Reference Architecture (RA) was already discussed in Chapter 4. We noted there that architectures simplify the characterization of the system's constituent functional blocks and the manner in which these functional blocks interrelate to each other. An RA facilitates the orderly partition of functions, typically in a hierarchical fashion. Such partition not only reduces functional redundancy, but it also promotes standardization with the possible definition of well-established layer-to-layer interfaces; this also allows the intermingling of products from an open set of vendors' products with the goal of layer-function cost optimization and/or usage of best-in-class technology for each layer. A number of formal RAs have been defined for basic communications, Internet of Things (IoT) environments, and application systems development, including the Open Systems Interconnection Reference Model (OSIRM) for communication; the Industrial Internet Reference Architecture (IIRA), the Reference Architecture Model Industrie 4.0 (RAMI 4.0), and the ETSI High Level Architecture for M2M for the IoT ecosystem, among many other; and the Open Group Architecture Framework (TOGAF), the Zachman International model, and the US Departure of Defense Architecture Framework (DoDAF), among many others for IT systems.

Unified Architecture A formal RA for a WiSNET is yet to be defined. Nonetheless, given that a WiSNET spans a large number of underlying communication technologies supporting a significant number of services and different media, mixed-media, and multimedia communications, it is important to have (i) one overall definition of the ecosystem, (ii) identification of its key functional blocks, and (iii) description of the input and output interfaces to/from the external world and between the discrete functional elements.

Figure 10.6 depicts a simple, unified hierarchical architecture that can be beneficially applied to the ecosystem with the goal of systematizing the underlying functionality framework (more elaborate models can be developed in the future). The figure depicts a large set of handheld, wearable, stationary, and mechanical/process control devices that populate the universe of things used to deliver services to the user (be it a human being or a mechanical system). A plethora of streams are generated (and/or needed) by these devices. Wi-Fi is the canonical discipline (technology) for local distribution. A data aggregation layer (possibly with some local processing – especially for flow or policy control) is implemented in support of its parent layer, the service-delivery engine layer. This service layer includes, for example, the connectivity services of the Wireless Service Providers (WSPs), the Internet Service Providers (ISPs), the

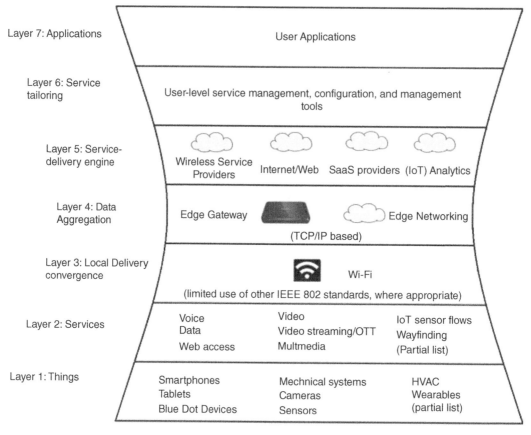

FIGURE 10.6 Simple, unified hierarchical WiSNET architecture.

Software-As-A-Service Providers (SSPs) (also sometimes known as Application Service Providers [ASPs] or Cloud Service Providers [CSPs]), and the IoT Analytics Service Providers (IASPs) (which increasingly utilize extensive AI/ML/DP techniques.). Cloud-based storage has also become prevalent in many application environments; additionally, there are content-providers. Above this layer, there is a user-managed layer for service configuration and tailoring. The ultimate user application resides at the next layer up (when the service is fully SaaS-based, this layer can, in some instances, be considered null).

Open, Integrated, Unified Platform, and Access Technology Consistent with the unified RA introduced in Section 10.6.1, a software/hardware platform to deliver WiSNET services is needed. The platform should be open in terms of using industry standards for all elements of the platform not only in the "user plane" but also in the "control plane" and in the "management plane"; it should be also open in the sense that any vendor that conforms to the RA can offer a network element product (traditional, or preferably conforming to SDN/NFV design) that supports the operation of the platform; it should also be open in the sense that any service provider (WSP, ISP, SSP, IASP, and so on) can contribute underlying services in a seamless manner (seamless in all three planes mentioned above). The platform should be integrated from a service, functionality, technology management, and usability perspective. The platform should also be unified (presenting a "comparable" view) to all stakeholders (users, providers, administrators) from a service, functionality, technology, management, and usability perspective; preferably a Service-Oriented Architecture (SOA) paradigm is utilized. User access should be consistent to the maximum degree possible, being predominantly Wi-Fi-based, or of not (and for some specific use cases) being at least based on IEEE 802.11 standards (e.g., Bluetooth Low Energy [BLE], Bluetooth Mesh, ZigBee, Wi-Fi HaLow). The higher protocol stack should be IP/TCP/UDP with appropriate application-support upper layers. Administrator access should also be based on a unified suite such as SNMPv3, HTTP, SOAP, JSON, REST, XML, open APIs, and so on.

SOA concepts, along with datacenter server virtualization of the 2000s -- first applied to application development and data centers -- have more recently influenced the networking arena in the form of NFV paradigms. NFV allows service providers and carriers to rapidly define software-based network functions on commercial, off-the-shelf hardware. Utilizing these Virtual Network Functions (VNFs) as the building blocks for creating Virtual Network Services (VNSs), a WiSNET administrator can change the way the network services are provided. NFV engenders independence from proprietary equipment and single-vendor dependence.

Wi-Fi/IP Centric   Consistent with the unified RA introduced in Section 10.6.1 and the open platform concept, the access discipline of a WiSNET should be Wi-Fi/IP centric. The core technology should also be IP centric, wherein the wired case (e.g., fiber-based with IP-over-SONET, IP-over-OTN, IP-over-DWDM, IP, MPLS), in the wireless case (e.g., IP-oriented 5G), or in the IoT/WSN case. This approach should apply not only to the "user plane" but also to the "control plane" and the "management plane." It should also be able to support IPv6, including MIPv6 where appropriate.

Consistent Core Technology   Consistent with the unified RA introduced in Section 10.6.1 and the open platform concept, the core networking technology (used in the MAN or WAN) should be preferably limited to a well-defined subset of communication technologies that make IP the cornerstone of the protocol suite at the lower layers.

Unified Administration   Consistent with the unified RA introduced in Section 10.6.1, the administration of the network should be unified, namely, similar (if not identical) procedures to manage all classes of resources (whatever they may be). Administration encompasses not only network management and monitoring functions to support the real-time delivery of services (more on this topic further in Section 10.6.3), but also, importantly, the service planning and provisioning interactions facing the service and equipment providers, as well as facing the ultimate user. This unified approach institutes discipline, simplicity, and quality assurance across all aspects of the WiSNET.

Secure Environment   Cybersecurity has become an ever-present, ever-pressing issue in modern environments. The WiSNET must support a high level of overall security. As discussed elsewhere in this text, cybersecurity has been minimally described as encompassing the following requirements, which are requirements across all services, environments, technologies, and management of the WiSNET ecosystem:

- Confidentiality (C) – making sure none of the data flows (be the voice, video, data, multimedia, sensor traffic, or management traffic) are not intercepted by unauthorized agents
- Integrity (I) – making sure that any information received over the WiSNET has not been altered, and that the resident data has not been changed in an unauthorized manner.
- Availability (A) – making sure that none of the devices used by stakeholders (smartphones, laptops, tablets, sensors, network elements, and so on) are not incapacitated and/or placed in a state where they are not properly performing their function; that the devices are not hijacked to become rough devices; that any of the analytics systems that might be used become flooded with spurious traffic; or that the communication channels or network element are not intentionally jammed.
- Authentication, Authorization, and Accounting (AAA) – making sure that only authorized users access the resources. Specifically, AAA is a set of mechanisms for controlling access

to computer resources and enforcing privileges and policies; AAA also enables auditing of usage and collecting the information needed to bill for services. Authentication and authorization are defined in RFC 2865, while accounting is described by RFC 2866. *Authentication* provides tools for identifying a user, typically requiring each user having a unique set of credentials and criteria for gaining access; AAA is normally supported by a dedicated AAA server that compares a user's authentication credentials with other user credentials stored in a database. Standards such as Remote Authentication Dial-In User Service (RADIUS) and Diameter are utilized. RADIUS is an AAA protocol for applications such as Network Access or IP Mobility. It is specified in a relatively large set of RFCs.[2] RADIUS is a client/server protocol that runs in the application layer; it can use either TCP or UDP at the Transport Layer (Layer 4). DIAMETER is an update to RADIUS defined in RFC 6733 and RFC 7075; it uses TCP at the Transport Layer and utilizes Transport Level Security (TLS) or IPSEC. It extends RADIUS by adding new commands and/or attributes, such as those for use with the Extensible Authentication Protocol (EAP). Related RFCs include RFC 4004, RFC 7155, RFC 4072, RFC 8506, RFC 4740; there are also definitions in the 3GPP IP Multimedia Subsystem. Following authentication, a user must be granted *authorization* for undertaking any number of specific tasks, including access to services or data; the authorization process determines whether the user has the authority to issue some commands or access some resources; namely, it is the process of enforcing policies: determining what types or qualities of activities, resources, or services a user is permitted. *Accounting* provides mechanisms, the measures, the resources a user consumes during access (e.g., amount of system time or the amount of data a user has sent and/or received during a session).

As a set, these are referred here as CIA[4].

Service Richness and Elasticity   The WiSNET must support a full range of contemporary services and be able to accommodate emerging services and use cases. Services must encompass full multimedia applications, such as voice, video, data, text, streaming. It must support the evolving IoT/sensor-based applications, including wayfinding and Real-Time Location System (RTLS)-type services. The services must be extensible and tailorable, providing an elastic, dynamic environment.

It must locally support all the services that have been identified for the 5G IoT as described in Chapter 4, as well as the other 5G services such as Ultra-Reliable Low Latency Communication (URLLC), massive MTC (mMTC), and enhanced Mobile Broadband (eMBB), described in Chapter 5 (specifically Figures 5.16 and 5.17). It must be able to properly interface to the core networks (Layer 5 of the Section 10.6.1 RA) (which include 5G) to provide end-to-end support. Naturally, the WiSNET by itself does not support mobile wide area networking, but it does support the panoply of evolving 5G-type services by providing local (campus) support of those services and properly handing off the IP flows to the core to achieve true global connectivity of all types.

IoT-empowered Ecosystem   Refer to the previous section. Furthermore, it is desirable for the WiSNET to support some of the IoT-specific RAs, under the auspices of the overall RA in Section 10.6.1.

---

[2]RFC 2058, RFC 2059, RFC 2138, RFC 2139, RFC 2548, RFC 2607, RFC 2618, RFC 2619, RFC 2620, RFC 2621, RFC 2809, RFC 2865, RFC 2866, RFC 2867, RFC 2868, RFC 2869, RFC 2882, RFC 3162, RFC 3575, RFC 3576, RFC 3579, RFC 3580, RFC 4014, RFC 4372, RFC 4590, RFC 4668, RFC 4669, RFC 4670, RFC 4671, RFC 4675, RFC 4679, RFC 4818, RFC 4849, RFC 5080, RFC 5090, RFC 5176, RFC 5607, RFC 5997, RFC 6158, RFC 6218, RFC 6421, RFC 6613, RFC 6614, RFC 6911, RFC 6929, RFC 7360, RFC 7585, RFC 8044.

Service Availability   Availability is quoted as the "number of 9s" for service uptime. For example, five (5) 9s means that the service is up 99.999% of the time. It should be noted, in contrast, that a system that is up 99.99% of the time is actually down about one hour a year (0.87 hours to be exact).

WLANs are subject to various "outages" due to interference from neighboring APs, or poor coverage, or inadequate backbone connectivity. The availability may also be related to the offered load: the higher the number of users (or the required throughput), the lower the actual availability as computed at the service level, not just at the system level. Although some WiSNET instantiations are directed to nonmission critical environments or applications, other instantiations may, in fact support quasi-mission-critical or mission-critical applications (particularly in the context of some IoT applications such as surveillance or building/campus support of mechanical systems).

Availability is not per se an intrinsic architectural issue, although some aspects of it may entail architectural considerations (for example, the use of a dedicated frequency spectrum instead of a publicly-shared frequency spectrum). Availability can be seen related to the engineering and provisioning of a system – for example, are redundant servers used, are redundant power supplied used, are redundant communication links used?

Availability can be defined end-to-end (including the service of the WAN providers), or it can be defined (measured) only over the local subnetwork. Naturally, it is desirable to have end-to-end availability, and the design can endeavor to achieve such end-to-end grade of service (for example, having redundant connectivity into the core, using reliable core service providers, having proactive/preventive monitoring of all elements of the network to anticipate or avoid outages). After all, what good is it if the local campus (e.g., airport) network is up, but connectivity to the Internet or the cloud is severed? However, the end-to-end availability can never be greater than the availability of the local subnetwork; therefore, it is important to properly design and engineer that local component.

Therefore, a WiSNET is defined as being a

- WiSNET-4, if it is designed and deployed such that all services enjoy a 99.99% end-to-end availability at expected traffic load
- WiSNET-5, if it is designed and deployed such that all services enjoy a 99.999% end-to-end availability at expected traffic load
- WiSNET-6, if it is designed and deployed such that all services enjoy a 99.9999% end-to-end availability at expected traffic load.

Simple Scalability   The WiSNET must be able to be easily scalable without requiring a major system or technology revamp or "fork-lift" to enable it to support new users, new sections of the local campus, or new services.

10.6.2   Technology Aspects of a WiSNET

This section discusses the technology aspects of a WiSNET. The innovative, all-encompassing, bleeding-edge design/engineering/deployment of the BWI airport developed by Slice Wireless in 2020/2021 brought into focus the need to support a comprehensive plethora of technologies in the WiSNET, as discussed in Section 10.6.2. To the degree possible, SDN and NFV principles and concepts should be employed in supporting and/or delivering the portfolio of constituent technology systems. The technologies listed in Section 10.6.2 represent a basic set of requirements; other and/or future technologies also should be supportable. Each of the subsections that follow describes important definitional aspects of the WiSNET. Table 10.2 depicts some of the key parametric capabilities intrinsically associated with a WiSNET.

TABLE 10.2   Technical Characterization of a WiSNET

Key Performance Indicators for Specific Venues of Interest	Key Performance Indicators	WiSNET Capabilities
Data/VoIP/video (OTT)/RTLS wireless connection density, for people on smartphones, laptops, tablets	Data/VoIP wireless connection density, for users on smartphones, laptops, tablets accessing the Internet, web services, or corporate intranets over a VPN	1 per 20 ft^2 throughout the venue
	User experienced wireless data rate	10–50 Mbps
	Peak wireless date rate	100 Mbps
	Traffic volume density	5 Gbps per individual zone throughout the venue; 100 Gbps aggregate handoff to backbone
	End-to-end latency	100 ms
	Wayfinding	Throughout the venue and in immediate adjacent spaces of interest
	Area of coverage	Entire venue and in immediate adjacent spaces of interest
	Security	Complete Confidentiality, Integrity, Availability, and AAA (CIA^4) mechanisms
	Availability	99.999% uptime
	Protocols	Wi-Fi 6 (and related standards)-focused at Layer 2; TCP/IPv4 and TCP/IPv6 at Layers 4/3; SIP for VoIP; MPEG-4 for video
Traditional telephony on DAS systems	Dialtone	50 Erlangs per zone
	Call length	10 minutes per call
	Availability	99.9999% uptime
	Protocols	3GPP, 4G/LTE, CBRS, 5G (including mmWave), VoWi-Fi, E911
Connection density, IoT devices and sensors	Wireless connection density, IoT devices	1 per 10 ft^2 throughout venue
	User-experienced wireless data rate	0.384 Mbps
	Peak wireless data rate	0.768 Mbps
	Traffic volume density	100 Mbps per 1000 ft^2 throughout venue and in immediate adjacent spaces of interest
	End-to-end latency	1–10 ms
	Area of coverage	venue and in immediate adjacent spaces of interest
	Security	Complete Confidentiality, Integrity, Availability, and AAA (CIA^4) mechanisms
	Availability	99.9999% uptime
	Protocols	Wi-Fi 6 (and related standards) focused at Layer 2 (including Wi-Fi HaLow, BLE, Bluetooth Mesh, ZigBee, M2M); TCP/IPv4 and TCP/IPv6 at Layers 4/3; MPEG-4 for IP-based surveillance cameras
	Operations	Secure over-the-air device updates

Voice-related Services   A WiSNET must be able to support the voice-related technologies listed below, although not all implementations need to deploy all platforms in all instances:

- VoIP
- VoWi-Fi
- DAS-based coverage for 4G and emerging 5G services
- 5G Network slicing (see Section 10.8)
- Near-term Private Long-Term Evolution/Citizens Broadband Radio Service (LTE/CBRS).

Appropriate backbone connectivity to the Public Switched Telephone Network (PSTN), to cellular networks, to the Internet and/or to appropriate clouds, and related service support is needed.

Data-related Services   A WiSNET must be able to support the data-related technologies listed below, although not all implementations need to deploy all platforms in all instances:

- Traditional wired LANs for various baseline application, including support of Power Over Ethernet (PoE)
- WLANs technologies, particularly IEEE 802.11ax
- Support new/evolving Wi-Fi bands, including the 60 GHz unlicensed band defined in IEEE standard 802.11ad, also known as WiGig.

Appropriate backbone connectivity to the Internet and/or to appropriate clouds and related service support is needed.

WAN-related Services   A WiSNET must be able to support VPN services that users may require while at the campus to access corporate intranets.

Appropriate backbone connectivity to the Internet and/or to appropriate clouds and related service support is needed.

Building Management-related Services   Campuses have requirements for mechanical support of systems. Although a BMS may already be deployed at the campus, WiSNET must support – or at least interface to -- BMS technology/functionality (occasionally replacing it). The WiSNET must be able to support a large number of ancillary sensor networks, whether IoT-based or non-IoT-based (for example, but not limited to, for connected lighting). Various ancillary support for DeviceNet, SOAP, XML, BACnet, LonWorks, and Modbus may be needed in some instances.

Various RTLS technologies must be supported, including Wi-Fi-, BLE-, ZigBee-, Radio Frequency Identification (RFID)-, and Ultra-Wideband (UWB)-based systems.

The plethora of IoT aggregations technologies must be supported, including, for example, NB-IoT, 5G IoT, LoRa, and Sigfox.

Cloud-supported Services   WiSNET must support access to all requisite cloud services, SaaS, and cloud analytics.

Video-related Services   Support of video platforms and technologies has become imperative in recent years. A lot of web content is video-based. Appropriate bandwidth, latency, jitter, and packet loss service goals (requirements) must be supported by the underlying technologies and platforms to deliver video content in a dense venue such as an airport, a stadium, or a convention center, to list a few. Technologies to support OTT and/or IPTV entertainment video that may be delivered over the WiSNET are needed.

Massive videoconferencing has become more prevalent in recent years (especially driven by the work/study at home paradigm shift engendered by the COVID-19 pandemic), spanning both the personal life and the business life of many. Technologies that support these services are needed by the WiSNET.

Security surveillance utilizing stationary IP cameras and more general (mobile) situational awareness capabilities (including, for, example face recognition) may be part of a WiSNET, including access to cloud-based analytics. Situational awareness is the ability to develop and deploy traditional or AI-based mechanisms to assess, recognize, anticipate, and intercept events specific to the use case of interest.

Artificial-related Services   The use of AI/ML/DL for various network-related/network-supported functions is becoming important, and a WiSNET needs to support and/or utilize these technologies.

ML predicts and classifies data using various algorithms optimized to the data set in question. ML is an example of AI, which itself is a subfield of computer science that focuses on the development of computer-based systems, applications, and algorithms that mimic cognitive processes intrinsic to human intelligence. ML is a mechanism used to implement AI concepts; it entails (complex) algorithms that parse data, learn from that data, and then apply what they have learned to make informed decisions. ML techniques are increasingly utilized to analyze, cluster, associate, classify, and apply regression methods to situational awareness data (as well as to many other IoT environments). ML techniques endeavor to examine and establish the internal relationships of a set of data collected from the plethora of input devices collecting visual, audio, and signal data from the field. Among other applications, ML techniques are applicable to image processing and analysis, computer vision, speech recognition, and natural language understanding. Although "machine learning" and "deep learning" are occasionally used synonymously, there are differences. DL as a discipline is a subset of ML -- thus, DL is a machine-learning mechanism (It is a subcategory of machine learning) and functions in a similar manner; however, DL uses a programmable neural network (an ANN) that enables machines to make more accurate decisions without help from humans. One application of interest relates to the correlation of network events for the purpose of more effective network management.

10.6.3   Management Aspects of a WiSNET

This section discusses management and administrative aspects of a WiSNET. The all-encompassing, bleeding-edge design/engineering/deployment of the BWI airport implemented by Slice Wireless in 2020/2021 brought into focus the need for a prescriptive set of high-usability management capabilities. Each of the subsections that follow describes important definitional aspects of the WiSNET.

Administration   Administration was already mentioned in Section 10.6.1 in the context of a unified approach to the management of service and technology providers. In this context, administration refers to the real-time oversight of the WiSNET to maintain it in optimal, secure working order.

Classically, the network management model included Fault, Configuration, Accounting, Performance, Security (FCAPS) tasks as described in the ISO network management recommendations, and in the ITU Telecommunications Management Network (TMN) M.3010/M.3400 recommendations [5–8]. Extensions to these definitions were developed by the TM Forum in the late 1980s. TM Forum is now an alliance of 850+ global companies, including the world's top 10 network and communications providers. Further along was the development of a service provider/telecom operator-owned framework of business processes related to network and service administration. The Telecom Operation Map (TOM) was extended in the early 2000s to the Enhanced Telecom Operations Map (eTOM), and in the 2010s to the "Business Process

Framework, the entire initiative now known as eTOM (the latest release being R19.0, 2019). eTOM is a comprehensive, industry-agreed, multilayered, hierarchical view of the key business processes required to run an efficient, effective, and agile digital enterprise, of which a telecom service provider is a premier example. It is promulgated by the TM Forum [9]. FCAPS can be seen as the predecessor of the newer Fulfillment, Assurance, Billing (FAB) model defined in the Business Process Framework/eTOM.

Fault management uses technologies and procedures to detect, address, and document faults that could interfere with network operations. These fault management capabilities report and record problems that administrators can analyze for trends of various types. The use of AI/ML/DP has become prevalent in advanced monitoring tools.

Configuration management uses a set of technologies and procedures to configure and setup routers, firewalls, switches, servers, APs, or other network devices. It offers tools to document the distribution and installation of new network element software release. It also can include tracking of any alterations to the configuration of the system by unauthorized agents.

Accounting management provides mechanisms to capture network utilization details; this can be part of usage-specific billing (although other types of billing may be instituted, at least for some services and/or some users).

Performance management provides mechanisms to monitor the status of the network to assure acceptable service levels. To support this function, the need exists for gathering statistics on network service quality on a consistent basis. Typical metrics include link utilization, packet loss rates, and network response times, over all segments of the network and for all types of services.

The WiSNET endeavors to use industry standards for network management, clearly including the Simple Network Management Protocol RFC 3413 Version 3 (2002). Network management capabilities embodied by the WiSNET include (i) having administrative control over the entire local network; (ii) having the ability to provision devices connected to a network; and (iii) having real-time logical/topological maps, along with the ability to discover new network connections.

Monitoring   Network monitoring capabilities embodied by the WiSNET include (i) the ability to constantly monitor the performance of an entire venue network; (ii) the ability to perform traffic management (monitoring the network and checking for the network overload); (iii) the ability to create a baseline for network performance metrics; (iv) the ability to alert administrators if the network crashes or varies from the baseline; (v) the ability to suggest solutions to performance issues when they arise; (vi) the ability to provide visualizations for network performance data; and (vii) the ability for load balancing, given that traffic would be distributed evenly on the network. Particular network management services of interest in WiSNET include the following additional capabilities (e.g., as described in [10]):

- Hardware diagnosis: capabilities to ascertain that all the hardware equipment is being monitored and in case of failure, the administrator is being notified.
- Hardware and software management: capabilities to automatically gather information about all hardware units in the network and about software installed.
- Data backup and restore: capabilities for automatic and scheduled storing a copy of configuration- and billing-data on backup servers and restoring information from them on demand.

Provisioning   Provisioning deals with (i) adding a piece of equipment to the network and activating it; and/or (ii) updating network elements to bring a customer online, based on some service-specific set of parameters, possibly along with some parameters that the user himself/herself would enter as service tailoring process, this being service provisioning. Among other capabilities, this could entail address management: managing network addresses and ensuring that there are no address conflicts in the network. Other capabilities include application-service management, business-service management, and mobility management.

Policy Enforcement   Users and other network entities typically need some kind of controlled throttling so that they do not monopolize the resources available to the detriment of other users. The WiSNET provides a number of tunable traffic/usage parameters to control the user behavior to a predefined "fair use." Other policy enforcement may be related to security and what the users can or cannot do when using the WiSNET.

Security Assurance   Security management seeks to ascertain that network is being currently secured and to notify the administrator in case of a breach; addressing security as part of managing a network was part of the original FCAPS management model. The main goal of network security management is to ensure that only authorized users and devices can access the network resources to which they have rights; unauthorized users or devices that are determined to have malware or some other malicious or harmful code are blocked. Functions that span security management cover network authentication, auditing, and authorization. A multilayered security process requires ongoing collection and a summary of critical information. A WiSNET will have rich VLAN and firewalling mechanisms to be utilized as appropriate. Some of the management functions include configuration and activation of network firewalls, intrusion detection systems, and VLANs; vulnerability management and unified threat management are also important. A roles-based function in security management software can also recognize if users should have access to specific resources based on their service options and service types [11].

Technology Vendor and Core Provider Management   As noted, the WiSNET makes use of several technologies and service providers. The value of the WiSNET to the venue property owner is that the administrator of the network provides a Single Point of Contact (SPOC) for the design, implementation, running, monitoring, and for expansion/scalability, and technology refresh. The tools available under the WiSNET ecosystem potentiate an efficient and cost-effective capability to transparently manage all providers of network resources.

Service Definition   The tools available under the WiSNET ecosystem enable an effective service creation mechanism that allows the administrators to introduce new or special-event services easily and rapidly. This covers not only the service itself as visible to the end-user but the monitoring, the performance management, the security, and the billability of such new service.

Technology Evolution   New technology now comes along at a rapid rate. The new technology may provide new features, be more economical, or increase the set of users that can be supported in a given area. The WiSNET allows the administrators to easily introduce new technology, as well as transparently remove obsolete equipment and/or technology.

User Support and Control   Users need to be controlled for a number of reasons, including for possible hoarding of network resources, for attempting to reach inappropriate sites, for (unwittingly) using infected devices, or for violating some security policy or terms of use. WiSNET allows administrators to easily monitor and control all users.

Revenue Generation and Collection   A variety of service plans may be in place, particularly for different classes of services (e.g., video, voice, and so on). WiSNET enables rapid and reliable collection of usage fees. It simplifies revenue generation. All fees are collected by an administrative system operated by the network administrator. The venue provider is able to get the revenue share without having to set up complex mechanisms. All service providers are paid in an expeditions manner by an(other) administrative system operated by the network administrator.

10.7 ECONOMIC ADVANTAGES OF A WiSNET SYSTEM

A review of the evolution of networks as depicted in Figure 10.1 progressing along the aggregated, integrated, and super integrated phases, along with the quick history described in Section 10.1 above, are a true testimonial to the economic, productivity, reliability, and functionality gains that can be achieved by "super-convergence" and "super-integration." To buttress these observations, it should be noted that the protocol/architecture convergence to the use of IP starting in the late 1980s already fostered a large degree of economic, productivity, reliability, and functionality gains in corporate and service provider networks. The introduction of a WiSNET protocol/architecture, with a Wi-Fi (and related standards) at the datalink layer for access support at the "last quarter mile," will further enhance these service and operational metrics.

In particular, WiSNET protocol/architecture will provide economic, productivity, reliability, and functionality gains on all three realms of the super-integrated network, specifically,

1. Gains obtained by the super-integrated architectural aspects of the WiSNET
2. Gains obtained by the super-integrated technology aspects of the WiSNET
3. Gains obtained by the super-integrated management aspects of the WiSNET.

The WiSNET architecture affords an open, integrated, unified platform, and access technology that is Wi-Fi/IP centric. As noted elsewhere in this text, as well in the IT/networking industry at large, such integration not only reduces costs by eliminating redundant physical infrastructure, equipment, and connectivity facilities but also greatly simplifies and optimizes costs associated with network administration, maintenance, monitoring, design, implementation, procurement, and system refresh. In addition, the use of standard protocols enables one to seamlessly inject best-in-class equipment wherever it makes sense, without requiring a design or implementation "forklift." Furthermore, such an approach enables the establishment of an ecosystem where service richness, service elasticity, service availability, and simple scalability are assured. Although it is difficult to make a broad generalization, one can typically expect operational savings in the 20–30% range for both capex and opex compared to an "as is" baseline (specific savings for particular venues and/or environments can be factually demonstrated).

The WiSNET technology suite further reinforces and invigorates the well-demonstrated economic and productivity improvements, along with reliability and functionality advancements, that have been documented by recent networking history, actually spanning a few decades. The integration of voice-related, data-related, video-related, and sensor-related information streams that support a large venue or business into one homogeneous bundle delivered by an integrated technology platform has clear advantages as measured by any number of Key Performance Indicators (KPIs), not the least being cost optimization. Although it is difficult to make broad generalizations, one can typically expect procurement and operational savings in the physical underlying technology platform exceeding 20% for both capex and opex compared to an "as is" baseline (specific savings for particular venues and/or environments can be factually demonstrated). It is of note that within the first year, the new network brought in over $0.5 million of new income to the airport.

The WiSNET management apparatus facilitates service monitoring and administration. FCAPS/FAB activities, includes traditional network monitoring, network provisioning, service definition, service provisioning, policy enforcement, user support and control capabilities, and security. The timely and judicious management of the technology vendors and core (WAN/ISP) providers itself establishes a beneficial cost-control function. Obviously, facilitating reliable revenue collection enhances the financial posture of the WiSNET provider. Although it is difficult to make broad generalizations, one can typically expect "run-the-engine" savings exceeding 30% for both capex and opex compared to an "as is" baseline (specific savings for particular venues and/or environments can be factually demonstrated).

10.8   5G SLICE CAPABILITIES

The WiSNET supports 5G slicing. Thus, a quick discussion on this topic is provided in this section.

10.8.1   Motivations and Approaches for 5G Network Slicing

Compared with prior generations of mobile and wireless networks, the 5G architecture is service based, meaning that wherever suitable, architecture elements are defined as network functions that offer their services to other network functions via common framework interfaces. In order to support this wide range of services and network functions across an ever-growing base of user equipment (UE), 5G networks extend the network slicing concept utilized in previous generation architectures [13]. In general,[3] as depicted in Figure 10.7 (right-hand side), a 5G system includes three subnetworks: an access subnetwork which generally comprises the physical resources at the network node, a transport subnetwork which generally comprises the physical resources at the one or more backhaul links, and a core subnetwork which generally comprises the physical resources at the communication service provider network(s). The left-hand side of Figure 10.7 illustrates an example network slices in a wireless communication system. Network slicing generally employs virtual networks created on top of a physical network by partitioning the physical network, where each partition (network slice) can be architected and optimized,

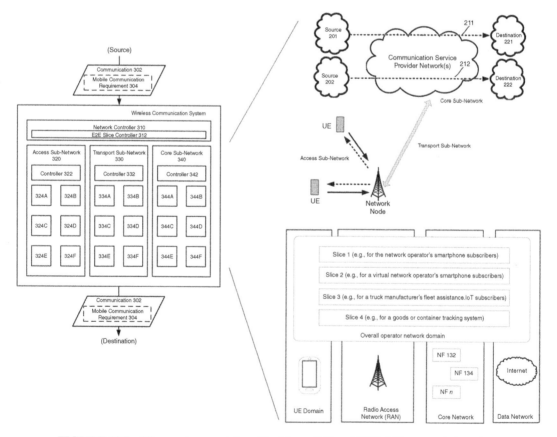

FIGURE 10.7   Slice environment (modified from [12] and [13]). NF: network function.

[3] Portions of this discussion are based on Reference [12].

e.g. for a user, application, or service. The network slice can be a self-contained network with its own virtual resources, topology, traffic flow, and provisioning rules. Conception, design, and implementation of E2E network slicing in 5G can expand over UEs, the access subnetwork, the transport subnetwork, and the core subnetwork. Network slices can have associated Service Level Agreement (SLA) commitments. For example, an example SLA commitment associated with a network slice can specify a total time for a 5G wireless communication system to deliver communication from the source to the destination. Another example SLA commitment associated with a network slice can specify reliability for delivering communication from the source to the destination.

As generally discussed in earlier chapters, the 5G core includes elements referred to as network functions, some of which can optionally be implemented software. The network functions include, for example: the Authentication Server Function (AUSF); the Core Access and Mobility Management Function (AMF); the Data Network (DN); the Structured Data Storage Network Function (SDSF); the Unstructured Data Storage Network Function (UDSF); the Network Exposure Function (NEF); NF Repository Function (NRF); the Policy Control Function (PCF); the Session Management Function (SMF); the Unified Data Management (UDM); the User Plane Function (UPF); the Application Function (AF); the UE; and the Radio Access Network (RAN).

The 5G wireless communication system of Figure 10.7 can support different network slices for different customer use cases, e.g. mobile telephones, smart home communications, IoT communications, connected cars, smart energy grid applications, etc. Each use case receives a network slice, which has a set of resources and a network topology and can provide certain SLA-specified properties, such as connectivity properties, speed properties, and capacity properties, to meet the needs of the use case. In another aspect, network slicing provides a virtual networking architecture which allows for better network flexibility through partitioning of wireless communication system into virtual elements. Note that network slicing for the plurality of network slices is implemented in E2E fashion, spanning multiple disparate technical and administrative domains, including management and orchestration planes. In other words, network slicing is performed from at least the enterprise or subscriber edge at UE domain, through the RAN, through the 5G access edge and the 5G core network, and to the data network. Moreover, note that this network slicing may span multiple different 5G providers. For example, as shown here, the plurality of network slices include Slice 1, which corresponds to smartphone subscribers of the 5G provider who also operates network domain, and Slice 2, which corresponds to smartphone subscribers of a virtual 5G provider leasing capacity from the actual operator of network domain. Also shown is Slice 3, which can be provided for a fleet of connected vehicles, and Slice 4, which can be provided for an IoT goods or container tracking system across a factory network or supply chain.

Network slicing can allow creation of multiple virtual networks, each virtual network supporting a network slice 211, 212, 213 within the shared physical infrastructure of wireless communication system 100. Logical partitions created through network slicing allow the capacity of wireless communication system 100 to be dynamically directed according to real-time needs. As needs change, so can the resources supplied to different network slices 211, 212, 213. Using common physical resources, such as storage and processors, network slicing permits the creation of network slices 211, 212, 213 which can be applied to logical, self-contained, and partitioned network functions. Network slices 211, 212, 213 can support providing wireless communication system on an "as-a-service" basis to meet the range of use cases present in communications from sources 201, 202, 203 to destinations 221, 222, 223. Network slice 211 can, for example, provide connectivity to IoT devices with a high availability and high reliability data-only service, with a given latency, data rate, and security level. Network slice 212 can, for example, provide very high throughput, high data speeds, and low latency for an augmented reality service. Network slice 213 can, for example, provide mobile voice telephony communications. Further network slicing use cases can include expanded mobile broadband with more video, higher speeds, and wide-scale availability; massive machine-type communications with transportation monitoring and control; mass market personalized TV with big data analytics; and critical machine-type communications with remote operation. Each of these use cases and others can use different configurations of requirements and parameters, and can therefore be provided with a tailored network slice. Network slices 211, 212, 213 can be optimized for different characteristics including latency and bandwidth requirements. In some environments, network slices 211, 212, 213 can be isolated from each other in control and user planes, allowing user, device, and application experiences of the different network slices 211, 212, 213 to simulate physically separate networks.

More specifically, the left-hand side of Figure 10.7 depicts a block diagram illustrating subnetworks equipped with various resources, wherein the subnetworks can be managed in connection with network slice management. The block diagram includes a communication 302 from a source, such as a UE or any of sources 201, 202, 203. The communication 302 has, or is associated with, a mobile communication requirement 304. The communication 302 is received at the 5G wireless communication system from the source, and the communication 302 is transmitted by the wireless communication system 100 to a destination, such as another UE, or any of destinations 221, 222, 223. The 5G wireless communication system comprises a network controller 310, an access subnetwork 320, a transport subnetwork 330, and a core subnetwork 340. Network controller 310 comprises an E2E slice controller 312. Access subnetwork 320 comprises a controller 322 and example resources 324A, 324B, 324C, 324D, 324E, 324F. Transport subnetwork 330 comprises a controller 332 and example resources 334A, 334B, 334C, 334D, 334E, 334F. Core subnetwork 340 comprises a controller 342 and example resources 344A, 344B, 344C, 344D, 344E, 344F.

There is a need to assign, e.g. by E2E slice controller 312, controller 322, controller 332, and/or controller 342, resources from the access subnetwork 320, transport subnetwork 330, and core subnetwork 340, respectively, to network slices in order to meet requirements (such as mobile communication requirement 304) associated with the network slices. In an embodiment, the E2E slice controller 312, controller 322, controller 332, and controller 342 can comprise 5G slice-aware controllers having any of the various tools and features presently included in 5G specifications and tools and features as may be later developed and incorporated into the 5G specifications. The illustrated resources from the access subnetwork 320, transport subnetwork 330, and core subnetwork 340 represent physical network resources and/or virtual network resources. Furthermore, there is a need for a mechanism to deliver SLA commitments by dynamically creating access subnetwork 320 slices, e.g. within the umbrella of Open Radio Access Network (O-RAN) in 5G networks and beyond. Some applications can use closed-loop auto-slicing through time-series inferences and statistical models, to successfully manage access subnetwork 320 slices. In some implementations, network slicing can be implemented in an E2E manner which can include slicing the access subnetwork 320, the transport subnetwork 330, and the core subnetwork 340. Aspects of this disclosure focus on the access subnetwork 320 in particular. There are a multitude of wireless access environments available, and moreover, access nodes can have numerous tunable parameters. Furthermore, there are numerous use cases requiring low latency, increased bandwidth, and priority SLA (e.g. FirstNet, Connect car, Industry 2.0, etc.) that can benefit from network slicing at the access subnetwork 320.

The Next Generation (NG) RAN supports resource isolation between network slices. NG-RAN resource isolation can be achieved by means of Radio Resource Management (RRM) policies and protection mechanisms that avoid shortages of shared resources in network slices, as such shortages can break the SLAs in other network slices. Solutions can optionally fully dedicate NG-RAN resources to a certain network slice. How NG-RAN supports resource isolation is implementation dependent. This presents a problem for wireless communication service providers, namely: how to provision and manage access subnetwork 320 slices based on E2E network slicing. That is, how to allocate the right resources of example access subnetwork 320 resources 324A, 324B, 324C, 324D, 324E, 324F, at the right time and in the right place, based on performance requirements in the access subnetwork 320. In another sense, the problem is one of selecting the right anchor points in multiple radio access technology (multi-RAT) access subnetworks.

Since network slicing is E2E, implementations can enable the access subnetwork 320 to meet network slice requirements in scenarios when another subnetwork (transport subnetwork 330 or core subnetwork 340) is not able to meet network slice targets. For example, suppose, an URLLC application requires an E2E latency of 18 ms or less between users. Core sub-Network Slice Instance (NSSI), transport NSSI, and access NSSI slices have delay budgets of 6 ms each. For some reason, "X," the transport subnetwork 330 slice is experiencing latency of 8 ms. But, the access subnetwork 320, with its available resources might be able to deliver service with 2 ms latency, thereby keeping overall E2E network slice latency within limits.

Advanced implementations can jointly address/enforce the SLA requirements of a network slice by exploiting near real-time states of access subnetwork 320, transport subnetwork 330, and core subnetwork 340. In other words, access subnetwork 320 can be tuned in real time or near real time to meet shortcomings of the transport subnetwork 330 and core subnetwork 340 slices.

Solutions to provision and manage access subnetwork 320 slices can separate management of access subnetwork 320, transport subnetwork 330, and core subnetwork 340. Factors for consideration to realize dynamically adaptive 5G access subnetwork slices include (i) performance (SLA achieved) of access subnetwork 320 slices for a UE at a given time; and (ii) performance (SLA achieved) of corresponding transport subnetwork 330 and core subnetwork 340 slices, corresponding to the access subnetwork 320 slice for the UE at the given time.

Some implementations can employ controller 322, e.g. a RAN Intelligent Controller (RIC) to manage network slices at the access subnetwork 320. Controller 322 can have a holistic view of access subnetwork 320 in a geographical location spanning across multiple radio access technologies. Controller 322 can assign resources of example resources 324A, 324B, 324C, 324D, 324E, 324F to network slices. Controller 322 can supervise, modify, and report data regarding access subnetwork 320 slices. Controller 322 can furthermore program-specific probes in connection with active testing to measure the performance of access subnetwork 320 and E2E network slices.

10.8.2 Implementation

As described in the previous subsection, in forthcoming 5G networks, slicing has been proposed as a means to partition a shared physical network infrastructure into different self-contained logical parts (slices), which are set up to satisfy certain requirements. In the future, organizations may choose to purchase a network slice from a carrier. Software will be able to provide a dedicated, virtual slice of the macro network, from the core, to the RAN, to the edge and through the transport.

10.8.3 Wi-Fi Slicing

In Wi-Fi, spectrum slicing allows multiple radios to operate in the same "band" within the same coverage area. This technology, can divide, or slice, both the 2.4 GHz ISM band and any of the three 5 GHz UNII bands into multiple concurrent channels.

10.9 CONCLUSION

It is expected that "super-integrated" wireless "supernetworks" such as the WiSNET will see increased deployments in venues that require high density, transparent ease-of-use, unified technology and business administration, service flexibility, and elastic scalability. As the world recovers from the COVID-19 pandemic, we already see increased economic and social activity, and some, [14], have predicted that society will greet post-pandemic life with a period of pent-up exuberance, including *inter alia*, increased travel, increased attendance to concerts and sporting events, and more emphasis on entertainment; business travel and attendance to trade conventions will see a clear resurgence. All of this could fuel the need for high-density wireless networks of which WiSNET is positioned to be the premier example. Further, we will witness a sea change in the telecommunications approach where users will just purchase a slice of the network and venues and concessionaire will be in control of their telecommunication and wireless networks. However, as a final epilogue to this textbook, make note that Moderna's CEO prognosticated at the start of 2021 – echoing comments from other public health officials – that COVID-19 is likely to become an endemic disease, meaning it will always be present in the population, but circulating at lower rates. "SARS-CoV-2 is not going away. We are going to live with this virus . . . forever" [15].

REFERENCES

1. www.wi-fi.org/value-of-wi-fi
2. Cisco, Cisco Annual Internet Report, 2018–2023.
3. Press Release. Slice wireless takes off with BWI Thurgood Marshall Airport (18 March 2020). https://www.slicewifi.com/blog/slice-wireless-takes-off-with-bwi-marshall
4. BWI Press Release. 31 January 2020 - Board of public works approves upgraded Wi-Fi service for BWI Marshall Airport. https://www.bwiairport.com/flying-with-us/about-bwi/press-media/january-31-2020-board-of-public-works-approves-upgraded-wi-fi
5. ITU-T. ITU-T recommendation M.3400 series M: TMN and network maintenance: international transmission systems, telephone circuits, telegraphy, facsimile and leased circuits TMN management functions, 2002.
6. ITU-T. ITU-T recommendation X.733: information technology - open system interconnection, systems management, alarm reporting function, 1992.
7. ISO. ISO 9595: information processing systems - open systems interconnection, management information service definition – part 2: common management information service (22 December 1988).
8. ISO. ISO 9596: information processing systems - open systems interconnection, management information protocol specification - part 2: common management information protocol (22 December 1988).
9. TM Forum. Business Process Framework (eTOM). https://www.tmforum.org
10. Fiberbit Staff. Types of network management services (June 2013). http://fiberbit.com.tw/types-of-network-management-services/
11. Hex64 staff. Managed network services: what are the different types of components of network management?.https://www.hex64.net/managed-network-services-what-are-the-different-types-of-components-of-network-management/
12. S. Jana, M. Tofighbakhsh, D. Gupta, D.D. Sharma, R. Jana, "Network slice management", U.S. Patent 11,012,312. May 18, 2021. Filed July 24, 2019. Uncopyrighted Material.
13. P. Patil; Prashanth, R. M. Ravindranath, N.K. Nainar, C.M. Pignataro, Blockchain-based auditing, instantiation and maintenance of 5G network slices, U.S. Patent 10,949,557. March 16, 2021. Filed August 20, 2018. Uncopyrighted Material.
14. Christakis, N. Apollo's arrow: the profound and enduring impact of coronavirus on the way we live (27 October 2020). Little, Brown Spark. AN/UPC 9780316628211.
15. Minkoff, Y. The world will have to live with COVID forever - Moderna CEO (14 January 2021). SA News. https://seekingalpha.com/news/3651349-world-will-to-live-covid-forever-moderna-ceo

INDEX

High-Density and De-Densified Smart Campus Communications: Technologies, Integration, Implementation, and Applications, First Edition. Daniel Minoli and Jo-Anne Dressendofer.
© 2022 John Wiley & Sons, Inc. Published 2022 by John Wiley & Sons, Inc.

Printed and bound by CPI Group (UK) Ltd, Croydon, CR0 4YY